T0360810

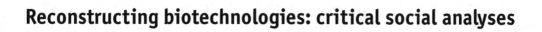

Reconstructing biotechnologies: critical social analyses

Reconstructing biotechnologies

Critical social analyses

edited by:
Guido Ruivenkamp
Shuji Hisano
Joost Jongerden

Wageningen Academic
P u b l i s h e r s

ISBN 978-90-8686-062-3

First published, 2008

Wageningen Academic Publishers
The Netherlands, 2008

Preface

On November 3 and 4, 2007, researchers associated to the Wageningen University in the Netherlands and the Kyoto University in Japan organised a conference about reconstructing biotechnologies for development. The conference was held in Kyoto and was attended by almost 40 scholars from 13 different countries. The aim of the conference was to discuss the possibilities of a reconstruction of biotechnology, one that contributes to the advancement of development and social justice, partly based on the achievement of our ongoing project of *Tailoring Biotechnologies* and its network (www.tailoringbiotechnologies.com). Many of the scholars, critically to the advancement of modern biotechnology, and its interrelatedness with corporate business and politics, had doubts about the idea that biotechnology could be reconstructed for the better, but at the same time were curious to consider the idea that not only another world, but also another biotechnology is possible. This idea may be traced back to the notion of Marx's critical dialectics, which is reflected in the following texts of a technology philosopher Andrew Feenberg and a Japanese agriscience philosopher Osamu Soda.

> 'To Marx, overcoming capitalism meant not just ending economic injustices and crises, but also democratising technical systems, bringing them under the control of the workers they enroll. This change would release technology from the grip of capitalist imperatives to a different development. Whatever our view of Marxism, a conception of technology open to a wider range of values remains essential to any real break with "technological thinking"' (Andrew Feenberg, 1999. Questioning Technology. London & New York, Routlegde: 224).

> '[A]lthough creation obviously entails a leap, it is not making something from nothing... In other words, creation is establishing new connections between existing elements, often elements that conflict and seem to be in a trade-off relationship with each other, differing from and conflicting with each other when placed in the same '*ba*' [place]. By putting two conflicting elements in the same place and considering them from a new perspective, the conflict may be sublated and synthesised, opening the world of creation. The German term '*aufheben*', the equivalent of 'sublate', includes both 'to eliminate, to deny' as well as 'to preserve' and 'to enhance', which are seemingly opposite meanings. In a sense, then, 'to sublate' means 'to preserve while denying'. While conflicting elements oppose and fight against each other and are exposed to criticism, they internally permeate each other and are thus synthesised. Criticism here means to position, and give meaning to, an object in the context of new developments, rather than breaking up and eliminating the object' (Osamu Soda, 2006. Philosophy of Agricultural Science: A Japanese Perspective. Melbourne, Trans Pacific Press: 278).

Most of the papers in this book were presented at this Kyoto conference. In the call for papers, we included diverse approaches and theoretical frameworks necessary to discuss the possibility to reappropriate the mode of biotechnology development and reconstruct biotechnologies for rural (endogenous and sustainable) development. The idea that technologies are products of agency and knowledge systems developed within particular social contexts is the object of study of 'science and technology studies' and 'social constructivism'. The analysis of technology within the context of uneven and contradictory power relations and the possibility of appropriation and democratisation is an important issue in 'critical theory' and 'political economy'. Furthermore, as any type of agricultural technologies is embedded in agrarian realities, we can rely on the contribution from 'rural sociology' and 'agrarian political economy' to understand the actual as well as potential contexts of development and application of technologies. We believe these critical social sciences provide us with diverse analytical points of view. The results of the attempt to rethink biotechnology from different perspectives but with common interests were thought provoking papers and a stimulating exchange of ideas. We thank the authors for their important contributions.

The Kyoto University Foundation as well as the Japan Society for the Promotion of Science (as part of Shuji Hisano's research budget) had provided funding for organising the Kyoto conference. In addition, graduate students supervised by Shuji Hisano were deeply involved in organising the conference. Without their contributions the conference would not have been possible. We have expressed our gratefulness for their involvement many times, but like to do this once again in this book. We would like to say a special word of thanks to Ms Ni Hui, Ms Sangeeta Jordan, Mr John Lambino, Mr Yoshifumi Ikejima, and Mr Hitoshi Miwa. We are also grateful to the editorial staff at Wageningen Academic Publishers for giving us an opportunity to publish the conference outcome as a book. Finally, we would like to thank our families for their enduring support and encouragement in this project.

The editors

January 2008

Table of contents

Part V. Potentialities of reconstruction: cases

Part VI. Regulating technologies

Introduction

Guido Ruivenkamp, Shuji Hisano and Joost Jongerden

For the past three decennia the public and scientific debate on biotechnology in the industrialised parts of the world has been dominated by an intensive pro/anti debate, diametrically opposed poles either accepting or rejecting biotechnology[1] as it has been developed.

Proponents have emphasised the advantages of the agro-industrial biotechnology, noting that the cultivation of 102 million hectares (equivalent to 252 million acres) by 10 million farmers in 2006 (James, 2006) represents the fastest adopted crop technology in recent history. Opponents refer to the social disadvantages and environmental risks of that technology, stressing that the fast adoption of the technology illustrates the power of dominant interest groups to implement (the latest phase of) the process of agricultural industrialisation. Beyond the pro/anti biotech debate is concealed a contrasting evaluation of modernity, in which optimists consider the development of science and technology as a progressive force for the further liberation and emancipation of mankind, while pessimists perceive the 'progressive' content of science and technology as a development of the forces and institutions antithetical to real freedom, and thus in reality masking new forms of oppression and domination.

The focus of this book is this scientific debate on the (im)possibilities of stimulating the emancipator potentialities of scientific progress in the reconstruction of biotechnology, focusing on agriculture (agro-biotechnology). In other words, it discusses (agro-)biotechnological developments as products of agency and knowledge systems, and explores whether and how the social relations and knowledge systems underlying the development of biotechnology can be reconstructed. In this sense, the book is strongly related to the long-running scientific debate over whether human beings are able to develop *a new way of thinking*, a new form of rationality, which offers them an unfolding capacity to unveil and deal with the radical modifications of the technologically advanced world. Heidegger expressed his doubts about this. Indeed, he stressed that '*what is disturbing is not that the world becomes dominated by technologies, but that man is not yet prepared to this modification and has not been able to develop an appropriate frame of reference to confront him/herself with that new reality*' (Heidegger, 1959).

This book challenges both the discourse of opposition, that one is either with it or against it when it comes to science and progress, and also the limiting conception of the human capacity for change implicit in Heidegger's analysis. The attitude of having to either accept or reject technology as it is is itself limiting, denying the potential of alternative models of progress and the reality that through technological innovation many new opportunities for human betterment do arise, and are arising. In fact, we see today, particularly in the non-industrialised parts of the world, a wide range of initiatives working to *reconstruct* technology. It is the objective of this book to present empirical evidence of various experiences of technology reconstruction, to reflect on these experiences and to develop a new theoretical frame of reference for dealing with the actual technology development.

---

[1] In the public discourse, biotechnology and genetic manipulation are treated as being the same. For reasons of clarity, however, the distinction ought to be made. Genetic manipulation involves the isolation, manipulation and reintroduction of *DNA* from one organism into another, usually aimed at *expressing new characteristics*, such as making a *crop* resistant to a *herbicide*. It generally implies that the process is outside the *organism*'s natural *reproductive process*. Biotechnology refers to the use of *microorganisms*, such as bacteria or yeasts, or *biological substances*, such as enzymes, to perform specific *industrial* or *manufacturing processes*.

1. Key concepts for developing a new theoretical framework

In view of this objective, the book presents a variety of perspectives, dealing with several issues. A *first* key issue for the development of a frame of reference for dealing with actual technology development is the debate on *reconstruction*. Reconstruction is a deliberate act of re-creation, applied at the level of knowledge systems and technical artifacts. It takes place by changing the social relations from which the artifacts emerge, as well as by modifying the material content of the artifacts. It is based on an appreciation of artifacts as ensembles of social and technical dimensions shaped by specific power relations among different actors. In other words, the concept of technology reconstruction is based on a socio-cultural critique and an awareness of alternatives. It takes as axiomatic the view that specific values are excluded from the original design (Feenberg, 1999); that technology has a specific code which is socially structured and biased, reflecting the unequal distribution of power (Noble, 1978); that technology has politics (Winner, 1986); and that social/techno activists networks are able to change the values, codes and politics of technology (Ruivenkamp, 2005). The reconstruction concept is strongly related to what Bertolt Brecht and Walter Benjamin have named the 'refunctioning of technologies' (referring to the efforts of engineers and workers to change the content of the technologies with which they were working), stressing that technology does not only take place at the stage of research but that technology may also be appropriated and redeveloped by its users. Indeed, the reconstruction concept restores *agency* in the actual technology development.

The *second* entry point for developing a new theoretical frame of reference on technology development implies that it becomes necessary to question whether a *social space of antagonistic agency* remains, whether there still is, in fact, a space from which antagonistic activities, such as techno-opposition, can be launched. The possibility of such a '*space for change*' is sometimes doubted, particularly by those who emphasise the socially manipulating force of new technological developments, as leading to new power relations and a reduction of the agency for change. In this respect, one might be mindful of Herbert Marcuse's analysis of advanced industrial society, in which technology is considered as the medium through which culture, politics and the economy converge into an omnipresent system that swallows up or repulses alternatives (Marcuse, 1964). Marcuse has stressed that the most singular achievement of this one-dimensional society is the containment of social change, paralysing antagonistic thought and behavior, a containment and paralysis realised through the specific manner in which the lives of the individuals are organised, i.e. the determination of their needs, aspirations and the content of socially required skills and occupations. In addition to this reduced space of agency in the one-dimensional society, David Noble (1978) described that a similar process takes place at the workplace in enterprises. Noble argues that – due to the use of particular production technologies at the workplace – workers loose their self-confidence and have difficulty with the idea of self-management. He also stresses that the workers' uncertainty and fear of exercising their own capacities in the workplace extends to the wider political arena where the workers tend to defer to the authority and prestige of their 'betters' rather than trusting themselves. The loss of self-esteem of the workers is – according to Noble – strongly related to the dominant perception of the prevailing production technology as a given. It is exactly this *false notion of technology as a given* that is symptomatic of the ideology of technological determinism – a perspective which is strongly challenged here. According to Noble, this false notion is an ideology shared by '*those lazy revolutionaries who proclaim liberation through technology and prophets of doom who forecast ultimate disaster through the same medium*'; an ideology that is highly contentious politically because it fosters passivity and rules out agency or human action. However, Noble indicates that this idea of *technology as a given* – which is based on the assumption that technology follows a self-generating, single course and has effects 'outside' it, usually referred to as 'social impacts' (Noble 1978: 316) – reflects the factual exclusion of workers from the realm of technological decision-making.

Some postmodern theorists even go a step further, arguing that *subjects are not only excluded from the decision-making process* regarding technology development, but that the knowledge and technology development are one of the multiple sites from which *new forms of domination and control are imposed on society*. This is the *third* issue to which the contributions in this book are strongly related, these postmodern perspectives on knowledge and technology developments as multiple sites from which new forms of domination and control operate. Such an approach to *power relations* has been developed by, among others, Michel Foucault. Indeed, Foucault stressed that the (post)modern era is characterised by an increasing and diffused domination of the individual through social institutions, discourses and practices. He described the modern era as a kind of 'progress' in the dissemination and refinement of techniques of domination (Best and Kellner 1991: 37), emphasising that knowledge, rationality and forms of subjectivity are multiple social sites through which domination is practiced. Best and Kellner, in their consideration of postmodern theory, emphasise that Foucault adopts a stance of hostile opposition to modernity, rejects the modern equation of reason with freedom, and attempts to problematise the modern forms of rationality as reductive and oppressive. This stimulated the vision – increasingly shared by social scientists today – of knowledge and technology as no longer the vehicles of development and emancipation, but rather as integral components of power and domination. Best and Kellner suggest that Foucault distances himself from totalising modes of theory and places himself in opposition to the rationalist myths of the Enlightenment, such as Marxism, emphasising that such totalising theories need to be superseded by a plurality of forms, forms of knowledge and microanalysis. Indeed, through microanalyses of 'madness', the 'clinic' and 'human sciences', Foucault aimed to reveal the assumptions, rules and ordering procedures of a discursive rationality – a discursive rationality that operates *beneath the level of intention* and represents the fundamental codes of a culture which constitute the episteme or configuration of knowledge, determining the social (and thus scientific, we would add) practices of a particular historical era. This archaeological approach fed further the developing suspicion of reason and emancipator schemes advanced in the name of that specific rationality, and in the process stimulated a pessimistic vision of knowledge and technology development.

The *fourth* core issue of this book is the debate over whether the focus on new power relations does not, in fact, underestimate the social potentialities of new technological developments; whether this focus on analysing Enlightenment rationality as the totalitarian force which eliminates all competing and antagonistic thoughts does not lead to a neglect of technologies as potential sources for emancipation and instead acts to reconfirm the ideology of technology as the fundamental organising principle of contemporary society, socially barely controllable and inseparably associated with the existing regimes of power. Foucauldian vision and postmodern analysis may lead to a passivity in respect of technological developments, which illustrates, indeed, and underscores the abovementioned social need to develop a new frame of reference. In this context, it is relevant to acknowledge that Foucault himself also gradually shifted from his archaeological towards a genealogical approach, from microanalysis in which he described the ways in which people are transformed (and become classified, excluded, objectified, disciplined and normalised), towards analysis of the ways in which people transform themselves – a gradual and strongly interrelated change from an analysis of the technologies of domination towards an analysis of the technologies of the self. Indeed, Foucault introduces a *redefinition of the concept of power*, developing the idea that power can no longer be envisaged just as a system of repression carried out by a sovereign state, but should also be seen as a force, as a capacity for individuals or singularities to transform, to intervene in the domain of the other's possible actions. This redefinition of power is not an analytical advance alone, for the intellectual development is itself indicative of a real shift in the balance of power, from the top-down (restrictive) system of unity to bottom-up (liberating) forces of diversity, a progressive dynamic that redefines the meaning of 'progress'. Power is also constituted from below, and seizes life and living being as the objects of its exercise. In the postmodern era, it is becoming clear that the grounding force of power can no longer be found on the side of the sovereign state, executing the dominion of a unilateral power relation, but is increasingly based on the dynamics of society, on the social relations between forces.

2. Multiplicity of critical perspectives

The objective of the present work is to work toward an unveiling of the social relations between forces in the domain of biotechnological developments, presenting a multiplicity of perspectives concerning the relationships between new forms of power and biotechnological developments and the different ways in which those relationships are challenged. In this sense, the book is related to the social constructivist approach to technology development but aims also to go beyond that and introduce other perspectives. Most of the papers gathered here share the vision of the social constructivists that technology is primarily social and thus politically neither neutral nor autonomous. Taken as a whole, moreover, the book shares the vision that the variety of local circumstances may have led to a plurality of different but comparable artifacts – in other words that there are viable technological alternatives. However, this book also emphasises that it remains necessary to address general societal developments and new power relations when analysing concrete biotechnological developments. It emphasises that it is insufficient only to carry out empirical constructivist studies focusing on the specific local groups involved in particular cases of technology development. Generally speaking, constructivist studies tend to lack any sense of the political context in which technology is developed. Moreover, social resistance is rarely discussed, with the result that the research is often skewed toward a few official actors whose interventions are easy to document.

The contributions to this volume challenge the political de-contextualisation of biotechnological developments and focus, in contrast to the social constructivist studies, on the *refusal to passively accept the actual biotechnological developments*. Indeed, three different types, or styles, of what might be termed 'active non-acceptance' with respect to biotechnology are presented: the (negative) act of refusing to accept technology (*rejection*), the (disruptive) response of opposing technologies that are disapproved of or disagreed with (*resistance*), and the (creative) attempts to develop new and different kinds of technologies (*reconstruction*). That is, the papers here include – as the subjects of their presentations and central to their arguments – examples of and case studies detailing an active non-acceptance of biotechnology-as-is (in its current state, in the context of the political status quo), examples and case studies which can be categorised into three, as different styles (negative, disruptive and creative), expressed in action through rejection, resistance and reconstruction. It should be emphasised also that despite their different perspectives and emphases, these three action styles are also closely interrelated. The act of reconstructing current agro-industrial biotechnologies in the movement of *tailor-made biotechnologies* implies the act of rejecting the actual form of biotechnology design as well as that of resisting the existing unequal power relations integrated in the design of biotechnologies. In other words, this book also aims to stimulate a reflection on the possibility of finding those new common conditions which may become – as Toni Negri and Michael Hardt (2004: 106) have formulated – *a multitude*: an emerging political project in which the innumerable types of labour, different modes of life and specific geographical conditions do not divide and prohibit but support and promote communication and collaboration into a new common political project. Importantly, these variables will always – and necessarily – remain, regardless of their changing specifications, and will remain as obstructions endlessly impeding the modernistic tendency to homogenisation. This new common political project is thus not based on sameness or unity, but rather on the recognition that the differences of nature or kind do not divide but enrich the multitude. Therefore, this book presents *a multiplicity of perspectives* on biotechnological developments in relation to new forms of power and rejection, resistance and reconstruction related to these developments.

The book aims to contribute to the development of a *multi-perspectival critical social theory*, what Marcuse has termed a multidimensional critical social theory, through the presentation of different views on technology development. As Best and Kellner (*ibid.*) have emphasised, the term 'perspective' presumes a specific standpoint, focus, position or even sets of positions that interpret particular phenomena. The term suggests that one's optic or analytical frame of reference never mirrors reality exactly as it is, that it is always selective and unavoidably mediated by one's position, one's pre-given assumptions, theories, values and interests. The notion of perspective also implies that no single view can ever fully illuminate the richness and complexity of any single phenomenon, let alone the infinite connections and aspects of social life. Thus, all knowledge of reality stems from a particular point of view, all facts are constituted interpretations and all perspectives finite and incomplete. An incompleteness, however, which may be reduced through the development of a composite combining different perspectives.

3. The organisation of the book

A multi-perspective critical social theory sees society as composed of a multiplicity of dimensions and potentialities for social transformation. The objective of developing such a theory is approached in this book by grouping the various perspectives into six parts.

In *Part I, The politics of biotechnologies*, two different perspectives are presented concerning the interrelation between politics and technology developments. The first paper focuses on uncovering the social space of antagonistic agency through an analysis of contradictions in the actual life-science technology development; the second focuses on the power relations of 'the Biotechnology Project' and the perspectives for an increasing techno-opposition.

In the opening contribution to this volume, Guido Ruivenkamp presents a manifesto, a *theoretical framework* to go beyond the factuality of the actual biotechnological development. A cornerstone of his critical-theoretical approach is the need to look for resisting and liberating forces that develop within established society and may offer perspectives for transformation. He refers to four historically contradictory trends in technology developments as potential entry points to a qualitative shift in the actual biotechnology development. The aim of this paper is to present these historical opportunities for agency, for the modification, re-appropriation and democratisation of life-science technologies through the organisation of specific strategic and tactical actions of resistance. Ruivenkamp emphasises the relevancy of developing 'situational politics' focused on creating new, alternative social relations from which another biotechnology trajectory may evolve, as well as on changing the material form of biotechnological artifacts so that an alternative biotechnology trajectory may come to function as a lever for location-specific developments.

In the second paper, Rachel Shurman and William Munro pay attention to the *resisting practices* of the anti-GMO (genetically modified organisms) movement. They refer to the concrete results of this movement in closing European markets to GMOs, in arousing consumer concerns about the safety of GM crops, putting pressure on governments to play a more serious regulatory role, and altering the political reception and trajectory of agricultural biotechnology. Shurman and Munro emphasise that these results have been realised by presenting specific narratives and an alternative interpretative framework, which directly challenge many of the claims and assumptions embodied in the dominant discourses of biotechnology. Therefore, they disagree with Ruivenkamp's assumption that the anti-GMO movement departs from the same epistemological frame of reference as what they term 'the Biotechnology Project', namely with the perception of biotechnology as an instrument in the hands of a specific group of actors, with negative (or positive) consequences for society. Shurman and Munro emphasise that the anti-GMO movement was not guided by a principle advocating the separation of technology and society but rather by the perception of the inextricability of technology from society and the possibilities inherent in existing social and power relations. In their vision, the anti-GMO movement has been counter-hegemonic in two important ways. First, it has challenged the political economic project of large multinational corporations, and second, it has challenged the cultural project of making a corporate-led, intellectual property-rights protected technology seem normal. By challenging these two basic tenets of the Biotechnology Project, the anti-GMO movement has helped expand our vision, in the direction of a quest for viable alternatives.

Despite the rather different perspectives presented in the first two contributions, the authors do agree with each other that politics and technology development can no longer be considered as two separate domains, and, moreover, that biotechnologies are socially constructed, with the social construction of biotechnology to be considered a complex and sometimes contradictory multi-actor and multi-agency process. In view of the actual power relations, the authors also concur that it is necessary to reflect on the possibilities for reintegrating agency in the re-construction of biotechnologies. In this respect, there is a combined emphasis on the perspective that regards as insufficient the focus on broadening the groups of actors involved in biotechnology design negotiations. Rather, what is stressed is the importance of reflecting on the ways in which interested parties can resist and modify the actual design of new biotechnologies in particular, and the Biotechnology Project in general. In the following parts, different perspectives are presented on the ways in which opposition to and/or participation in the development of the Biotechnology Project can be organised.

In *Part II*, these different perspectives are discussed as *Opposition and participation* in the shaping of biotech policies and techno-scientific issues. Franz Seifert describes the resistance to GMOs in two European countries, Austria and France, arguing that the farmers' resistance to GMOs there has had a crucial impact on national policies of countries averse to agro-biotechnology as a rule. These two cases, however, also show conspicuous differences. In Austria, organic farmers are striking in their GMO-opposition. However, they take a backseat in public protest, while Austria's national biotechnology policy simply bars GMOs from its territory in order to protect organic farmers from 'GMO-contamination'. Franz Seifert calls the Austria stance a national NIMBY (*Not In My Back Yard*) approach. By contrast, the movement of farmers' protests in France, led by the Farmers' Union *Confédération Paysanne*, has become the leading anti-GMO voice among the French public. French farmers employ media-savvy protest tactics of symbolic politics and civil disobedience, which are part of a much wider ideological agenda than in Austria, denouncing modern capitalism and globalism. French GMO critique envisions a NIABY world: GMOs should not be grown *anywhere* (*Not In Anybody's Back Yard*). Seifert's analysis traces the causes for these different conflict patterns to *dissimilar contextual factors* and *opportunity structures* in Austria and France, specifically, to differences in national political cultures, agricultural policies and politics. As to the impact of these two types of criticism on the development of modern biotechnology, the two national cases also provide different lessons, which nevertheless have in common a general rejection of modern biotechnology for symbolic reasons.

In the following paper, Les Levidow argues that European policy-making on techno-scientific issues has anticipated or responded to European public concerns over agro-biotechnology, and to this effect state bodies have sponsored participatory exercises in technology assessment (TA). These exercises, however, have encountered public suspicion and legitimacy problems. This is because government policies have promoted specific technologies as if they were objective imperatives and efforts to govern societal conflict. Although public participants in TA sought to open up technological decisions vis à vis alternative futures and normative choices, such efforts were marginalised. Technology-induced TA was favored, as opposed to problem-induced TA. Questions about social problems in relation to agro-biotechnology were displaced and channeled into regulatory issues and risk control measures. In designing and managing the TA exercises, *boundaries were imposed* – between biotechnological imperatives versus alternative options, between scientific versus policy issues, and between expert versus lay roles – thus closing down issues. Some participants challenged the boundaries, opening up issues for a broader, lay expertise. By contesting lay/expert boundaries, participants introduced different models of the public. As a result, on-going deep-seated tensions have arisen between discussing a 'common' problem – how to make agro-biotechnology safe or acceptable – versus containing conflicts around problem definitions of societal needs. To some extent, participatory TA exercises have helped to hold governments accountable for regulatory criteria, but not for innovation choices. These participatory TA exercises generally internalised assumptions about agro-biotechnology as societal progress. *Levidow* concludes that despite aspirations to democratise technological choices, the exercises tended to biotechnologise democracy.

In *Part III, Potentialities of reconstruction: critical reflections*, three different perspectives on reconstruction are presented. In 'First the Peasant', Joost Jongerden offers a reflexive and critical analysis on the reconstruction of modern technology. The adjective 'modern' should be emphasised here, since the major concern of this paper is with a critical treatment of modernity. Jongerden argues that the destruction of peasant production systems was target and outcome of the modernity project. In addition, he builds the argument that modernity has been usurped by an authoritarian, instrumental approach to development, aiming at a more efficient and effective use of technology. A reconstructionist approach, he argues, is essentially a reflexive and questioning approach, critically assessing means and ends systems and ultimately the nature of modernity itself.

Wietse Vroom's 'Resistance and Reconstruction' considers concrete technical objects and technological development trajectories in relation to different views on agricultural modernisation. Vroom refers particularly to the efforts of the International Potato Centre in Peru to develop appropriate biotechnologies for potato farmers. The discussion of three case vignettes demonstrates how alternative modernisation trajectories are feasible, trajectories in which biotechnology development empowers informal local innovation and production systems. This leads to an elaboration of three distinctive, conceptual levels on which technological development is being contested, and can be reconstructed: (1) a pragmatic level, (2) a social level related to a restructuring of the roles in innovation and production, and (3) a level related to identity construction and a fear of cultural erosion. Finally, the implications of this for deconstructing the ongoing public debate and controversy on GMOs are discussed.

Given the potentially crucial role of scientists in the process of social transformation of challenging the technological hegemony, Shuji Hisano argues that it is worth asking if and how these scientists can be critically reflexive in regard of the social meaning of their research activities. As the issues of technology and related public concerns are increasingly and actively being framed in ethical terms, the possibility and limitations of morality as a sort of nexus of science and society, a discursive tool to stimulate dialogues among scientists as well as between scientists and the public at large, or a platform for scientists to reflect on the nature of their practice need to be examined. Hisano coins the term '*ethicisation*' to refer to a blind tendency to introduce ethical considerations into an institutional setting or a decision-making discourse. Referring to several experiences of the '*institutionalisation of ethics*', Hisano argues that what should be expected from ethics is not just an instrumental tool for judging things according to values couched in abstract terms, and that there is an urgent need to politicise the ethics of biotechnology, to include values and interests embedded in biotechnology research, and to involve scientists in the issue of the sub-politics of biotechnology. By looking into several instances of ethical remarks made by agricultural scientists, Hisano also examines the possibility of bringing scientists into a process of critical reflections on the social meaning of agricultural science and technology by the use of critically reconstructed ethical terms. This paper concludes with an argument for the need for interdisciplinary approaches in the context of scientists' critical reflection.

Part IV, Quality agriculture and networks, focuses on descriptions of possible social spaces for technology reconstruction in Europe, related to quality agriculture. European policies have promoted capital-intensive technological innovation as an essential means for more efficient production methods, more lucrative products, and thus a competitive advantage in the global economy. This agenda setting, Les Levidow argues in his second contribution, corresponds to a series of policy slogans such as the 'Biosociety' in the 1980s, the 'Knowledge-Based Economy' in the 1990s, and the 'Knowledge-Based Bio-Economy' (KBBE) in the current decade. The latter narrative contrasts with the 'quality' narratives of Alternative Agri-Food Networks (AAFNs). The KBBE emphasises agricultural weaknesses that need technological, laboratory-based solutions, while AAFNs emphasise local resources as bio-vitalities to be promoted. Each narrative has different ways of defining agricultural problems, devising solutions, relating producers to consumers, adding value, assigning the extra value and commoditising resources. As defined through AAFNs, quality agriculture implies an alternative type of knowledge-based bio-economy. These networks create, add and capture market value, especially for the benefit of farmers, by contrast to the techno-economic treadmill of agro-industry. Quality agriculture valorises on-farm resources through diverse cultural meanings of agricultural inputs, outputs and wider societal benefits. Through regional authorities, a Europe-wide 'GM-free' network brands regions as sources of quality agro-food products. Quality agriculture has also been turned into an antagonistic narrative, portraying agro-biotech as a multiple threat of symbolic contamination, globalisation, economic competition and political domination over government policy. This counter-narrative challenges the supposedly objective imperatives of global competition for bulk commodity production. As an alternative bio-economy, then, quality agriculture can play an oppositional role.

Although agriculture and food production are pushed towards traditional industrial models, Ezio Manzini describes a stratified reality where different behaviors, agricultural ethos's and food cultures exist side by side. He differentiates between five different sub-systems: the traditional, classic and experiential systems, the advanced agro-business system and, finally, social experimentation. In this general framework, Manzini examines the role that design can, or could, play, and criticises the current notion of food design as aiming to transform food and food preparation into a kind of show, into a sensory theatre where everything has to be thrilling and exceptional. This paper highlights the need for a design culture capable of raising more profound questions about the sensory nature of the food and food preparation experience. By way of a concrete example of where this reflection could lead, the Slow Food organisation is offered as an excellent instance of strategic design in the agriculture and food system. Manzini goes on to consider the opportunity for designers to look at new food networks built by creative communities all over the world, and concludes with an outline of how designers can enhance these new forms of organisation, bringing their specific skills into play to improve visibility, making communication channels more fluid, and implementing enabling platforms that encourage the spread and development of the new approaches of these communities to agriculture and food design.

The third paper in this section focuses on a concrete experience in the Netherlands to strengthen quality agriculture by challenging specific institutional arrangements and their technological artifacts. Joost Jongerden and Guido Ruivenkamp discuss the development of alternatives to the mechanised, large-scale agricultural production systems in agriculture in relation to the development of alternatives in technology development. They argue that the increase in the number of locally embedded food production and consumption networks, so-called 'regional initiatives' for the production and distribution of high quality food products, is an important indicator of the emergence of alternatives for modern industrialised agriculture. However, analyses of such alternatives in agriculture – referred to as 'regional agriculture' or 'quality agriculture' – tend to focus mainly on distribution and consumer/marketing issues. Jongerden and Ruivenkamp aim to show how these alternatives make new demands on the development of technology and require institutional (re-)arrangements. Referring to the functioning of the Dutch Variety List, the Genebank and participatory plant breeding, the authors describe ways in which attempts were made to reconstruct these arrangements, which are revealed as strongly related to modern agricultural practices. Jongerden and Ruivenkamp advance in the form of a proposition the argument that the development of alternatives in agriculture goes hand-in-hand with the development of alternatives in technology, and that alternative technology developments require institutional re-arrangements enabled by a modification of the power relations between actors.

Guido Nicolosi emphasises that the development of food technologies is closely interrelated to the local social space (termed 'territory'), and that food has to be considered as a crossroads between territory and technique. However, he also stresses that this relationship is mediated through material objects and immaterial or symbolic objects (beliefs, codes). Indeed, Nicolosi argues that food is a central symbolic and communicative issue, and needs to be studied from the perspective of its embedment in a socio-cultural matrix. Nicolosi's contribution aims at showing the changes that the advent of (bio)technological modernity has brought about in the communication flows that involve eating. Despite the fact that modernity has activated processes of removing nutrition from the social and cultural context of appurtenance, human beings continue to feel the need for food to have meaning, to be constructed, embedded in a symbolical frame. Any kind of new intervention, aimed at the introduction of new practices, new technologies or food production, must be able to deal with this aspect. Nicolosi emphasises that with the great transformation that the alimentary underwent with the advent of modernity, food (un)safety became the prior semantic issue. He characterises late-modern society as 'an orthorexic society', that is, a society founded on an alimentary hyper-reflexivity in its various meanings: dietetic (fitness), ethical (critical consumption), aesthetic (food design), symbolic (slow food), psycho-pathological (eating disorders) anxiogenic (alimentary fears), etc. This is a hyper-reflexivity, moreover, in which new hegemonic discursive forms emerge.

Part V, Potentialities of reconstruction: cases, begins with a contribution from William Munro, who analyses conflicts over the implementation of the Biotechnology Project in South Africa by focusing on the politics of biotechnology regulation. Through a brief case study of South-African smallholders who plant GM cotton on the Makhathini Flats, Munro shows how these conflicts have been catalysed in dueling discourses of rights, risks, and authoritative knowledge. The paper then shows how the *logics* of regulation shape the politics of rights and risks in two particular ways. First, the location of agricultural biotechnology regulation by the state within the framework of science and technology policy pushes conflicts over regulation into the realm of 'boundary making', which involves attempts by various interested parties (regulators, experts, activists, citisens, farmers) to draw the epistemic and normative boundaries of authoritative knowledge, more or less broadly or narrowly. Second, these logics are rooted in a meta-narrative of economic modernisation and development that makes the South African state sympathetic to the Biotechnology Project. An examination of the politics of boundary making in the construction of South Africa's regulatory framework for agricultural biotechnology shows how a small group of scientific experts was able to 'capture' the regulatory process, defining authoritative knowledge narrowly and marginalising social constituencies such as farmers. As such, the regulatory framework for biotechnology sustains the Biotechnology Project, through its grounding in the logic of modernisation. Nevertheless, the paper shows that the political conflicts generated around agricultural biotechnology have opened up spaces for new actors and directions in agricultural development, and perhaps reduced the control of the biotechnology industry by creating new opportunities for marginalised groups such as the Makhathini farmers to take 'independent participatory action'. In sum, biotechnology is there to stay, but it may not represent the Biotechnology Project that anyone envisaged.

In the second paper in this section, George Essegbey examines the relevance of biotechnology policies in Sub-Saharan Africa, reviewing the general trends in policy in six selected African countries, and drawing out the similarities and contrasts. The content analysis of the biotechnology and biosafety policies of the countries (Ghana, Nigeria, Kenya, Rwanda, Namibia and South Africa) highlights characteristics in terms of, among others, the stated objectives, policy guidelines, and mechanisms. The premise of these policies is that they define the framework for the development and application of biotechnology. The paper discusses the validity of the premise by taking into account the context of the formulated policies, and assesses the orientation and thrust of the policies. Some fundamental constraints in biotechnology policy formulation and implementation are indicated, such as the relatively weak capacities for application and innovation, particularly at the grassroots of society, and the limited potential for public investment. Above all, there is the greater challenge facing the transformation of biotechnology of the need for tailoring, to be orientated to the specific societies and socio-cultural conditions of these countries. In the process of transformation, argues Essegbey, there must be an ownership of the technology by the people, precipitating in the products of biotechnology an enhancement of sovereignty in broad terms.

Finally in this part, Elibariki Emmanuel Msuya argues that Genetically Modified (GM) cotton Controlled Field Trials (CFTs) have started in the southern part of Tanzania, as an initial stage before embarking on GM food crops, and this despite great concern over whether Tanzania should promote GM crops and initiate their commercialisation. The overall goal of Msuya's paper is to conceptualise ways in which it may be possible to develop feasible agro-biotechnologies that address the problems of the smallholder farmers in Tanzania, who often have limited access to high potential land and capital resources. Based on a Tanzanian cotton sub-sector case study and literature review, the paper concludes that unless agro-biotechnologies are reconstructed, the existing technologies in their present form are likely to have only a minor impact on improving the livelihood of resource-poor farmers in Tanzania.

Concluding this presentation of different perspectives on relationships between new power forms and challenges in the sphere of (agro-)biotechnological development, *Part VI, Regulating Technologies*, concentrates on common property, open source and regulatory issues. Eric Deibel's 'Life in Common' looks at ownership and 'common property relations' in life sciences. Deibel takes Foucault's concept of a 'classical space of representation' as his point of departure, in order to reconsider the historical relationship between science and society. As such, he approaches ownership in the life sciences as analogous to the early modern state-of-nature theories of Hobbes and Rousseau, as well as a 'Locke-Marx relationship'. Such a definition of natural foundation of society, literally, assumed a 'world in common', with its proposed counterpart here, of a 'life in common', regarded as reemerging in terms of a shift from the patentability of a 'biodiversity market' to the sharing of a 'free market for genetic information'. It is proposed that a principled open source-like approach is needed, one that would aim at a recoding of the biological potentiality of a post-modern space of representation – as a translation of genetic materials into genetic information (technologies) and *vice versa*, as a fact of our communal existence, our living together.

The issue of common property and open source is also discussed by Kate Milberry, although here in the context of information technology. Milberry discusses the disconnection between modern technology and social values, mainly focusing on the internet. Her focus on tech activists having appropriated Internet technology, instilling it with the goals and concerns of the global justice movement, is highly relevant for others concerned with the appropriation of technology, i.e. biotechnology. Through their development of free software – in particular their customisation of wiki technology – tech activists have created a space and tool for communication in cyberspace. This in turn has enabled the realisation of new communicative practices offline, establishing a dialectical relation between the technological and the social, and restoring technology's transformative aspect. The desire for a more just society prefigures democratic practice online, actualised as democratic interventions in the development and use of technology, which then manifest in alternative modes of social organisation in the 'real' world.

The final contributor, Niels Louwaars, argues that the debate on tailoring biotechnologies has concentrated mainly on conceptual and technical aspects, ignoring right regimes. A reconstructive approach, however, needs to take into account the rights systems that determine whether, how and by whom the technologies and the derived products (e.g. plant varieties) can be accessed. Louwaars analyses the impact of rights systems, and also comes up with a proposal. At the policy level, he argues, due attention should be given to the development-related rather than trade-related aspects of intellectual property rights ('DRIP's rather than 'TRIP's), even when the policy space is reduced, due to relatively strong demands from trade negotiations. In the field of licensing strategies, examples of broad humanitarian licenses and open source strategies deserve due attention.

A variety of perspectives emphasising different aspects of the social constitution of agro-biotechnological developments is presented in this book. These perspectives range from the politics in biotechnology, to the resistance practices of the anti-GM movements, to the management of TA participation practices, critical reflections of potentialities of reconstruction, alternative agricultural practices for alternative biotechnological developments and location-specific cases of biotechnology reconstruction and re-technology regulation. Moreover, the contributions here have been made by writers from various disciplinary fields and disparate backgrounds, presenting a range of different and sometimes contrasting perspectives. As editors of this book, we hope that the wealth in the diversity of the contributions may assist in a further surge in the expansion of opportunities for resisting, rejecting and reconstructing agro-industrial biotechnological developments, and, most particularly, for reconstructing them as and into people-based development projects.

References

Best, S. and Douglas K. (1991). *Postmodern Theory: Critical Interrogations*. New York, The Guilford Press.

Feenberg, A. (1999). *Questioning Technology*. London & New York, Routledge.

Hardt, M. and Negri, T. (2004). *Multitude, war and democracy in the age of empire*. New York, Penguin Press.

Heidegger, M. (1970). *An introduction to Metaphysics*. Anchor, New York.

James, C. (2006). International Service for the Acquisition of Agri-biotech Applications. Available at: http://www.isaaa.org/default.html

Marcuse, H. (1964). *One Dimensional Man, Studies in the ideology of advanced industrial society*. Boston, Beacon press.

Noble, D. (1978). Social Choice in Machine Design: the case of automatically controlled machine tools, and a challenge for labor. *Politics & Society* 8 (3-4): 247-312.

Ruivenkamp, G. (2005). Tailor-made biotechnologies: between biopower and subpolitics. *Tailoring Biotechnologies* 1 (1): 11-33.

Winner, L. (1985). Do artifacts have politics. In: *The Social Shaping of Technology*, (Eds. D. MacKenzie and J. Wajcman), Philadelphia, Open University Press.

Part I. The politics of biotechnology

Tailoring biotechnologies: a manifesto[2]

Guido Ruivenkamp

1. Introduction

During recent decades the scientific and public debate on biotechnologies has arrived at a pro/anti deadlock. The debate has been characterised by two contrasting positions, that of an unqualified *acceptance of* agro-industrial biotechnological developments, in the name of progress and the free market, or else a complete *rejection* of them, as a form of (self)protection. Notwithstanding their different evaluations of the social impacts of the agro-industrial biotechnological developments, the supporters and the opponents of the agro-industrial biotechnological developments both underestimate the opportunities for *modifying* these developments. Each group tends to consider the development of agro-industrial biotechnology, and especially its related gene-technology, as an inevitable, almost autonomous, science-driven development, a given which may be used for good or for evil, for the enhancement or the destruction of individual and social life.

In contrast to this vision – expressed by various different interest groups and non-governmental organisations – fluid social-technical networks have been formed which no longer position themselves within the intensive pro/anti biotech debate, but are interested instead to develop concretely a *third approach*, that of re-establishing new, co-creative relations between biotechnological and endogenous developments. By reconnecting biotechnological developments to human agency, these fluid social networks aim to transform the agro-industrial biotechnological developments into tailor-made biotechnologies (TMBT) attuned to sustainable, location-specific (endogenous) developments[3].

In this manifesto, this third approach will be elaborated *conceptually*, reflecting on the activities of those techno-activist networks in different regions of the world, on their struggle to modify, re-appropriate and democratise biotechnological developments. The reflection of these activities is based on an elaboration of the scientific debates on the critical theory of technology development, enriched by the critical-constructivist analysis on the practices of tailor-made biotechnologies and genomics, carried out in the Critical Technology Construction (CTC) research unit of Wageningen University.

The aim of this article is to consider, at a conceptual level, some additional opportunities for the fluid TMBT networks to transcend the established state of affairs and reconstruct technologies. These opportunities are sought from within the tension between four historical trends of technology development, leading to the *apparently autonomous technology* development and the *increasingly social character* of that technology development. It is this tension between historical trends in technology development (shaping that which is) and the agency or praxis character of society (shaping that which may become) which is considered as an important entry point for discussing and indicating additional opportunities for a reconstruction of life-science technologies

---

[2] This manifesto is part of a continuing debate that takes place within CTC in collaboration with members of the Athena Institute.

[3] An international network was formed in 1999, called 'Tailor-Made BioTechnologies (TMBT) for endogenous developments', to stimulate an international information exchange concerning the experiences of six regional units to develop such a 'third approach'. In 2005 it was decided to reflect on these experiences through the establishment of a journal, Tailoring Biotechnologies (www.tailoringbiotechnologies.com).

2. The tension between what is and what may become

According to Feenberg (2005: 87) critical theory elaborates the notion that 'what is' is fraught with a tension between its empirical reality and its potentialities. Applying the description of 'what is' to the development of biotechnology and genomics, one sees an enormous growth of specific agro-industrial biotechnological applications – such as the increased cultivation of herbicide and pest-resistant crops (from 1.7 million hectares in 1996 to 81.0 hectares in 2004 – James, 2004: 20) – with millions of dollars and euros being spent on research to develop biotechnology and genomics. The other side of this empirical reality, however, is that despite these efforts of applying bioscience and biotechnology to industrialised agricultural practices the number of hungry people in the world remains shamefully and shockingly high. Estimates refer to a total of 800 million undernourished people globally, and of one child dying every 15 seconds due to undernourishment – while at the same time there is a sufficient total food supply in the world and a continuing scientific and technological 'progress'. Characteristic of the actual empirical reality (i.e. of what is) is this tension between the enormous amounts of research efforts and money spent on developing new technologies and the failure of these new technologies to meet the demand for food and shelter from people 'beyond the market'. The following questions, therefore, are crucial:

1. Why has the technological rationality – as it has developed – not been able to reduce social inequality? Or: Why have the affluent technical means been so marginally connected with addressing the large societal issues of sustainability and equality, and in general with respecting life?
2. Why, moreover, has the developing technological rationality become a major factor threatening the continuity of life on 'Mother Earth'?

Referring to the second question, in the face of global warming it has become clear that the technologies we use today cannot secure our lives, due to the massive pollution produced by modern industry in its present form[4]; and it cannot be denied that a monstrous indifference to life prevailed throughout the 20th century in wars, genocides, tyrannies and exploitation (Feenberg, 2005: 106) rendered possible through the development of all kinds of technologies.

The tension inherent in the actual empirical reality of the development of generic technologies, such as biotechnology, is that these generic technologies are designed according to a path of development that threatens human survival, while potentialities yet exist for these very technologies to also affirm life. This tension between a technological rationality that inclines towards an instrumentalisation and a denial of the complexity of life, and one which inclines to the affirmation of life as fundamental, this tension between so-called *life-denying and life-affirming technologies* is especially striking when it concerns the development of technologies dealing directly with bios, with life and living things themselves[5]. This article thus enquires into opportunities to develop life-science technologies – no longer within the societal context of a developmental path that threatens human survival, but, on the contrary, –within a societal context in which technology development is explicitly reconciled with the value of life.

---

[4] A development now compounded by the economic development of India and China, which is causing a rapidly rising world demand for oil and other raw materials (Feenberg, 2005: 107-108).

[5] Although in livelihoodstudies one also speaks about livelihood-threatening or -denying strategies and impacts, in this article I use the concepts of life-denying and life-affirming technologies from another point of view – referring neither to impactstudies nor to personal choice in research objectives, but to two different forms of rationality from which technology may evolve, or, as Habermas has put it, to two different knowledge-directing interests, differentiating between an instrumental and communicative reason: an instrumental and emancipator technological rationality (cited by Hoy and McCarthy, 1994: 179-180).

By considering four historical trends in the development of life-science technologies, the possibility for reconciling life-science technologies with the value of life through living labour will be discussed (Hardt and Negri, 2004: 146)[6]. These historical trends put technology development under pressure and, through the internal contradictions of this development, also offer perspectives on reconstructing technology. This will be illustrated by referring to some actual practices that are currently effecting a reconstruction of agro-industrial biotechnological developments. In order to present a conceptual discussion, however, I first revisit the history of the critical theory. The question of why mankind has failed to reduce social inequality – despite the ever growing development of technical means – has been a core issue in the debates among critical theorists. Returning to these theoretical discussions, we start with a brief overview of some of the key elements of critical theory which may still be usefully applied to a macro-sociological analysis of the life-science technologies we have today.

3. Analytic perspectives deduced from critical theory: macro-sociology

Traditionally macro-sociology was defined as referring to a specific level of the object of study, differentiated, that is, from the meso- and micro-levels. In our vision, however, the macro-sociological approach is rather different, specifying not the object but the focus of study. It incorporates a specific way of looking at society, based on an explicit relation with critical social theory. A macro-sociological analysis can thus investigate issues related to micro-products, such as seeds and enzymes, as well as broader areas of technological development, such as genomics or nanotechnology, and the development of biotechnology in general. Characteristic of a macro-sociological analysis of micro-products, macro-techno-developments and local efforts to reconstruct technologies is its explicit relation to critical theory, which also implies a specific research methodology. We will first lay the ground by outlining four basic characteristics of the critical theoretical frame of reference to be used in these macro-sociological analyses.

3.1. The dialectical-critical approach

A core aspect of a macro-sociological analysis is that it follows a *dialectical approach*. This implies that an empirical sociological analysis has to be related to and based on a critical reflection of, among others, the various contributions of the founding fathers of the dialectic approach, such as Hegel, Feuerbach, Fichte and Marx, and members of the Frankfurt School, along with more recent philosophical currents. In general terms, dialectics can be referred to as 'a means of reaching truth through reason, built around internal contradictions' (MacLennan, 1998)[7]. As well as the proposed effort to reconnect empirical sociological analysis to philosophical thought. A dialectical approach concretely signifies that actual (bio)technological developments are analysed from *their internal contradictions* and particularly from the tension between the used and unused capacities for improving the human condition (Marcuse, 1964: xliii)[8]. A macro-sociological analysis of technology perceives its analytical object from within this

---

[6] Hardt and Negri refer to Marx's *Grundrisse* in using this concept to indicate the human ability to engage the world actively and create social life.

[7] Sometimes the definition is extended with the addition that 'the dialectic approach proceeds in a linear and automatic fashion'. However, this Hegelian approach to dialectics has been criticised by Bhaskar's dialectic in which Hegel's triad of identity, negativity and totality has been substituted by the quadruplicate of non-identity, negativity, totality and transformative agency. We are particularly interested to elaborate further this transformative agency in technology development, and strongly criticise linear historical interpretations of dialectics (see the discussion of critical-realism at MacLennan, 1998).

[8] The dialectical approach is, indeed, the epistemological basis of the critical theory, and, as defined by Marcuse, characterised by the aim of studying contemporary society in the light of its used and unused capacities for improving the human condition (Marcuse, 1964: xliii).

tension between the existing 'is' and that which is possible, i.e. that which can be realised from within the existing 'is'. Therefore, a second and strongly interrelated characteristic of the macro-sociological approach is that it is principally *critical*, in the sense that it constantly confronts existing society and its social relations with the inherent possibilities of emancipation. For example, it confronts society with those additional needs which society at that moment cannot (does not) satisfy. The point of criticism, however, is not voluntarily and arbitrarily formulated, but is based on a concrete relationship with the existing reality that is criticised (and particularly with what that reality can become, its potentiality).

The criticism is dialectic in the sense that the antithesis of the existing reality (its potentiality) is not an abstract construction but *is already present in the actual described reality*. The dialectical criticism – upon which critical theory and the macro-sociological approach are based – always *departs from* the existing, from the confrontation of the existing 'is' with its potential becoming. Methodologically, this implies that macro-sociological analysis is based on an empirical research of the existing, but does not limit itself to that: it does not aim to reproduce the existing, but, on the contrary, focuses on criticising the existing in order to show what may become. Adorno has stated it thus: 'Thinking about society is only possible when we purposely intend to criticise that society; otherwise social sciences will be reduced to be a pure ideology of the existing social relations' (Adorno, 1969).

3.2. The historical-empirical approach

In view of this dialectical and critical content, a macro-sociological analysis necessarily also contains a *historical characteristic*. A basic assumption of the dialectical theory is that the essence of Being is considered dynamically, in terms of change, as movement, or Becoming (Marcuse, 1964)[9]. This implies that being is understood not so much in its actual form, which is provisional, as in its becoming. A macro-sociological analysis takes actual society to be just one provisional phase in a historical developmental process, determined by its previous history and referring to its further developments. The actual society is perceived as a junction of historical development and potential future possibility, that which may evolve from the existing actuality. By applying such a historical approach, one can prevent what is empirically observed from being assumed as the only possible existing actuality[10].

In order to go beyond the existing and to relate the actual situation to its historical development (from the past and into the future), it is necessary to perceive the historical development from a coherent and specific frame of reference, applying a philosophy of history – a philosophy of history which explains that different societal forms or phases are more or less logical consequences deriving from each other, that history has a structuration, through which it gradually realises itself, realises its sense. Indeed, another aspect involved in applying a historical approach is that *historical development gets a sense*, a meaning. A historical approach does not only refer to continuities and discontinuities in historical developments, but also refers to the notion that history has an end; has a meaning which gradually

---

[9] One of the most central discussions in the long history of philosophy concerns this debate about being and becoming. It started with the debate between Heraclitus, who emphasised that becoming, movement, change is the essential characteristic of reality, and Parmenides, who emphasised contrarily that just being (*Sein*) is the essence of reality, and that becoming, change and movement are just appearances of that being (Plato *et al.*, 2006).

[10] It is precisely through this confrontation between 'what actual is' with 'what may become' that the actual factuality becomes criticised, instead of taken as a absolute. In the dominant practice of case-by-case studies carried out to research biotechnological innovations, however, such a confrontation, between what is and what may become, is often missing – which implies that despite, even through, criticism of the *casus*, the reality of that *casus* is made absolute, leading to either its wholesale acceptance or complete rejection, and thus ignoring the potentiality of its transformation.

reveals itself through historical developments. A crucial question, of course, is where this sense or meaning comes from. Idealism assumes that 'what is' exists in thinking, as a metastructure of the human mind (Kant's ontological idealism), or that 'what is' emerges from and returns into thinking (Hegel's historic idealism). Marx inspired by Feuerbach – adapted Hegel's philosophy of history and the idea that history can be made by men – subsequently reversed Hegel's historic idealism into a historical materialist approach. The main characteristic of this historical-materialist (historical-empirical) approach is that labour is seen as the crucial element in history, that history gets its sense through labour, labour perceived as the elaboration of nature to enable man to satisfy his needs and be liberated from the domain of necessity (Marcuse, 1964).

3.3. The praxis-oriented approach

For Marx and the Frankfurter School the sense or end (direction) of history is the gradually *emancipation of humankind*, which implies that man realises an increasing level of freedom in respect to the surrounding nature. Indeed, a crucial aspect of critical theory is this *praxis character of history*, which has man as the subject of history: historical developments are the result of human actions and the sense of history is the (increasing) emancipation of man, realised through action, as labour. Praxis has two specific aspects. First, it relates to labour – i.e. to those forms of human activities that aim, through manufacturing nature, to satisfy primary human needs through the development of production forces. Second, as Marx emphasised, through labour humans establish specific social relations with each other. From labour, that is, specific societal relations emerge. In the analysis of his contemporaneous society, Marx stressed that the development of production forces and that of production relations enters into a dialectic tension. Marx posited the idea that the subjective process of elaborating nature (to satisfy needs) becomes reified, historically objectified, taking a specific historical form and becoming a power into itself that controls men and hinders further development of the production forces. It is *within this contradictory development* between production forces and social relations that Marx and the Frankfurters search for moments of emancipation (below, Section 5). By analysing the material circumstances, Marx intends to denounce the alienating character of history and show how the lack of historical consciousness reduces subjective opportunities for change. Indeed the category *consciousness* is another crucial element of critical theory and the Frankfurters repeatedly emphasise that whenever the historical material circumstances seem to reduce the opportunities for consciousness this should be perceived as the moment for changing these circumstances by strengthening the contradictions within these circumstances.

3.4. The totality-oriented approach

Another important aspect of dialectical theory is that contradictions are seen as a totality. Within a dialectical approach, the two opposites – such as production forces and production relations – determine as well as negate each other. The dialectical opposites cannot exist apart from each other, and therefore no solution to a dialectical contradiction can be found within the totality that they form together nor created through a kind of compromise (adapting one or both the sides toward the other, to some common, central ground). It is necessary to change the quality of the totality. Indeed, another basic assumption of the critical theory is that history develops and determines itself through its contradictions, and that at a certain historical level *quantitative changes transform into a qualitative change* – a discontinuity which has sense because it offers mankind the opportunity to liberate itself from coercive forces and attain a higher level of development and freedom.

The totality concept also implies that social phenomena cannot be studied as isolated facts but are perceived as part of a totality. The dialectic character of the phenomena stands in the fact that each phenomenon is an expression of a totality, but at the same time masquerades that totality and manifests itself as an autonomous event. Indeed, the Frankfurters reproach other social scientists and particularly empirical sociologists for focusing only on empirical facts as isolated phenomena. The macro-sociological approach is characterised by its effort to define social phenomena in relation to the social totality (which is primarily praxis).

Adorno summarises it thus: '...isolated facts are only apparently facts; breaking through their isolation gives their factuality a real significance'. In his work *Minima Moralis*, Adorno (2005) shows concretely how it is possible to interpretate isolated facts as expressions of social totality.

3.5. A synthesis: the dialectics of emancipation

Society seen as historical praxis requires a *specific research methodology*. Research has to focus on essential contradictions within 'the existing is' by confronting this with its potential becoming. Indeed, the above (Sections 3.1-3.4) may be summarised by stating that a macro-sociological approach is both shaped by and shapes *the dialectics of emancipation*. It views society as historical praxis which means that man is considered as being increasingly able to create the good life through a progressive development of factors of production – constrained, of course, by contradictions present in the existing reality. A macro-sociological approach will constantly confront existing technical developments and social relations with their own potential to emancipate, and, moreover, plays a role in the realisation of this by investigating and supporting the social praxis[11]. It orientates itself to describing those mechanisms which repress the realisation of emancipation within the actual society as well as to the opportunities for realising that emancipation by formulating concrete programs for a social praxis to strengthen specific opportunities. Therefore, the key study object of a macro-sociological approach is the tension between social control and freedom. It aims to clarify the ways in which efforts to realise emancipation, which technically is possible, are socially constrained. Actually, a core constraint in realising socially that which is technically possible concerns the lack of historical consciousness of the potentialities of emancipation through technology developments. In actual society, therefore, emancipation primarily means a reduction of the beliefs in the so-called objective laws of history, including the belief that social changes may be the unconscious results of human actions, as illustrated in the actual discussion about the so-called 'unplanned side-effects' of technical developments on society (see also page 42).

This lack of *consciousness* about the potentialities of emancipation through technological developments has been analysed by Marcuse. Marcuse stressed that the technical development of the advanced industrial societies is producing a qualitatively different social structure, through which it becomes possible to contain antagonistic social forces. He argues that in the medium of technology culture, politics and economy merge into an omnipresent system which swallows up or repulses all alternatives (Marcuse, 1964: xlviii). However, it is not correct to read Marcuse's analysis of advanced industrial society exclusively as a description of the emergence of a new monolithic power system, able to absorb all opposition and to control thought and action indefinitely (Kellner, 1984). On the contrary, Marcuse's analysis has even contributed to the social turmoil continuing from the early 1970's to the

---

[11] This can be considered as the key difference between a critical theoretical and an empirical sociological approach. The critical theoretical approach aims to contribute, to 'go beyond the actual factuality', by uncovering the societal potentialities for emancipation.

present, strengthening the crisis-tendencies and contradictions of the advanced industrial societies. Indeed, at the end of his book Marcuse emphasises that 'the unification of opposites in the medium of technological rationality must be, in all its reality, an illusory unification' (Marcuse 1964: xxxiii). In the following sections of this paper, I will describe some historical trends which put technology development under pressure, and offer opportunities for various social-technical networks to work for its transformation (Section 4), and then enumerate a few concrete efforts by different social-technical networks to use these historical trends as entry points for reconstructing technologies (Section 5).

4. Historical trends in technological developments

Critical analysis of technology strives to identify those trends within technological development that put that development itself under pressure and offer opportunities for it to generate its own transformation. In this section, I describe four historical trends within technological developments that may offer social-technical networks opportunities to re-appropriate, modify and democratise (bio)technological developments. These trends are described in general terms and illustrated with concrete examples taken from analyses carried out in biotechnology.

4.1. Lack of consciousness due to an instrumentalist approach

An important constraint to the social realisation of what is technically possible concerns the *lack of consciousness* of the potential use of (bio)technology as a force of emancipation, i.e. for the real possibility for (bio)technology to be used to improve the human condition. This may seem a rather strange statement in view of the many programs that exist to relieve poverty, for example, through attuned biotechnological developments[12]. However, most of these programs do ignore the presence of a specific knowledge-directed interest of technology development (Habermas, 1973). Indeed, there are no doubts about the instrumental rationality of technology, and no discussion takes place about whether the instrumental reason should be substituted by an emancipator-directed interest. In this sense I refer to a general lack of consciousness of the potential use of biotechnology as a force of emancipation, a lack based primarily on the widespread belief of *technology as a value-free, neutral instrument* at the disposal and applicable for the benefit of human mankind. In this vision, technology is seen as something without meaning in itself; as something from which its own objective content has been dissolved; which has become just a tool, a tool of which the power to use it is in the hands of the users. This vision of technology as a value-free, neutral tool is largely shared by supporters as well as opponents of agro-industrial biotechnological developments. One may even suggest that an important social impact of the pro/anti biotech debate is the anchoring of this vision in society at large. It is

---

[12] I wish to thank Wietse Vroom of CTC who emphasised in his comments on the draft of this paper that there is no general lack of consciousness regarding the positive potential of biotechnology – on the contrary there are many efforts to link biotechnological development to, for example, the Millennium Development Goals (MDG) – e.g. the UN Millennium Project Task Force working toward the first MDG and to eradicate extreme hunger and poverty, states that 'Genetically superior crops, vegetables, trees, and animals can greatly increase the productivity of small farms. The task force supports... transgenic research...' (Sanchez *et al.*, 2005: 14). Rather, as Wietse noted, there are differences of opinion as to how one can, and should, attune biotechnology to these objectives – e.g. see the UN-sponsored paper on concerns related to large companies dominating biotechnology research in respect of the failure of poor farmers to benefit from it (Pray and Naseem, 2003). Indeed, the current historical material circumstances clearly do illustrate that there is no lack of consciousness as such. In my response to these comments, however, I emphasise that this discussion of the use of biotechnology for specific societal objectives takes place within the context of an instrumental vision of technology and that I prefer to downplay this discussion, focused on applying the instrumental rationality in a socially positive way, and try to insist, instead, on the relevancy of reflecting on the possibility of applying the emancipator-directed interest in technology development, as indicated in this section (4.1) (see also footnote 33 for a concrete example related to this issue).

exactly this vision of an instrumental, value-free technology to be used as a tool for the good (or evil) of humankind that forms the first constraint for a reconstruction of technology – particularly because the anchoring of this vision has also been *historically* shaped, enabling this vision to become part of the cultural horizon of society[13].

In his 1947 work *'Eclipse of Reason'*, Horkheimer emphasised that in relation to the emergence of industrial society and culture *a qualitative shift* has taken place in the balance between *objective and subjective reason*, which influences the way technology is used. Indeed, technology does not only contain those various instruments that form the whole of a technical apparatus, but also *a specific rationality*, a specific theory of Reason, that determines its functioning (Galimberti, 2003: 34). For a long time in Western thought, according to Horkheimer, it was the *objective theory of reason* that prevailed. A long line of philosophical systems were constructed and different doctrines formulated following a tradition in which Reason was considered as a force present in the objective world – in nature, in social relations, in institutions – and, we add here, in technologies[14]. It was a force living in each man and enabling each man to realise Reason in all his practical and theoretical activities, appreciating, however, that a distance still remained between the actual practice (what is) and the realisation of Reason in that practice[15]. In this Western tradition, the idea of Reason expresses the antagonistic structure of reality and thought, in which thinking aims to understand reality. It also signifies that the world of immediate experience – the world in which we find ourselves living – must be comprehended, transformed, and even subverted in order to become that which Really is[16]. Reason was thus perceived as the subversive power, the power of the negative, establishing the conditions in which men and things become what they Really are (Marcuse, 1964: 123). The aim of Reason was to unveil an all-embracing or essential structure (of Being in Reality) from which a conception of human destination could be derived, thereby shaping our relations with other human beings and nature. Leading principles of man's activities, such as justice, self-interest, equality, happiness and tolerance, were supposed to be inherent in or sanctioned by Reason. Indeed, Reason was conceived as an instrument for understanding, determining and/or *revealing the ends* inherent in reality (Horkheimer, 1947: 7).

Through various historical developments, Reason became dissolved from its objective (but historically shaped) content and gradually acquired a new connotation as the search for reason in all practical and theoretical activities was weakened and reason became exclusively related to the subjective faculty of

---

[13] The cultural horizon of technology development in a society refers to the general accepted assumptions about technology development which are present in that society. It refers to visions that are broadly shared (see Feenberg, 1999).

[14] Horkheimer refers to philosophical systems as founded on an objective theory of reason ranging from those of Plato and Aristotle through to German Idealism (Horkheimer, 1947: 4).

[15] Thus the term 'objective reason' refers to an essence in reality, accessible to anyone who takes upon himself the effort of dialectic thinking, differentiating between what is and what may become (Horkheimer, 1947: 8).

[16] Marcuse summarises this way of thinking with two short formulations: 'that which is, cannot be true' and 'what is real, is rational' (Marcuse, 1964: 123) – indicating that behind the 'factual is' (which is not real), the 'real is' needs to be found, a vision which also urges the development of a research methodology opposed to the prevailing empirical sociological research focus on 'facts as they are'. The critical theoretical analysis refuses to accept the given universe of facts as the final context of validation, and aims to develop a 'transcending analysis of facts in the light of their arrested and denied possibilities which must be definable goals of practice in the realm of the respective society' (Marcuse, 1964: xliii).

the individual mind[17]. This *subjective reason* – originally regarded as one of the many universalistic principles rooted in the objective structure of the universal rationality – gained the upper-hand during the industrial age (Horkheimer, 1947: 14). It attached little importance to the question of revealing ends in reality, and if concerned with ends at all, took for granted that the ends were reasonable in the subjective sense, i.e. that they serve the subject's interests in relation to self-preservation, be it that of the individual or of the community (Horkheimer, 1947: 3). In the subjectivist vision of reason, thinking can no longer be of help in determining the desirability of any goal in itself. Goals such as equality, justice, and property are supposed to be matters of choice, made to depend upon factors other than reason. While objective reason was conceived as playing a leading role in regulating our preferences and in shaping our relations with other human beings and nature, the subjective idea of reason relegates it to a subordinated position of *adjusting means to conform with reality as it is*[18]. The operational value of reason becomes its sole criterion, and the leading principles of activities aimed at goals that were originally sanctioned by reason – equality, justice etc – now lose their intellectual roots. The goals are still perceived as ends but there is now no rational agency authorised to appraise and link them to an objective reality (Horkheimer, 1947: 16). They still may enjoy a certain prestige, embedded in national and international law for example, but they lack any confirmation by reason in its modern sense (Horkheimer, 1947: 16).

It is the disconnection of Reason from its objective content and subsequent reduction to an instrumentalist and subjectivist rationality which makes this emptied rationality vulnerable to social processes. Sharing Horkheimer's historical analysis of the transformation of objective Reason into a subjective or instrumental reason, Marcuse adds that the operational value of instrumental reason in the historic actuality of advanced industrialised societies reveals a rationality of domination and social control. Indeed, Marcuse's '*One Dimensional Man*' can be read as an analysis of how the historical trend towards a dominancy of instrumental reason becomes a specific technology development, which repress other historical alternatives and projects, *enabling technology to become directly political*[19].

4.2. Technological rationality as political rationality

After the lack of consciousness of the potential use of (bio)technology as a force of emancipation due to an instrumentalist approach, the second constraint to the social realisation of what is technically possible concerns exactly this *instrumentalist technological development – revealed to be a development of social control*. Marcuse emphasised that the technological rationality of the advanced industrialised

---

[17] The search for objective reason was weakened due, among other reasons, to the divorce of reason from religion, which undermined its grounding in universalistic principles. Indeed, Horkheimer emphasised that Catholicism and European rationalist philosophy were in complete agreement regarding the existence of an objective reality. Philosophers did not intend to abolish the belief in an objective truth, and were attempting only to give it a rational foundation. Indeed the assumption that such a reality existed from which right modes of behavior and activities could be derived was the common ground on which their conflict took place. The increasing neutralisation of religion, however, contradicting its claim to incorporate the objective truth, paved the way for an ultimate abolition of the concept of such an objectivity altogether, also strengthened by some specific pragmatic currents in philosophy (Horkheimer, 1947: 12).

[18] Horkheimer also emphasises that through the instrumental aspect of subjective reason of adjusting ideas to reality as it is, concepts have been reduced to summaries of characteristics that several specimens have in common (Horkheimer, 1947: 15). Concepts are seen as mere abbreviations of the items to which they refer; no longer thoughts with a meaning of their own. They are considered things, machines (Horkheimer, 1947: 15).

[19] Marcuse also introduces the concept of *project*, referring to the element of freedom and responsibility in historically (relatively) determined situations, a concept that links autonomy and contingency (Marcuse, 1964: xlviii), and may therefore be helpful in debate about a reconstruction of technology.

societies has become a *political rationality*; in the sense that technology becomes the medium which unifies and disempowers the antagonistic forces and tendencies of the advanced industrialised societies for the defense of that society[20]. He stressed that the containment of social change – i.e. of qualitative change which would establish essentially new modes of human existence – is perhaps the most singular achievement of advanced industrial society (Marcuse, 1964: xliv).

Due to this specific political content rendering oppositional forces impotent as it swallows up as well as repels, the traditional vision of a neutral technology can no longer be maintained. On the contrary, Marcuse emphasised that the instrumental technological rationality is primarily political, and that this shift has been anchored to and strengthened by various historical developments. He refers, among others, to the overpowering productivity of the technological apparatus that enables the advanced industrial societies to reduce the social space of oppositional forces and even to modify their antagonism to a positive force for the maintenance of the advanced industrial society. Marcuse stressed that the political content of the industrial technical rationality concerns the specific way in which the advanced industrial society *organises the lives* of its members – determining their needs and aspirations as well as the content of the socially required skills, attitudes and occupations – which makes the productive apparatus a totalitarian system[21].

A system of domination is emerging that also reduces the social space for antagonistic thought and behavior, giving advanced industrial society its one-dimensional character. The containment and paralysing of antagonistic thought is also related to the transformation of objective reason into subjective reason. Especially, Marcuse refers to the reduction of Reason as the subversive (negative and critical) power of reality. Just like Horkheimer, Marcuse also suggests that within industrial society and culture the concept of Reason has been changing. He argues that Reason was originally perceived as a way of thinking to indicate the conditions in which men and things *become* what they Really are – beyond the appearance of what they actually are. Reason was based upon an elaboration of this antagonism between appearance and reality, between actuality and potentiality. This concept of Reason has increasingly been substituted by another version of reason in which concepts are formulated to describe *reality as it is*, assuming that no other truth exists beyond actual reality. Indeed, Marcuse emphasises that the one-dimensional society is shaped by this specific way of thinking in which transcending elements are neglected and thought is reduced to describing, in a sense copying, reality as it is. He stresses that if the experience of the world – antagonistic in itself – no longer guides the development of thought, and the categories of social science only identify what is real, what actually is – whenever no distinction is made between that which appears to be (the appearance) and that which Really is (beyond appearance), when social science focuses exclusively on reality as it is – then the technological rationality will become increasingly separated from the ends of realising Real Being, giving the instrumental vision of reason its absolute technological and status-quo confirming position.

The dominance of instrumental reason enables technology to become a means to conform to reality as it is, and that within technological developments reference can no longer be made to other values or ends

---

[20] In the introduction of 'One Dimensional Man', Marcuse describes how technology literally becomes 'the medium through which culture, politics and the economy merge into an omnipresent system which swallows up or repulses alternatives' (Marcuse, 1964: xlviii).

[21] It is important to emphasise that Marcuse has not only described the emergence of a new (one-dimensional) social order able to control thought and action, but also referred to new contradictions within changed social order and perspectives of multidimensional thinking and behavior, referring especially to technological developments

than the instrumental efficiency. At the same time, the dominant societal application of instrumental reason implies that values that can no longer be derived from scientific reasoning, such as justice and equality, become mere ideals. Values and ends – being separated from an objective theory of Reason and no longer related to a Reality that exists beyond what actually is – become increasingly elevated above and beyond reality (Marcuse, 1964: 147-148); they are placed in an ethical or metaphysical atmosphere from where these values may 'look down' on technical practices, which are developed separately from these values[22]. These values are considered to be mere matters of choice, having no relation to an objective vision of Reason and a (historically shaped) reality of which only a part of its potential has been realised[23].

The emergence of technological rationality as political rationality is not only related to the reduced way of thinking about what actually is, excluding, that is, a critical reflection on what 'the existing is' may become. It is also based on the *reduction of the social space for critical behavior*. Marcuse emphasised that the roots of individual protests are increasingly being whittled down and even reformulated as forces to expand the instrumental technological rationality (see Section 4.4). It is this continuous expansion of the instrumental technological rationality that leads to another technological development, known as the reversal of means and ends.

4.3. Technology as an abstractum: the reversal of means and ends

The historical trend towards an instrumental development of technology may expand quantitatively so much that the technology itself may undergo a qualitative change. The quantitative expansion of the instrumental technological development now manifests itself in the affluent and ever-increasing presence of technical artifacts in all aspects of daily life. Our conservations involve telephones and computers, our traveling cars, trains and airplanes, our food refrigerators, cooking-rangers and microwaves, our free time television, video and audio-apparatus, and our being born, getting ill and dying all kinds of medical artifacts (Verbeek, 2000). People have become part of a technological culture in which objectives and ideals, means and ends are also *technically defined*. Worldwide, a shift is taking place, and has taken place, from a society with technology as still a specific means to realise a specific end, the meaning of a technology determined by the end for which it is used, towards one in which the expansion of the instrumental means determines the ends to be realised. Technology as a means is thus reversing, as it were, the means-to-ends relationship and becoming an end in itself – and not because technology presents itself as an end, but, on the contrary, because 'man' only searches for those ends which can be realised technically.

In his historical-materialist approach, Marx referred to the possibility of this reversal of means into ends when he spoke about automation (see Marcuse, 1964: 36) and the new period of technology which he described as 'the civilisation of the machine' (Galimberti, 2003: 42). He also described the reversing of means into ends concretely when he discussed the functioning of money. As a means, money serves men in the production of goods to satisfy needs, but whenever all goods and needs are mediated by money then the moment arrives when money itself becomes the objective. Producing goods may then

---

[22] The dominance of this approach manifests itself in the establishment of various ethical committees that, separated from the ongoing research, have to approve or reject biotechnological research projects and field trials.

[23] The integration of specific social objectives in the design of technologies seems to be just a matter of choice; there is no effort made to relate these choices to a social theory, and it seems to be that one can no longer identify criteria on which these efforts may be evaluated. The only criteria that counts is that which concerns the technical efficiency of the tool.

serve the objective of acquiring money and the satisfaction of needs may even be postponed to save money. Money is no longer the intermediate instrument that brings products and needs together, but has become the object in itself (Galimberti, 2003: 37). The same is occurring now with the quantitative growth of the instrumental rationality.

Due to the expansion of instrumental rationality, technology is no longer the intermediate instrument for realising specific objectives, but has become the object in itself from which other specific objectives are subsequently deduced. Indeed, a continuous transformation process of technology takes place. Technology as an object in itself becomes exclusively re-connected to its own technical organisation (Barcellona, 2007: 13). It is no longer society which leads the use of the instrument to realise specific objectives; it is no longer the social organisation of technology which determines its development. This reversal becomes circular as technology becomes increasingly influenced by the way the development of technology is technically organised. *The technical organisation* has become the decisive aspect that determines the development of technology and which other (additional) objectives will be pursued. This also implies that technology becomes an *abstractum*, an object for the development of itself, realised from within. The abstractum is the development of a specific technology, its growth and expansion: For example, it will be the expansion of genomics compared to, say, a reduced growth of biotechnology that will decide which specific societal problems will be tackled through genomics instead of through biotechnology. No longer will values or social ends, but the comparative lead of one of the technological fields determine which social objectives will be tackled first. The reversal of technology from means into ends also changes the relation between people and technology development.

In the society in which technology is still a specific means to realise a specific end, man can be considered as the subject that decides how technology as a means is to be handled. Through the expansion of the technological means, however, this relation is reversing. The abstract development of technology – disconnected from values and ends and therefore applicable to any and all ends whenever they can be realised, technically – is no longer a domain separated from human beings. On the contrary, man has become an integral part of the technological apparatus. Indeed people's lives are characterised, both directly and indirectly, by enabling and furthering the functioning of this apparatus, into which everybody delivers a specific input, changing life as an end-in-itself into life as a means for the maintenance and development of the new abstractum[24].

4.4. The technologisation of life

The historical trends outlined above come together in the actual development of the life sciences, or, as they are in actuality, the life-science technologies, including biotechnology, genomics, nanotechnology and synthetic biology. These life-science technologies represent and strengthen the historical trends towards:
1. An increased dominance of the instrumental, value-neutral vision of technology (Section 4.1).
2. The politicising functioning of the (dominant instrumental) technological rationality (Section 4.2).

---

[24] Galimberti (2003: 599) speaks about the ontology of the technological apparatus in which he refers to Heidegger stating that only what can be qualified as being part of the apparatus can be considered as representing Being (*Seiende*). And because the apparatus looks at the world exclusively in terms of raw materials to be used, all that can be used becomes real (being) compared with that which cannot be used, which becomes non-being, thus transforming being into means.

3. A reversal in technology as a means into an object in itself (Section 4.3), enabling the technology to evolve into an 'abstract technology development' and become a key subject for historical transformations.

These trends all materialise in the design of new life-science technologies, resulting in the emergence of a fourth trend that can be defined as an increasing technologisation of life.

The emergent historical trend towards a technologisation of life refers to a *new system of domination*, which is based on an interiorisation of power within the inner space of each individual. Marcuse emphasised that advanced industrial and technological society changes the way in which the lives of the individuals are controlled. He stressed that the technological society is organised in such a way that specific needs can be continuously satisfied, and that the plethora of images produced by the ubiquitous media industry lead to a new situation in which the inner space of individual life, originally separated from external influences, is gradually vanishing. This implies that the control of the individual by society no longer takes place through a process of internalisation, 'introjection', in which the individual adjusts him/herself to external influences, because that inner space for self regulation is disappearing; the antagonism between inner and outer space fades away. Therefore, the control of the technological society on individual life takes place through a process of 'mimesis', a direct identification of individual and societal needs (Marcuse, 1964: 10). This shift implies that individual psychology is changing: with a reduced, and reducing, inner space of the individual, the psychology of the individual has become part of the psychology of, and for, technical development. There is an increasing technologisation of individual life, a *technologisation of the psyche*.

The technologisation of the psyche – one of the most important transitions of (and expressions of the dialectic inherent in) the contemporary phase of historical material development – reduces the social space for critical behavior and dispels the opportunity for individual protest at its roots. Marcuse insists that individual protests are not only repelled, and with the increasing efficiency afforded by the technical apparatus, but also reformulated in such a way that the protests themselves become forces to expand the instrumental technological rationality. Indeed the defense of the instrumental technological rationality is increasingly based on the assumption, an interiorisation, of the ideology of the instrumental technological rationality (as the only rationality that may bring happiness in life, excluding all other kinds of historical projects)[25].

The technologisation of life also refers to a *changing character of praxis*. We have considered the trend whereby technology is reversed, moving from a means for the realisation of specific objectives, defined by people, towards an abstract object in itself for realising ends which are determined by the actual state and potential development of technologies. This reversal of technology from a means into an end also changes qualitatively the relation between people and technology development. In the societal context in which people still decide which technologies are used for which ends, one can argue that people are the *subjects* of technological development and that *their actions are goal orientated* (Galimberti, 2003: 685). In the context of the advanced technological society, however, it is the expansion of technological means as a whole – the actual state of and potential for technological development – that decides which social objectives will be pursued through the realisation of which technologies (and when and where and

---

[25] Clearly this development urges us to reflect on the meaning of participatory developments. It will become necessary to investigate ways in which a participatory design process of technology is able to disconnect itself from the dominance of the instrumental vision of technology.

how, and even why). In this context, the continuously expanding technological development becomes the key player in historical development, *reducing people's activities to a functional dealing* with the constant adaptation of the instrumental, technological rationality (and its concomitant maintenance, administration, etc.).

Indeed, this shift in the character of praxis is already announced by the ever-growing debate on the 'unplanned side-effects' of technology development. In general this debate is focused on the issue of whether and how man can reduce the unplanned side-effects of new technological developments, accepting that a complete control of these side-effects and their risks is not possible. What this debate does not do is to raise the more basic question of whether the unplanned side effects and risks indicate that man no longer plans technology development, but rather has to adapt himself to that technology development, which can no longer fully be planned or controlled.

The historical trend towards a technologisation of life is not only strengthened by the quantitative shift from goal-oriented to functional acts for maintaining, developing and adapting to technologies. It is also reinforced by an orientation of life towards the *potentialities* of (life-science) technologies, which again transform life into a means for technology development. This trend becomes understandable if one takes into consideration Marx's concept of labour power, in which a crucial difference is made between *labour power* and *effective labour*, between the generic undetermined potential of any kind of labour and the concrete execution of specific acts, a concept that can also be applied to the difference between potential and actual technological development.

To understand the relevancy of these notions, it is important to stress that the purchaser of labour power buys the *generic capacity for producing as such*, in other words buys, as Marx has described it, 'the aggregate of those mental and physical capabilities existing in the physical form, the living personality, of a human being' (Marx, Capital Volume 1: 270, cited by Virno, 2004: 81). This implies that the core of the exchange between the buyer and the seller of labour power is exactly *this potential*, this something *unreal and non-present*, which is, however, bought and sold as any other commodity. The crucial aspect of the labour-power concept is precisely the fact that it deals with the non-present, the non-real, and deals with it, moreover, as reality. The same accounts for technology development, which also refers to the non-present and non-real, to what can be imagined in terms of fields of applications, and again, dealing with this as though it were already real[26].

The trend towards a technologisation of life is based on using *life as the centre of potentiality* – for the further development, that is, of the instrumental technological rationality. Life becomes no longer pure and simple bios, but acquires a specific importance insofar as it becomes this centre of potentiality, the centre of realising the unreal, treated as reality. However, just like labour power – in which the generic capacity for producing as such cannot be separated from the living person, selling its effective labour – so also is it not possible to separate the potential of life, as centre for realising the unreal, from the effective life itself[27]. The effective life itself is still the tangible sign of a yet unrealised potential to develop an as

---

[26] This happens particularly in the field of nanotechnology, in which the non-present, the invisible is rendered visible through the specific design of specific instruments, which realise the possibility of seeing at nanoscale. The social implications of this development are investigated by Martin Ruivenkamp at the University of Twente.

[27] As Marx writes, in respect of labour power: '[T]he use value which the worker has to offer to the capitalist, which he has to offer to others in general, is not materialised in a product, does not exist apart from him at all, thus exists not really, but only in potentiality, as his capacity (cited by Virno, 2004: 82)

yet untechnologised life, of developing, as Marcuse has emphasised, *life-affirming technologies*[28]. In the following section we will discuss how the four historical trends considered here offer new perspectives for life-affirming technologies, specifically, for the modifcation, re-appropriation and reconstruction of life-science technology developments.

5. Reconstructing life-science technologies

The reference to a critical theoretical framework (Section 3) implies that a debate about the potential for reconstructing life-science technologies is necessarily built upon a reflection on the *historical trends in technology development and its internal contradictions*, as entry-points for defining concrete activities that challenge and go beyond these trends. In the effort to transcend the apparent factuality of the actual reality, a critical research approach aimed at reconstructing technology may be focused on the following aspects:

1. It may go beyond the historical trend towards a dominance of the instrumental technology rationality and look for opportunities to challenge the lack of consciousness for using life-science technologies as a force for emancipation. This may be done by *reintegrating values and ends* in the material development of life-science technologies.
2. It may go beyond the historical trend towards the politicisation of technology and challenge the actual politicised character of the instrumental life-science rationality by a *re-codification* of that technology.
3. It may go beyond the historical trend towards an abstract technology development as an object in itself and to look for concrete opportunities *to localise (glocalise) the abstract life-science technology developments*.
4. It may go beyond the historical trend towards a technologisation of life and realise a shift away from technology as an end in itself, in which life becomes a means for technology development, towards *technologies in which the affirmation of life becomes the core object*, a shift summarised here as a move away from the instrumental reason and life-denying technologies towards life-affirming technologies and a *communicative reason*.

I have looked at a reintegration of value into the material development of life-science technologies elsewhere (Ruivenkamp, 2005), with a consideration of the possibilities of re-coupling ethics to technical potentialities. In the final sections of this article, therefore, I will limit myself to a discussion of three other aspects of a reconstructive strategy for life-science technologies.

5.1 The re-codification of life-science technologies

A characteristic of the critical-theoretical approach is that it looks for those (historically shaped) forces that may give direction, point the way forward for the 'going beyond' of the actual reality. As Marcuse explains, a cornerstone of the critical-theoretical approach is to look for the liberating historical forces which develop *within* the established society (Marcuse, 1964: 23). For a technological society this implies that the historical trends within technology development may also be perceived as announcing perspectives for the realisation of a qualitative shift in the way in which technology development is embedded in society. The critical-theoretical approach, and particularly its totality concept, emphasises

---

[28] I refer to a discussion by Paolo Virno emphasising the importance of Foucault's concept of biopolitics, in which Virno emphasises – referring to Marx potentiality issue – that although life lies at the center of biopolitics, the living body is still the tangible sign of a yet unrealised potential, of labour not yet objectified, as Marx has said of labour as subjectivity (Virno, 2004: 83).

that such a modification of the political rationality of the instrumental technology development presupposes a qualitative shift in the *technology structure itself* as well as in *its material content*. Moreover, critical theory is characterised by its effort to find those subjects, those protagonists, that will become the social carriers for such a transformation. To this end, therefore, we will refer to some concrete experiences of social-technical activists networks that aim to change the technical code or script of technology.

The technical code or script concept refers to Winner's (1985) argument that technology development takes place within societal contexts, and that these contexts inscribe a social code into the design of an artifact and in particular its unequal social order[29]. The design of technological development mirrors the social order, and technologies are developed under a *cultural horizon* which refers to the general cultural assumptions that form the unquestioned background to all technology (Feenberg, 1999).

It has been argued here that the instrumental, value-neutral vision of technology forms an important aspect of the unquestioned background of the actual development of life-science technologies (Section 4.1). It has also been emphasised that it is exactly this value-neutral and instrumental perception of technology that gives technology its political content (Section 4.2). This implies that a re-codification of technology starts with *the unraveling* of the political particularity of the actual technology.

5.1.1. The deconstruction of technologies

The first step towards a re-codification of technology is to analyse the kind of *social organisation of society* to which technology contributes. In this way the script concept is first used to un-do or deconstructs technology. A deconstruction of a technology is based on a critical analysis of those historical trends that are structuring the development of that particular technology. Importantly, attention needs to be paid to discussing the contradictions within these historical trends, which opens perspectives for change. Indeed, the critical theoretical approach reveals itself in the effort made to uncover the cultural horizon of the technology development, and to demystify the illusion of the actual technical necessities. Critical theory exposes the relativity of the prevailing technical choices. In other words, the deconstruction of the actual code is characterised by 'taking apart' those concepts which serve as the axioms for the development of the existing technology.

5.1.2. The reconstruction of technologies

The second step of a re-codification of technology is based upon the effort to use the technical code or script concept as a reconstructive tool, as an effort 'to rewrite' its content. This can be achieved through *strategic actions* focused on changing the social structure of technology, as well as through *tactical actions* orientated to re-writing the (im)material and politicised content of the technology.

The *strategic actions* focus their re-writing activities on challenging the institutionalised, unequal balance of power. They search for concrete opportunities to change the *social organisation* of the society from which the technology emerges. The actions challenge those social relations which are the constituent elements of the actual technology development, in order to create space for alternative technology projects. The strategic actions aim to identify contradictory developments among the actual

---

[29] Feenberg emphasises that technical codes define the object in strictly technical terms in accordance with the social meaning it has acquired. These codes are usually invisible because, like culture itself, they appear self-evident (Feenberg, 1999: 88).

technological developments so as to articulate the antagonism between the technical developments as they are and as they could be.

Tactical actions are particularly focused on changing the *material form* of a technology through a combination of resisting, subversive and redesigning activities. The tactical actions are characterised by efforts at enlarging the rules of the game: they attempt to involve other players in the game of innovation and to make room for maneuver, modifying the dominant technical code by introducing other, alternative social dimensions into the material form of the technical artifacts. The aim of these tactical actions is to create opportunities for various subjects to become (again) protagonists in technology development, and even to apply an orientation of what Beck *et al.* (1994) has called sub-politics in the design of new technologies. These general remarks about strategic and tactical actions – which are also closely interrelated praxis aspects – are further advanced by referring to some practical examples of the re-codification of agro-industrial biotechnology into tailor-made biotechnologies.

5.1.3. De- and reconstructing agro-industrial biotechnology

The *deconstruction* of the technical code of agro-industrial biotechnology is initiated with the uncovering of the *historical trends* that are structuring the development of the political particularity of that technology. *Appropriation* and *substitution* have been indicated as two crucial historical trends from within which the particular development of biotechnology has taken shape (Goodman *et al.*, 1987). Appropriation here refers to the gradual takeover of biological activities, from farming practices by external institutions, especially industry, while substitution refers to the gradual replacement of the agrarian origin of food sources by products derived through an industrial-biochemical methodology (Goodman *et al.*, 1987; cited by Pistorius and Van Wijk, 1997: 17). Appropriation and substitution are historical trends that shape the development of biotechnology, while also having been quantitatively strengthened by biotechnological developments in such a way as to lead to a contemporary and ongoing *qualitative transformation of the social organisation* of food production. It is crucial that the deconstruction of technology focus on this qualitative transformative force. In other analyses of agro-industrial biotechnological development, therefore, it has been emphasised that appropriation and substitution are no longer the main characteristics of the social organisation of global production – these have now been shifted to the trends for an increased *control at a distance* from the seed-supplying industries and an increased *interchangeability* of agrarian food sources as inputs to food components aggregating companies, supported by a third trend, that of an increasing *privatisation of the politicising products* of the reorganised food production system (Ruivenkamp, 1989).

These historically evolving trends take place against the background of *three separation processes* that are characteristic of the social relations between various actors within global food production chains – indeed, these three separation processes form the real core of the technical code of agro-industrial biotechnology (Ruivenkamp, 1989, 2005). By disconnecting agriculture from its ecological environment (generally, through the complete technologisation of the farming process, and specifically, through distancing farming as producer from both its markets and its suppliers); and by disconnecting agriculture from direct food production (primarily, through the widespread employment of farming as food component input provider for industrial production); and by disconnecting agricultural products even from their food-orientated applications (for example, the massive transition underway to land use for ethanol), the agro-industrial biotechnology reinforces very specific social relations and becomes representative of the unequal power relations in the socially reorganised food production system.

Indeed, the *technical code of the agro-industrial biotechnology is formed by a changing labour organisation of food production*, which is still, and in many aspects increasingly, characterised by inequality, not to say iniquity. It is precisely these social relations that are challenged by the various social-technical activist networks in regions worldwide aiming to re-write the agro-industrial biotechnology code and begin to develop the approach of tailor-made biotechnology.

Tailor-made biotechnology constitutes a *reconstructive* approach that begins with a variety of *strategic actions* aimed at changing the societal context from which the technology design emerges. No longer is reference made to the social organisation of global food chains with their characteristic separation processes. The societal contradictions within that trajectory are so immense that another frame of reference is sought, and another trajectory of technology design required. Rather, concrete efforts are made to develop *biotechnology from within other social relations* (other, that is, than those relations which maintain the three separation processes upon which the hegemonic position of the biopower system is based). Instead of referring directly to the separation processes – which characterise the social organisation of global food chains – efforts are made to establish new re-connections, between, for example, the production of seeds and the cultivation of crops at a local level. It has been argued elsewhere (Ruivenkamp, 1989, 2005) that the unequal power position of seed supplying industries in respect of farmers is based on a dual process of uncoupling the reproduction of seeds from the constraints of the natural environment and re-coupling it to specific scientific information. Strategic actions, therefore, challenge both these aspects. Farmers in different regions have acted communally to establish new social relations between themselves and with specific groups of researchers for the purposes of introducing scientific information that either offers them opportunities for a re-appropriation of the (re)production of seeds or for realising location-specific, sustainable developments. Concrete outcomes of these strategic actions include, for example, the development of early mature, dual purpose and disease-resistant varieties of sorghum in India, of a lysine- and tryptophan-enriched maize variety (Obatampa, literally 'good mother') in Ghana, and the Sol da Manha maize variety in Brazil (Ruivenkamp, 2005).

Concerning the reconstruction of social relations between agricultural produce and food production it is emphasised that strategic actions focus on the re-linking of agricultural produce to food production and consumption at local level. On the ground, new social relations are being established between food scientists and consumers, through which location-specific food products (e.g. lupine in Ecuador) are no longer seen as symbols of backwardness, but, on the contrary, as potential levers for local development. By recognising the symbolic and cultural significance of local food habits, scientists, farmers and consumers are able to work together in local food networks and introduce specific scientific information into enzymes and other biocatalysts to promote a regional enhancement of food sovereignty.

One might also refer here to the multi-diversity of *tactical actions* to reject, resist, attune, re-appropriate and modify the material content of the technology. Multiple forms of so-called micro-politics are emerging that are beginning to attempt a re-codification of location-specific biotechnological developments. These micro-politics will be considered in the next section, as illustrative of ways in which various networks challenge the historical trend of a life-science technology development towards an abstractum. However, I wish first to conclude this section by emphasising that it will be *the combination of strategic actions* – challenging the social structure of technology – *and multiple tactical actions*[30] – challenging the material content of technology – that will make possible the re-

---

[30] Feenberg also refers to these tactical actions as 'micropolitics' (1999: 104) – 'situational politics based on local knowledge and action'.

codification of biotechnology, transforming it to a supportive force for those social-technical networks which challenge the emerging biopower system. A critical theoretical approach may strive towards a strengthening of these combined praxis potentialities for re-codifying life-science technologies.

5.2. Localisation of life science technologies

Alongside the argument regarding the politicised character of the instrumental rationality of life-science technologies, the idea has been developed (Section 4.3) that the quantitative expansion of the instrumental means enables technology to function as an abstractum and to become a core subject of historical transformations. It is no longer exclusively society or people that determine for which specific ends which specific technical means will be developed. It is increasingly the presence of an ever-expanding range of technologies that will influence and even determine man's choices in selecting specific technologies for solving specific societal problems. This reversal of technology from a concrete means to realise a specific end towards an end in itself, able to solve all social problems under the condition that these problems are technically formulated, implies that technology development becomes disconnected from any local context – albeit that the abstract technology development inevitably re-conceptualises local developments within its own abstract technology trajectory.

It is the dual process of a politicised instrumental rationality of life-science technologies which at the same time functions as an abstractum determining ends that can be (and is being) challenged by various social-technical activist networks. Indeed, the *deconstruction* of the historical trend towards an 'abstract technical code' begins with the uncovering of the specific ways in which the abstract technological development re-conceptualises the social space of locality; how it re-shapes the local social relationships.

It has been argued elsewhere that the development of agro-industrial biotechnology rearranges existing local social spaces and produces new ones (Ruivenkamp, 1989, 2005). Concretely, we see that, through the development of new enzymes and microbiological production processes, the agro-industrial biotechnology creates increasing possibilities for the extraction and production of food components from various farming and biochemical raw materials at regional level. Each region (locality) may be able to aggregate its own desired combinations of food components from various global/local flows of food components. An increased autonomy of regional food supply may arise. This is dependent, however, on the use of globally developed biocatalysts: the re-conceptualisation of local space – currently conceived in terms of the production of food components – is directly linked to the re-production of an abstract and virtual space of scientific and technological developments. And it is exactly within the reformulated local space that new social relations and particularly new forms of political control emerge. Indeed, it has been argued that international competition between the food components aggregating companies may become increasingly based on using the advantages of the specifically regionally (and socially) organised forms of producing, converting and assembling food components. These regionally diverse forms of social cooperation may effectively be transformed into interchangeable production units (of the food components assembling companies). This implies that a qualitative shift is realised in which socially diverse, organised forms of production at local level (small- or large-scale, publicly or privately owned) become interchangeable local units within global networks that are able to de-empty the original political content of the socially diverse forms of production[31].

---

[31] This also indicates the need to reflect critically on what it actually means for increasingly participatory approaches to be followed within this new social context of an expanding competition between regionally organised forms of cooperation.

This development of, on the one hand, an apparently increasing autonomy of food component production at the regional level, and, on the other, an increasing dependency on scientific and technical research networks, also indicates that the identity of locality can no longer be defined in purely geographical terms, in relation to the local food produce, farming methods, etc. It will increasingly be formed in relation to these new global/local flows of enzymes, biocatalysts and other biotech products. However, the spatial organisation of the development of these 'politicising products' is already characterised by a very complicated and fluid organisation of research networks, in which research results and products flow continuously back and forth between multifarious public to private research institutions. And it is exactly the fluid, non-transparent – and particularly the abstract – nature of these research networks that is challenged by multiple tactical actions, spread over different regions and leading to efforts aimed at re-constructing technologies to the local context.

The *reconstructive* approach challenging abstract technological development is indeed based on efforts to re-link technology development to location-specific objectives. After unraveling the specific way in which the abstract technology development re-conceptualises local space, people organise themselves, purposefully, to re-conquer the locality (Magnaghi, 2005). Various forms of 'situational politics' of technology development are formulated, ranging from rejecting to re-appropriating and redesigning the material culture of technology.

These politics are sometimes the work of *individuals* who are directly affected by a particular technical decision and/or by the re-conceptualisation of the local space. Individuals, for example, may simply *refuse* new food products due to the incorporation of what to their perceptions is 'alien' scientific information. *Groups* of persons, moreover, can *resist* technical transformations because of their anticipated territorial impact, perceived as harmful to local culture and resources. It can also happen that individuals and groups of persons aim to *re-appropriate* some specific aspects of a technology. For example, user groups such as farmers may aim to cross modern, foreign varieties with their own local, traditional landraces in order to improve their established strains with some specific traits from the exogenous varieties. It may also happen that user groups and scientific researchers open 'innovative dialogues' that work towards a *redesign* of the material culture of a technology. Limiting myself in this article to a discussion of these redesigning efforts, I shall refer to recent 'innovative dialogues' in Ghana, Ecuador and Cuba[32].

In Ghana, the combined efforts of farmers and local researchers have resulted in the development of protein-enriched maize and cowpea varieties. One may seek to downplay the local, and radical, aspect of this, arguing that attention is paid within global biotechnological networks also to the development of similar nutritionally-enriched food crops – e.g. vitamin-A enriched ('golden') rice. However, I would stress that what appears to be the same kind of technical research – developing nutritive-enriched crops, in this case – may be socially dissimilar. Such a distinction tends to be important whenever research is strongly embedded in local social relations and local food habits. In Ghana the aim of developing protein-enriched maize varieties represents a location-driven research approach. This implies that additional research targets are constantly redefined in consultation with local social groups. The farmers in Ghana, for example, emphasised the importance of maintaining certain other (non-nutritional) traits in the protein-enriched maize crops, such as high yield and resistance to specific pests and diseases. Measures have also been taken to ensure that the farmers cultivating these protein-enriched maize varieties are able to sell their products at the local markets. Indeed, the enriched maize and cowpea varieties are

---

[32] In four documentaries on 'Reconstructing biotechnologies for development' (www.ctc.wur.nl/UK/Education) these redesigning efforts are shown.

bought by school centres and used in meal preparation for the school children (in local dishes, such as *kenkey* and *waakye*). A two-way process is thus evolving around the new crops: the nutritional and health status of school children improve, while farmers earn additional income selling their products to the school centres. In this way – due, that is, to the attuning of the technical plant breeding means to the potentialities of the local natural and social resources for development – the protein-enriched crops are functioning as dynamic forces for a small-scale but hugely significant social and economic upheaval, a shift, in fact, of the locality[33].

There is a similar story in Ecuador where groups of farmers and scientists are developing new lupine, quinoa and amarantha varieties, location-specific crops that had been perceived in the global re-conceptualisation of the Andes-area as symbols of backwardness. Small-scale farmers and scientists are now working together, however, to develop early maturing and high-yielding varieties of these crops, varieties that need less pesticides, have good nutritional qualities and are adapted to the harsh conditions of the Andes and the resource-poor conditions of small-scale farmers[34]. Through the attuning of the technical means to the location-specific characteristics, concrete improvements have been realised which have rendered it possible for these crops to function as *potential levers for local development*. Through small and multiple specific choices in the research activities farmers and scientists have been able to change these location specific crops from worthless resources and symbols of backwardness into catalysts for location-specific developments.

Finally, we can mention the efforts in Cuba to link biotechnological developments to small-scale urban agriculture. Alongside the development of high-tech biotechnology, which is strongly interwoven with large-scale agriculture in Cuba, an urban agricultural orientated biotechnology approach is also being developed. This approach is characterised by the effort to tailor traditional and modern forms of bio-techniques to the social objective of improving the biological food supply in urban quarters. It demands a decentralised model of science and technology development in view of the decentralised model of food production and consumption in urban quarters. The move towards attuning science and technology to a decentralised model of food production within urban quarters implies that both the social organisation of food production and of science and technology development are currently undergoing huge change (see Manzano, 2007). It is, indeed, through a gradual, accumulative process of small steps and generally low cost innovations that a technical and social re-functioning of technologies begins to take place.

---

[33] Al-Hassan and Jatoe suggest that Ghana's achievements here – since the late 1950s yet relative failure to import the Green Revolution model – is characterised by 'low levels of adoption and poor performance' of improved varieties due to 'a number of factors, including the physical environment, farmers' socio-economic conditions and the generally poor rural environment' – which attests to a historically unsuccessful program driven by a primarily top-down, non-location specific approach, and inviting the conclusion that 'the disabling nature of these local conditions must be recognised by policy makers for immediate redress' (Al-Hassan and Jatoe, 2002: 1,3). The more recent rectification of this approach – through redesign and re-codification – is exemplified by the work since the 1980s of the regional (north-west Ghana) project, the Nyankpala Agricultural Experiment Station/Savanna Agricultural Research Institute, with its emphasis on local farming needs in a social context. By raising production but without damaging the environment and with addressing local consumption preferences, this project does, in fact, begin to approach an emancipator-directed technology (see: http://www.csir.org.gh/index1.php?linkid=117&sublinkid=73).

[34] The objective of reduced pesticide use is especially pertinent in the light of the 2002 reported results from research conducted in the Ecuadorian Andes, in which high pesticide levels in the heavy highland soils there were found to be particularly harmful (nearly twenty times more so than had previously been estimated) (see: http://web.worldbank.org/WBSITE/EXTERNAL/COUNTRIES/LACEXT/0,,contentMDK: 20046695~menuPK: 258569~pagePK: 2865106~piPK: 2865128~theSitePK: 258554,00.html).

Viewing technologies as an ensemble of technical and social dimensions, it becomes manifest that a re-functioning of technology implies changes in its social and material structure. The re-codification involved in a re-functioning of abstract technology development – into location-specific technologies – implies a strong cooperation of different actors within each locality, focused on the discovery of opportunities for local, sustainable developments strengthened by an attuned use of location specific technical means. In contrast to global biotechnological developments in which new global products are invented by scientists in anonymous labs and sold worldwide, this location-driven biotechnological research is based on the cooperative efforts of farmers, scientists and other people of local communities, participating in local food networks, to enhance local developmental perspectives through an attuned use of technical means. It is importantly based on combining *different life experiences* in selecting local natural and social resources. Therefore, it becomes increasingly important to understand in the ways in which such a variety of life experiences may find room for manoeuvre in order to challenge the (fourth) historical trend of the technologisation of life (as outlined in Section 4.4).

5.3. Life-affirming technologies

The technologisation of life is closely interrelated with the other historical trends, such as the trend towards a dominant vision of technology as a value-free instrument which becomes directly political and strengthens an apparently autonomous technology development in which technology becomes increasingly abstract and disconnected from any location-specific context. These trends – which certainly should not be understood as linear developments, but on the contrary, as fluid, interconnected processes – are increasingly contrasted by the efforts of scientists, farmers and villagers to create social life and challenge the technologisation of life with the construction of life-affirming technologies, or 'technologies of the self' (Foucault, 2004: 113-147).

An important stage in the process of challenging the trend of a technologisation of life starts with an unravelling of the changing praxis character of society. To speak of the technologisation of life is to refer to the continuously expanding development of technologies, which enables a social transformation in which technology, not man, becomes the key player of historical development; it also refers to the trend in which people's activities are increasingly shifting from goal-orientated actions towards a functional dealing with the maintenance, adaptation and development of the instrumental rationality of technology (Galimberti) – a trend in which life is no longer pure bios, a domain separated from technology development, but in which life itself becomes an integrated part of the development of technology and of the governing of life through technology (Lazzarato, 2005).

This historical transformation already reveals itself in the changed social organisation of labour (Hardt and Negri, 2004), in which there is no longer a clean, well-defined threshold, separating labour-time from non-labour-time (Virno, 2004: 103). Pertinently, this occurs in the domain of science, especially scientific research. For the scientific worker his/her work is his/her life, and his/her life his/her work. There tends to be a personal involvement on the part of scientific workers in their research activities, implying that we can talk about an expansion and intensification of working hours for the producers of these politcising products. There is also a change in the relationship between life- and labour-time of the individual researcher in the sense that the personal life-style of the labourer becomes the basis for his/her work. For the scientific labourer, all aspects of his/her individual human life become a potential to be used in the production of knowledge-intensive (immaterial) products (Hardt and Negri, 2004: 109). The ability to learn, to adapt to different networks, to integrate different disciplines, to be accustomed to mobility, to be emotionally involved, effective and communicative – these are all elements, inseparable from the person, that contribute (increasingly) to the weight of the individual contribution (outside as well as inside the workplace) to the process of and final production of the politicising products (Virno, 2004: 84). One practical consequence of this is that the personal life of the researcher and the hegemonic biopower system are intimately interrelated (Ruivenkamp, 1989, 2005). Moreover, this intimate relationship is reinforced by the specific social organisational form from within which the politicised biotech products are produced.

Products such as informationalised seeds and enzymes are developed within numerous fluid and hybrid networks of public/private, fundamental/applied research institutions. Behind this diversity, however, there is a strong concentration of power. The global production of seeds and enzymes as information-intensive biotech products is actually in the hands of a very limited number of companies. Indeed, the social organisation of these products is characterised by a dual process of a concentration of economic power and decentralisation of research networks (Ruivenkamp, 1989, 2005). Within this concentrated-decentralised production schemata, the effective labour of the scientific worker is increasingly replaced by his/her generic capacity to develop and produce politicised, knowledge-intensive products. This transformation of effective labour into living labour (Hardt and Negri, 2004: 146), as a social base for the social organisation of producing knowledge-intensive products can be (and is being) challenged by the strategic and tactical actions of various 'alternative' networks.

The *deconstruction* of this trend towards a technologisation of life is carried out by those researchers – working within local-orientated networks – who are aware (rationally or emotionally) that they live through or even within a sharply contradictory development[35]. This concerns the complex organisational form of fluid networks of private/public, fundamental/applied research institutions in which their work as living labour power engaged in the production of informationalised seeds and biocatalysts is carried out. This socially complex organisational form implies that the individual researchers – particularly those who dedicate their lives to their work – loose their grip on and insight into their own contribution to the politicising content of the end-product, even though the end-products are clearly characterised by their politicising content (of creating new social relations). It is the non-transparency of the complex production system which causes researchers to become completely alienated from the social significance of their work, which is in fact the core aspect of their work. There develops an intellectual and emotional estrangement from the true purpose and the heart of the scientific enterprise. It is this contradictory development that disconnects the research work of the individual researcher from its social significance that may explain why the developing technological rationality is not able to reduce social inequality (see question 1). Indeed, it is the highly specialised form of work within international, hybrid networks that leads the technical potentials of the knowledge-intensive production to be increasingly separated from critical-reflexive ethicality – notwithstanding the fact that a broad ethical content is exactly the main characteristic of the products that are delivered by the scientific labourers! In other words, it is precisely because of this specific social organisation, from within which the life-science technologies evolve, that we may expect the developing technological rationality to fail to reduce social inequality. This, of course, implies the need for a reconstruction of that social organisation, which is exactly what activist networks are aiming for.

---

[35] It would be nice to reflect further on the statement of Marcuse (1964: 127) that the established ways of life can be destroyed by the exigencies of thought (Logos) and the madness of love (Eros), and that the reconciliation of these two contrasting rationalities may offer opportunities to challenge the trend in which the rational society subverts the idea of Reason (Marcuse, 1964: 167).

The *reconstructive actions* challenge this disjunction in which the social organisation of research offers opportunities to individual researchers to technically develop their work, while at the same time setting constraints on with regard to critical reflection on its social content. Therefore, the reconstructive approach here is based on the efforts of researchers to re-establish a direct connection between ethicality and technical potential – not only in general terms but also, and particularly, in all the different sub-segments of the complex networks of the knowledge-intensive products. By making concrete choices about the specific information to be incorporated in knowledge-intensive products, individual researchers can challenge the hegemony of the bio-power system and become actively involved in the reconstruction of other, new informationalised seeds and biocatalysts within location-specific networks. It will be these divergent and multiple efforts of local networks to apply strategic and tactical actions and make specific choices to include ethicality in new technical means which may ensure that life itself becomes reconfirmed as an object instead of a means for technology development.

6. Concluding remarks

The article started with the questions of why human mankind – despite the ever growing development of technical means – is still not able to reduce social inequality, and why the survival of mankind and 'Mother Earth' seems even to be threatened, precisely because of this growth of technical means. These questions have always been core issues in the analyses and debates of critical theory. Therefore, we return to the ideas of the founding fathers of critical theory – particularly to the analysis by Marcuse of the advanced industrial, technological society – in order to see how they approach these issues and whether they may still be usefully applied in a contemporary analysis of life-science technologies.

The conclusion of this article is that dialectic, critical, historical-materialistic approaches continue to offer important analytical frames of reference for dealing with the actual, contradictory developments of life-science technologies. We consider the method of 'going beyond' the actual factuality of technology development relevant, and particularly the approach of uncovering historically contradictory trends as entry-points to a qualitative shift in the actual technological development. Indeed, the critical theoretical approach is primarily relevant because it investigates the (historical) opportunities for transformation (from within the described contradictions).

Concerning the questions posed of why the technological rationality, as it has developed, is not able to diminish social inequality and why it is, furthermore, that technological development even seems to be so environmentally and ecologically damaging, possibly disastrously, the critical theoretical approach does not limit itself to a description of these phenomena and thereby recopy that reality. On the contrary, its primary aim is to search for opportunities to change this situation and contradictory developments that may lead to its ending. Therefore, critical theory explores the dual process of an uncoupling of ethicality from the technical potential (a historical trend) and a re-coupling of technical means to social objectives (a re-integration of values) in order to investigate the potential of this contradiction to create a social space for the transformation of technologies.

The critical theoretical approach is sometimes characterised as an *utopian-realistic approach*, in which the aim is to realise social change (utopism) based on the concrete opportunities that the existing reality affords (realism), real possibilities for transformation offered by historical contradictory developments that come together in the actual reality, enabling people to change that reality. Indeed, the central aspect of the utopian-realistic or critical-theoretical approach is exactly this tension between *historical trends* (shaping that which is) and the *praxis character of society* (that which may become) in which people can grasp the opportunities to realise historical transformations. It has been argued that this praxis character of society is in fact itself challenged by the contemporary development of technologies. Still, it is just the trends outlined, towards an ever-expanding development of technology encompassing even human life, which form the concrete preconditions for the transformation of that technology.

In Table 1 I have summarised the way in which this tension between four historical trends in technology development and the praxis character of society is discussed in this article, expressed, that is, in terms of the dialectics of emancipation, i.e. an emancipator-directed technological development. In the columns, we present the four evolving *historical trends*, headed by the (black) Boxes I-IV[36]:

- *Box I* represents the development towards the dominant position of the *instrumental, value-free notion of technology*. In Section 4.1 it is argued that a qualitative shift has taken place in the balance between objective and subjective reason related to the emergence of industrial society, leading to the dominant vision of an instrumental, value-free technology. The instrumental, value-neutral technology development, it is argued, is historically shifting into the development of a politicising technology.
- Therefore, *Box II* represents this trend towards: The development of the *instrumental, technological rationality* that becomes directly *political*. The content of technology is political because the dominant instrumental vision reduces the opportunities for antagonistic forces to change the technology trajectory. Moreover, people's actions are no longer primarily goal-orientated but increasingly rearranged into a functional dealing with the maintenance, adaptation and development of that instrumental technological rationality. Therefore, a third trend emerges concerning the apparently autonomous development of technology, seemingly disconnected from any specific societal context. Thus:
- *Box III* represents the development of *technology as an abstractum* in which there is reversal of means and ends. Technology development may increasingly become an object in itself and the expansion of specific technical means the determining factor in choosing which means will be used for handling which social problems. Moreover, it is argued, even people's lives may become a means for realising an apparently autonomous technological development; human life integral to the smooth and optimum functioning of the technological development. Thus, a fourth trend:
- *Box IV* represents the development of a *technologisation of life*, which refers to the trend in which life is perceived as the centre of potentiality for the further development of the instrumental technological rationality. However, this trend also strengthens the contradiction within the praxis character of society in which man aims to govern technological development. Indeed, it is emphasised that the apparently autonomous technological development contrasts sharply with its essential social character and its specifically political content. Therefore, we have also argued that the trend in which a person's life becomes a means for the efficient functioning of the technological system is increasingly put under pressure by the appearance of living labour in the production of knowledge-intensive products. This living labour creates opportunities for social life to be re-established, for people to become protagonists (and subjects) in social transformation and develop the potentialities for reconstructing technological developments.

---

[36] I want to stress again that these trends should not be interpretated as indicating a linear development. They refer to developmental tendencies that may be strengthened or weakened according to the presence of political and social struggles in different regions. Moreover, it may happen that some trends may shift to another trend and/or a so-called previous trend may become again dominant. The schemata should be interpreted as representing as discrete trends that are in reality a constantly changing, fluid mix.

Table 1. The dialectics of emancipator technological developments.

Praxis (A)	Historical trends (I–IV)			
	I: Instrumental, value-free, technology	II: Technological rationality as political rationality	III: Technology as an abstractum	IV: Technologisation of life
A1: Deconstruction (unravelling of)	I.A1: Uncoupling ethicality and technical potential	II.A1: The social organisation of production	III.A1: Re-concep-tualising local space within global networks	IV.A1: Governing of life
A2: Reconstruction	I.A2: Reintegrating ethical values into technology	II.A2: Re-codification of technologies	III.A2: Re-linking technologies to localities	IV.A2: Life-affirming technologies
A2.1: Strategic actions	I.A2.1: Social cooperation in local networks	II.A2.1: Re-organising immaterial (knowledge-intensive) production	III.A2.1: Strengthening location-specific natural and social resources	IV.A2.1: Transforming effective labour into generic labour
A2.2: Tactical actions	I.A2.2: Selecting specific scientific information to incorporate into artefacts	II.A2.2: Designing sub-political means as micro-politics	III.A2.2: Developing situational politics	IV.A2.2: Strengthening living labour
B: Manifestations of emancipation[1]	I.B: Sub-political products, *challenging* social relations	II.B: Social dynamic products *strengthening* social transformations	III.B: Products as *levers* for potential local developments	IV.B: Creation of social life

[1] The explication here of manifestations in terms of products is not intended to imply that emancipator-directed technology development is necessarily orientated towards products in the narrow sense of physical artifacts. On the contrary, products should also be understood in the wider sense as representing social relations. Regarding Table 1 here, the B row, for example, might be instantiated by e.g. localised developments of new varieties, but it may also be interpreted as including e.g. innovations in complementary early-planting techniques – or, more generally, the establishment of regional researcher-farmer initiatives; equally, the A2.2 row may refer not only to the scientific information used in the creation of biotech artifacts, but also to the use of environmental risk factors in determining optimum levels of inorganic fertiliser use – or, to the types and specifications of community-based considerations informing the establishment of local projects; A2.1 boxes need not only involve e.g. new business partnerships including local communal bodies as well as private companies and public institutions, but also methodological principles of researchers and farmers/farming communities working collaboratively together – and the involvement of NGOs and activist organisations in scientific establishments and political life at all levels; and while we may relate the A2 and A1 rows to e.g. the (lack of) attention paid to local consumption needs when introducing new varieties to raise productivity, we may also relate them to (a response to) the failure of regional initiatives when their planning and operation are impoverished by the lack of a major, guiding bottom-up element – or, more generally, to a recognition and understanding of (and struggle to revolutionise) the amorality embedded in the technology-related policies and practices of national governments and international agencies and organisations. It should go without saying that Table 1 is not necessarily limited solely to biotechnology, either.

In the (light shaded) rows headed A, we present four aspects of praxis through which people may become again the real protagonists in the development of life-science technologies. The following four aspects are distinguished:

- *Box A.1*: People can *de-construct* the actual technological developments, by unravelling the societal context from which the actual technology emerges, indicating which key social relations are core constituent elements in the shaping of a particular technology.
- *Box A.2*: People can also *re-construct* technologies, by changing the social structure of technology as well as by modifying the material (politicised) form of a technology. Indeed, we have emphasised that these two orientations can be taken up by strongly interrelated actions.
- *Box A.2.1*: People can carry out *strategic actions*, orientated to challenging the institutionalised unequal balance of power by transforming the social cooperation within which technology design takes place.
- *Box A.2.2*: People can also carry out various *tactical actions,* focused on changing the material form of technology.

The different boxes in Table 1 indicate ways in which the tension between historical trends (columns) and the praxis character of society concerning technology development (rows) is engaged and struggled with by various regional initiatives. The discussion of these tensions in general terms in this article is summarised in the heading boxes as follows:

- *Box I.A1*: refers to the argument that a de-construction of the instrumental, value-free notion of technology can be carried out by unravelling the historical trend of uncoupling ethicality from the technical potentiality.
- *Box I.A2*: summarises the argument that the reconstruction of that trend may be carried out by efforts to reintegrate ethical values into technology, through the setting up of specific cooperative networks (Box I.A2.1) and/or by choosing specific information that will change the material form of a technology (Box I.A2.2).

In other words: This table summarises the dialectics of emancipator technological developments based on the tension between the historical trends (what is) and people's capacities to transform these developments (what may become) from within these trends.

In order to further clarify this, I will also describe in brief the content of the Boxes III as well as B, in which I summarise the trend towards an abstract technology development contrasted by various forms of praxis to localise that technology development:

- In *Box III.A1* I refer to the general argument that de-constructive activities may be focussed on the unravelling of the social relations from which the specific political characters of new technologies emerge. Particularly, it refers to the de-construction and unravelling and the re-conceptualisation of the social space of locality within global production chains.
- *Box III.A2* summarises the argument that the restructuring or re-writing of the technical code of abstract technology may be based on efforts to strengthen an elaboration of a redefined concept of locality, and that concrete strategic actions (*Box III.A2.1*) may be carried out for the linking of technology to the local natural and social resources as potentialities for location-specific developments. In order to strengthen the reconstructive strategies, furthermore, it is relevant to elaborate 'situational politics' (*III.A2.2*). Also, it is emphasised that the combination of *strategic actions* – aiming to create new, alternative social relations from which other technologies may emerge – and multiple forms of *tactical actions* – aiming to change the material form of a technology – may

result in the development of specific artefacts that may function as levers for location-specific developments (*III.B*).

- The final (dark shaded) horizontal line (*I.B–IV.B*) represents provisional target-points on which an utopian-realistic approach of life-science technologies may be focused. It summarises the conclusion that, at this particular stage of the historical material development of technology and praxis character of society, the dialectics of emancipator technological development come together in concrete material forms (*Box B*). Indeed, the research unit Critical Technology Construction (CTC) aims to investigate further the ways in which specific life-science products may challenge the dominant social relations (*I.B*), strengthen social transformations (*II.B*), function as levers for location-specific development (*III.B*), and create new social life (*IV.B*)37.

By way of conclusion, I would like to emphasise that the historical trends incline towards an apparently autonomous technological development, including even human life as becoming an integrated part of that development. However, it has also been emphasised that the presence of this apparently autonomous technological development is contrasted with its essential social character and its historically, concrete politicised content. It is argued that this basic contradiction manifests itself within the work of the scientific labourers. These labourers may increasingly experience that their way of life, dedicated so completely to their research work, represents the transformation of effective labour power into generic labour, which stands for the generic labouring capacity of human beings to produce, develop and transform, a capacity that goes beyond the limits of productive labour. Indeed, it is through this existing trend of transforming effective into living labour (Hardt and Negri, 2004: 146) that oppositional forces to the actual technology development are appearing. This manifests itself through the social cooperation of panoply of farmers, villagers, scientists, etc. raising specific social-technical networks in different regions throughout the world, trying to deconstruct and reconstruct the social structure of technology as well as its material (politicised) content. There are a myriad of hopeful signs indicating a growing awareness of the potential uses of life-science technologies as forces for emancipation.

A manifesto, however, such as this paper declares itself to be – albeit limited just to a conceptual discussion of the opportunities for raising the praxis content within actual trends of technology developments – is duty-bound to sound a warning note as well as clarion call. The smattering of concrete examples in the field of agro-industrial biotechnology mentioned here might indicate that reconstructed seeds and enzymes can play a socially dynamic, emancipator role in society, and some may indeed see the current developments as signs of the crumbling and immanent collapse of the hegemonic edifice. A darker reading, on the other hand, might note how the popular 1960s dialectic of the counter-culture seems to have been usurped and transformed recently into the mass phobia of the War On Terror, in which even antipathetic forces are carolled into alignment with one side or the other in the Clash of Civilisations. There is a continual struggle for control, even of the definition of the defining dialectic of our age.

We at CTC offer our services in the name of a transformation of what we see as the deeper processes at work in human society, hoping to contribute to a sense of common humanity, valuing people and communities as ends in themselves. We continue offer a platform for the critical constructivist project in the field of biotechnology, supporting a host of activities, including analyses of the multiple, location-specific forms of strategic and tactical actions in which the social structure *and* material/political content of life-science technologies are challenged. It will be the analysis of the location-specific combination of these two fronts,

---

37 For example, three new CTC-sponsored doctoral theses, considering ways in which genomics products may also play dynamic, sub-political roles, will be described by Wietse Vroom, Daniel Puente and Eric Deibel as results of their PhD research projects.

social structure and material/political content, that will indicate how well regional (bio)technical activist networks are proceeding in their efforts to re-conquer the praxis character of local societies regarding (bio)technology development. Macro-sociological approaches based on a critical theoretical research methodology may be helpful to strengthen these multiple forms of 'situational politics', reconfirming the specific position of critical social scientists of not only interpreting and (re-)copying reality (as it is), but aiming also to contribute to a transformation of that reality (as it may become).

Acknowledgements

For this article I wish to thank particularly Joost Jongerden and Andy Hilton for their stimulating comments.

References

Adorno, T.W. (1969). Zur Logik der Sozialwissenschaften. In: *Der Positivismusstreit in der deutschen Soziologie*, Neuwied und Berlin.
Adorno, T.W. (2005). *Minima Moralis, Reflections on a damaged life*. Verso, New York.
Al-Hassan, R. and Jatoe, J.B.D. (2002). *Adoption and Impact of Imroved Varieties in Ghana*. paper presented for the Workshop on the Green Revolution in Asia and its Transferability to Africa, organized by the Foundation for Advanced Studies in International Development (FASID), Dec. 2002, Tokyo, Japan. At: http://www.fasid. or.jp/chosa/forum/fasidforum/ten/fasid10/dl/2-7-p.pdf
Barcelona, P. (2007). Modernity between Abstraction, Logos and Globalization. *Tailoring Biotechnologies* 3(1): 9-17.
Beck, U., Giddens, A. and Lash, S. (1994). *Reflexive Modernization. Politics, Tradition and Aesthetics in the Modern Social Order*. Cambridge, Polity Press.
Feenberg, A. (1999). *Questioning Technology*. London, New York: Routledge.
Feenberg, A. (2005). *Heidegger and Marcuse. The catastrophe and redemption of history*. Oxon, Routledge.
Foucault, M. (2004). *Breekbare Vrijheid. Teksten & Interviews*. Amsterdam, Boom, Parresia.
Galimberti, U. (2003). *Psiche e techne. L'uomo nell'eta della tecnica*. Milano, Giangiacomo Feltrinelli Editore.
Goodman, D., Sorj, B. and Wilkinson, J. (1987). *From Farming to Biotechnology. A theory of agro-industrial development*. Oxford, Basil Blackwell Publications.
Habermas, J. (1973). *Erkenntnis und Interesse*. Suhrkamp verlag, Frankfurt am Mainz.
Hardt, M. and Negri, A. (2004). *Multitude: War and Democracy in the Age of Empire*. New York, The Penguin Press.
Horkheimer, M. (1947). *Eclipse of reason*. Oxford, Oxford University Press.
James, C. (2004). *Global Status of Commercialized Biotech/GM Crops*. New York, Ithaca, ISAAA Briefs, no 32.
Hoy, D.H. and McCarthy, T. (1994). *Critical Theory*. Cambridge, Massachusetts, Blackwell Publishers.
Kellner, D. (1984). *Herbert Marcuse and the crisis of Marxism*. Berkeley, University of California Press.
Lazzarato, M. (2005). Biopolitics. Available at: http:// www.generation-online.org/c/cbiopolitics.htm
MacLennan, G. (1998). Begin the dialectic: the fruits of diffraction. Available at: http://archives.econ.utah.edu/ archives/critical-realism/1998m11/msg00004.htm
Magnaghi, A. (2005). Local Self-Sustainable Development: Subjects of Transformation. *Tailoring Biotechnologies* 1(1): 79-102.
Manzano, A.R. (2007). Socialization of science and technology: An exploration of the decentralized research and production units within Cuban urban agriculture. *Tailoring Biotechnologies* 3(2): 49-78.
Marcuse, H. (1964). *One Dimensional Man, Studies in the ideology of advanced industrial society*. Boston, Beacon press.

Pistorius, R. and Van Wijk, J. (1999). *The Exploitation of Plant Genetic Information. Political strategies in crop development.* Amsterdam, Print Partners Ipskamp.

Plato, Aristoteles, Seneca, Erasmus (2006). *Filosofie: Van oudheid tot renaissance*, Amsterdam, Uitgeverij 521.

Pray, C.E. and Naseem, A. (2003). *Biotechnology R&D: Policy Options to Ensure Access and Benefits for the Poor.* ESA Working Paper No.03-08, FAO. Available at: http://www.undp.org/biodiversity/biodiversitycd/key4.htm.

Ruivenkamp, G. (1989). *De Invoering van Biotechnologie in de Agro-Industriele Productieketen. De overgang naar een nieuwe arbeidsorganisatie.* Utrecht, Jan van Arkel Uitgeverij.

Ruivenkamp, G. (2005). Tailor-made biotechnologies: between biopower and subpolitics. *Tailoring Biotechnologies* 1(1): 11-33.

Sanchez, P., Swaminathan, M.S., Dobie, P. and Yuksel, N. (2005). *Halving Hunger: It Can Be Done.* Summary version of the report of the Task Force on Hunger, New York: The Earth Institute at Columbia University. Available at: http://www.unmillenniumproject.org/documents/HTF-SumVers_FINAL.pdf.

Verbeek, P.P. (2000). *De daadkracht der dingen.* Amsterdam, Boom Uitgeverij.

Virno, P. (2004). *A Grammar of the Multitude. An analysis of contemporary forms of life.* Cambridge, Massachusetts, Semiotext(e).

Winner, L. (1985). Do artifacts have politics. In: *The Social Shaping of Technology*, (Eds. D. MacKenzie and J. Wajcman), Philadelphia, Open University Press.

Local activism and the 'biotechnology project'

Rachel Schurman and William Munro[38]

1. Introduction

> 'The concept of hegemony refers to domination which is so strongly embedded in social life that it is not queried anymore; not even by those who are subject to it. It concerns a domination that has the force of culture behind it (Feenberg 1999). One can speak of *hegemonic biotechnology* when the bio-power system is based on assumptions which seem so natural and obvious that these assumptions lie below the threshold of conscious awareness' (Ruivenkamp, 2005).

In the opening article of the first issue of *Tailoring Biotechnologies*, Guido Ruivenkamp offers a highly provocative theoretical argument about the emerging system of bio-power and its central characteristics. Ruivenkamp sees this emerging system of bio-power as being fundamentally rooted in the development of biotechnology, 'a social-technical ensemble' whose products take the form of seeds, enzymes and biocatalysts, and whose signature quality is that they enable 'control from a distance'[39]. Even though this social-technical ensemble is still relatively embryonic, Ruivenkamp suggests, its power derives from the transformative pressure it exerts on the global food system. As the global food system changes in response to these pressures, the new bio-power system is presumed to become hegemonic, as its underlying assumptions (and material practices) become so taken for granted that they are no longer subject to questioning.

In the course of making this argument about the emerging bio-power system, Ruivenkamp also provocatively suggests that the diehard proponents and opponents of biotechnology – those whom he collectively refers to as 'the splitters', or alternatively, as the 'pro's' and the 'anti's'[40] – play the critical ideological function of strengthening this system. According to Ruivenkamp, they do this in several ways. First, as the splitters carry out their polarised debate on biotechnology, their arguments invariably focus on biotechnology in its most extreme forms, ie. 'gentechnology,' or genetic engineering, which involves the production of 'novel' organisms (GMOs) through gene splicing. By doing so, they foreclose the possibility of thinking about other forms of biotechnology (non-GMOs, hybrid forms). Second, both those who believe fully in the need for the technology and those who are diametrically opposed to

---

[38] Both authors contributed equally to the preparation of this paper.

[39] Ruivenkamp identifies five key tendencies in this emerging system. At risk of oversimplification, they include: (1) a shift of the political system into the social organisation of production, as researchers in the life sciences industry embed new knowledge and production control directly into the seed, enabling 'control at a distance'; (2) the associated creation of new social relations as farmers take up these new seeds (and other new biotechnologies); (3) the devaluation of the traditional knowledge held by farmers and its replacement by the external – and commodified – knowledge of agricultural researchers, creating a new balance of power in the agroindustrial food chain; and (4) the declining social-cultural sovereignty of food producing regions, as (a) biotechnology/seed firms offer (locationally tailored) patented seed technologies that displace location-specific production methods and (b) food processing companies reduce their dependence on specific locales and break agricultural output down into its interchangeable components out of which they (re)construct various processed foods. The fifth characteristic relates to the changing nature of the locality as a food production unit, which becomes reconstituted and redefined in terms of its relationship to global sources of enzymes, biocatalysts and other biotech products (see Ruivenkamp, 2005: 14-16). In our view, it is the first of these – control at a distance – that is most central and revolutionary; all the rest are derivative from this new source of power.

[40] '[T]he splitters are divided into two subgroups, the optimistic and the pessimistic, those who think that biotechnology might solve all problems and those who emphasise its negative impacts' (Ruivenkamp, 2005). In this article, we focus on one group of those pessimists, namely, those who comprise the anti-biotechnology movement.

it hold a linear and socially disembedded model of technological change which sees (bio)technology as existing outside of society, and exerting 'effects' upon it from this external position. Consequently, both supporters and critics alike hold an absolutist view of the technology such that it can only be accepted or rejected *tout court*. On this account, both groups take as given the motives of technology developers and the implied inflexibility of the technology, thereby foreclosing an understanding of the alternative ways in which the technology could be used[41]. Third, by taking what is happening on a very small proportion of agricultural land (that planted with GM seeds) and extrapolating from that to the rest of the agricultural system, the splitters add to the immanence of a hegemonic bio-power[42]. Fourth, in the course of carrying out their struggle, the pro/anti-biotech forces create a complex around their own interests which effectively closes off the space for ideas that do not share their same basic assumptions. And finally, through all of the above processes, the splitters mask the possibilities for change in the form of a creative or productive redeployment and redesign of the technology through the constitution of a different social-technical ensemble.

In this paper, we want to engage with some of Ruivenkamp's claims about the role played by the splitters in the emergent system of bio-power, or what might also be called the 'Biotechnology Project', to signify the efforts of the large biotechnology companies and supportive states to promote a *particular sort* of biotechnology[43]. We use the term 'Biotechnology Project' to emphasise the point that this system of bio-power is a political, economic, and ideological undertaking designed to produce a global agricultural order based on corporate control of agriculture, strong private property rights, technologies that facilitate 'control from a distance', and supportive regulatory frameworks. This Project is advanced by a complex of interests, including agribusiness, states, key agricultural research institutions, some research scientists, aid agencies, and private philanthropic foundations. As Jack Kloppenburg Jr. has argued, control in the system is constructed around two essential vectors, namely, technology and intellectual property rights (Kloppenburg Jr., 1988). *Technological control* is exerted by breeding certain capabilities (and incapabilities) into the seed[44]. *Private property rights* are exerted through the legal system, which enables those who possess these rights to determine the way in which these technologies get used[45]. Moreover, as an effort to institute a 'control from a distance' regime, the Biotechnology Project is characterised by a desire on the part of industry and GMO-producing countries to establish a modular approach to GMO regulation which is science-based and internationally harmonised, thereby reducing the space for adaptation and choice by end-users or recipient countries (cf. Gupta and Falkner 2006).

It is important to note that the Biotechnology Project also rests upon and expresses certain ideological assumptions. One is the classic liberal tenet that public good comes from private gain; thus, the best way to achieve the optimal societal good is to let the private sector seek to maximise its utility (typically

---

[41] This mechanism, as well as few others, is inferred by the present authors, since Ruivenkamp's argument doesn't make entirely clear how these processes work to strengthen the emerging bio-power system.

[42] Along similar lines, one could also argue that both groups underscore this immanence when they stress the immense power of gene technology to solve the world's agricultural problems (in the case of biotechnology supporters) or to take over world agriculture (in the case of critics).

[43] We use this term similarly to the way that Philip McMichael and others use the term, 'development project'. This is of course a central element in Ruivenkamp's notion of a system of bio-power.

[44] An example of one such incapability is the infamous 'Terminator' technologies, which build a sterility trigger into the seed.

[45] As Smith and Rajan (Smith and Rajan, 2007) point out, however, not all property claims are equally powerful in the sense that some patent holders, eg., multinational companies, have resources and reasons to protect those rights against violation, while others, such as research institutions in poor countries, do not. Furthermore, not all countries have equally developed and enforceable property regimes.

defined as profits, or shareholder value). A second assumption is that technology exists apart from society, and as such, does not reflect or reinforce particular political-economic interests and values. The third (and perhaps most critical) ideological assumption of the Project is that 'there is no realistic alternative' to solving world agricultural problems or for meeting human needs. Taken together, these assumptions produce a social-technical ensemble in which the private sector plays the leading role in technology production, dissemination, and control.

Given that our area of expertise is on the anti-genetic engineering movement, or the most politicised subset and face of those whom Ruivenkamp calls 'the *anti's*', the focus of our analysis will be on these activists and the specific role they have played *vis-à-vis* the Biotechnology Project. Contrary to Ruivenkamp, we are not so ready to dismiss this group of 'splitters' as simply moving us toward a hegemonic system of bio-power. Rather, we think a focus on the *effects* of the actions undertaken by these politicised technology opponents reveals that they have played a much more complex and variegated role than Ruivenkamp allows. Through their activism, they have destabilised the Biotechnology Project and forced open the future of agriculture, rather than ensuring that a hegemonic system of bio-power will obtain. Nevertheless, we recognise that because of the way in which these activists have conceptualised, framed, and mobilised around agricultural biotechnology, they may not be capable of moving society toward a biotechnology for the social good. In other words, anti-GM activists may not be able to play a central role in reconstructing agricultural biotechnologies for human-centered development. Others will have to take on this task.

In the rest of this paper, we examine the ways in which the anti-GM movement has sought to subvert the Biotechnology Project and the success it has had in achieving this goal. We argue that in the late 1990s/early 2000s, the anti-GM movement had a profoundly destabilising effect on the Project, most notably in Europe but also in other parts of the world (including Africa and Asia). Through its mobilisation of a powerful and highly resonant set of oppositional discourses, the anti-GM movement threw a significant monkey wrench into the gears of the corporate/state biotechnology nexus, temporarily slowing down the Project and setting it off in new directions. As a direct result of this activism, agricultural biotechnology became intensely politicised and the level of public scrutiny of its chief developers (large multinational corporations) and its chief supporters (states) dramatically intensified. New actors came into the picture, new systems of biotechnology regulation and governance were established at the national and international levels, and, at least in the global south, increased attention came to be paid to the question of who the technology will be useful *for*. Today, the direction in which the Project is going is pregnant with ambiguity. From our perspective, its future will depend upon a range of factors, including: (a) the degree to which these new regimes of technology regulation (e.g. the Cartagena Protocol on Biosafety, national regulatory systems) will actually be mobilised to *constrain* the power of large biotechnology companies and their ability to exert 'control from a distance', rather than legitimating and facilitating them; (b) the speed with which alternative agricultural options that increase the relative power, autonomy and economic resources of farmers are forged (one might imagine the development of non-proprietary technologies which are broadly attractive to farmers, the growth in fair trade or forms of 'quality agriculture' that give farmers more market options and power, or the growth of farmer/distributor cooperatives that have the same effect); and (c) the extent to which farmers and states simply refuse to respect and enforce the companies' intellectual property rights, thereby rendering profit-making from these technologies more problematic for corporations. Another important determinant of the future will lie in the extent to which critics and other actors (research scientists, farmers, development analysts and practitioners) are able to effectively counter the discourse that a private-sector driven, biotechnology-based future is the only one imaginable and capable of meeting our needs for food, fuel, and environmental sustainability, and to reconstruct technologies that advance alternative models of agricultural development. Herein lies the challenge.

Before moving to a discussion of the anti-GM movement's actions and their effects, we would like to advance two brief comments about the movement itself. First, in contrast to Ruivenkamp's characterisation, we do not believe that many in the anti-biotech movement see technology as existing apart from society and independently exercising 'an impact' upon it from outside. In fact, as we show elsewhere in our work, one of the ideas that motivated many early biotechnology resisters to get involved and to take the actions that would later form the basis of a movement was their belief in the *inextricability* of technology from society, and from existing social and power relations (Schurman and Munro, 2006; personal interviews). This was particularly true of those who came to the issue from the perspective of science studies or critical Marxism, but it was also true of the movement's more 'organic intellectuals' who were (and still are) keenly aware of the social/political contexts in which technologies evolve[46]. For many of these activists, the technology's intimate, and perhaps inextricable, association with the particular institutionalised patterns of power and authority embodied in the Biotechnology Project made it unacceptable. Second, even though both Ruivenkamp and we write about the 'anti's' as if they were a unified and undifferentiated actor, in reality there is significant diversity within this group in terms of people's views and understanding of gene technologies and their social potential, liberatory and otherwise. For some, 'another biotechnology' is indeed imaginable, while for others, it is not. In this sense, we prefer to think of the anti-GM activists as 'spoilers' rather than 'splitters.' That said, let us turn to the ways in which activists have resisted the development and deployment of agricultural biotechnology through their technology-oriented political practice.

2. Constructing an alternative framework of interpretation

As we have shown elsewhere in our work, a vitally important part of what the anti-GM movement did in the 1980s and the first half of the 1990s was to develop an in-depth and multifaceted critique of biotechnology, or an 'alternative interpretative framework' (Schurman and Munro, 2006). Working together across tables, telephone lines (and later, the Internet) and continents, these individuals developed a critical interpretation of genetic engineering and of the biotechnology industry that was profoundly opposed to the one advanced by the mainstream. Indeed, this alternative interpretative framework directly challenged many of the claims and assumptions embodied in the dominant discourses on biotechnology[47], and raised a set of questions about its safety and value (including the values embedded in biotechnology) that its chief proponents would have preferred to have kept off the table. As the public, consumers, and government officials around the world became exposed to the discourses associated with this critique through their networks and through the mass media, a not

---

[46] Of course, few activists use the language of science studies, or go as far as STS community does in terms of thinking of society and technology as mutually constituitive. But that does not mean that they are unaware of the many forces that shape the emergence and use of particular technologies.

[47] See (Gottweis, 1998; Jasanoff, 2005; Wright, 1998) for a discussion of these dominant discourses. These authors show how genetic engineering was framed in the 1980s as a revolutionary new technology, one that could raise agricultural productivity by shortcutting the slow process of traditional plant and animal breeding; improve health care and lower health care costs by providing cheaper and better medicines; and generate a new wave of economic growth based on science, high technology, and innovation, rather on than the mass production of widgets, as was the case for industrial production. For an individual country, the development of the biotechnology sector was framed by policymakers as a virtual imperative, in the absence of which no country could possibly hope to remain competitive in a rapidly changing global economy.

insignificant number of people found these ideas persuasive. Another, even larger group simply began to wonder about the value of this technology in improving the quality of life in light of the risks it appeared to offer.

As is by now well known, one of the central discourses the movement mobilised against the new biotechnologies had to do with the very serious potential health and environmental risks associated with the re-engineering of plant and animal (and human) genomes. Whereas the dominant discourses on agricultural biotechnology portrayed genetic engineering as being associated with increased precision and control in plant and animal breeding, the critics of gene splicing (and of the industry behind it) vociferously challenged this assumption, and pointed instead to the uncertainty of the science. They suggested that much was still unknown about how various genomes and biological organisms work, and that these new gene-manipulating technologies could lead to health and environmental problems that even the best 'scientific experts' could not foresee. Their discourses identified specific risks that GMOs potentially posed to ecological systems and biological diversity, and emphasised the notion that the effects of releasing such organisms were likely to be irreversible and uncontrollable[48]. Although these discourses never gained real legs among the American public, they were immensely powerful in Europe, where the public has a much stronger environmental sensibility. They also had considerable resonance in the global South, where many view their country's genetic diversity as a vitally important national resource, and one that requires careful stewardship in order both to promote economic growth and to secure the livelihoods of poor agrarian populations.

Another, closely related set of narratives the anti-GM movement advanced concerned the state and its responsibility to stringently regulate these new technologies. Virtually as soon as scientists discovered rDNA techniques and the petrochemical, agrochemical, and pharmaceutical industries began applying them, many governments jumped on the bandwagon. The United States government was particularly supportive, and created a range of policies designed to facilitate the industry's and technology's development[49]. The British, German, and French governments, among others, were also very excited about the prospects of biotechnology and the development of this new area of the economy (Gottweis, 1998; Jasanoff, 1995). Deeply disturbed by this enthusiasm, critics mobilised a narrative that questioned the motives and responsibilities of governments with respect to biotechnology. They harshly criticised governments that had rushed to embrace the new biosciences and biotechnology industry in the 1980s and 1990s, and that had sought to support the sector rather than to regulate it. In short, activists painted governments as being in bed with the industry and shirking their responsibilities as protectors of the public interest.

Anti-GM activists also vigorously attacked the state on the grounds that it was not doing enough to secure the public's right to know – and choose – what it was eating. This discourse became particularly salient in Europe in the 1990s, when a number of serious food scares and catastrophes swept through the continent, and public concern about food safety was at an all time high. In Britain, for instance, activists frequently charged that the government was not doing enough to inform the public, and that most of its actions were aimed at trying to protect the industry. Throughout Europe, the anti-biotechnology

---

[48] Les Levidow has shown, for example, how the movement described GMOs in terms of genetic pollution, genetic contamination, and 'Frankenfoods' to emphasise the risks of gene flow, loss of genetic diversity, and an uncontrollable technology (Levidow, 2000).

[49] For instance, the product-based regulatory system the US government created in the late 1980s, called the 'Coordinated Framework', was openly designed to facilitate the movement of GM crops from laboratory to field trial to market.

movement argued that it was a basic democratic right of consumers to know what was in the food supply, and to be able to make their consumption choices on that basis. This discourse had a profound effect on the policy sphere, where it helped to compel both European governments and the European Union to develop policies that would ensure this 'right to know' through GMO labeling.

As part of their counter-discursive work against the Biotechnology Project, anti-GM activists mobilised a powerful critique of the motives and behavior of the multinational corporations at the forefront of developing and commercialising these new technologies. A central part of their narrative focused on the Herculean and imperial efforts these corporations were making to push these new crop technologies on countries and consumers around the globe, whether or not they wanted them. Activists also criticised the biotechnology industry for trying to induce other countries to adopt US-style intellectual property rights (IPR) so that it could protect its economic interests, and for engaging in 'biopiracy', or the theft of genetic information from the global South. A third element in their anti-corporate discourse laid bare the 'true motives' of the industry. Whereas the biotechnology industry was quick to point out that their technologies would 'feed the world' and 'reduce world hunger', the anti-GM movement countered that the motives of these corporations were not to improve the lives of the poor and hungry, but rather to make profits for their shareholders. As evidence for their argument, they noted that the biotechnology majors (or the 'Gene Giants', as one activist group cleverly named them) had not spent their resources on developing GM cassava, millet, or other staples for developing countries, but had instead invested in crops that are planted heavily by North American farmers. As further evidence of these companies' true intentions, suggested the activists, they had also bought up scores of seed companies around the world. The leader in this 'gene grab' was Monsanto. Taken as a group, these counternarratives put the motivations of the industry under suspicion and countered its claims it was doing something useful for the world.

Thus, in sharp contrast to the dominant discourses which framed genetic engineering in terms of its economic growth- and productivity-enhancing potential in the areas of agriculture, medicine, and energy, the anti-GM movement's alternative discourses highlighted the wide range of 'bads' and 'unknowns' that could be associated with the technology: the uncertainty of the science, its health and environmental risks, the nefarious motives of the industry, and the institutional and structural changes these technologies were helping to bring about. As these counter-discourses found resonance with parts of the public and various political actors around the world, they profoundly politicised the biotechnology issue and drew attention to what scientists, multinational corporations, and governments were doing. Next we look at some of the material effects these discourses had and the ways these social activists concretely affected the Biotechnology Project.

3. The effects of anti-GM activism in Europe

Nowhere did the discourses described above – and other actions taken by the anti-GM movement – have a more obvious and powerful effect on the deployment of biotechnology and the fate of the biotechnology industry than they did in Europe. During the decade of the 1990s, European anti-GM activists managed to turn public opinion against agricultural genetic engineering, press the region's leading supermarkets and food processors into denouncing the use and sale of GMOs (Schurman, 2004), and help push several formerly supportive governments to reverse their positions on new

GM crop approvals (Carr, 2000)[50]. The ultimate effect was to close European markets to GMOs. One could also plausibly argue that it was primarily (although certainly not only) because of social activism that European governments and the EU as a whole adopted a strong precautionary approach to biotechnology.

Space limitations do not permit us to go into the details of how the anti-GM movement managed to achieved such remarkable results here (though see (Schurman, 2004)). We also want to emphasise that it wasn't *only* activists that pushed Europe into shutting its doors on GMOs; many factors obviously contributed to Europe's remarkable turn-around. But there is little question that activists were the key instigators of the shift. They set things in motion, got their messages out to the public and to the press, and maintained intense and continuous pressure on governments to regulate the technology and to respond to changing public sentiments about it. If they had not accomplished these tasks, a worldwide controversy over genetic engineering never would have erupted.

A few key points about what happened in Europe are worth noting. First, anti-GM activists were highly effective in arousing consumer concern about the safety of GM food. It was here that their discourses about unsafe and untested foods, 'Frankenfoods', and the public's right to know what it was eating, had their largest impact on the consuming public. Reeling from the impact of Mad Cow disease and several other serious (but less catastrophic) food scares, European consumers shifted their position on GMOs quite dramatically. Whereas the vast majority of Europeans were generally agnostic about agricultural biotechnology in the early part of the 1990s, by the end of the decade, 'widespread public ambivalence about GM foods...[gave] way to widespread public hostility' (Gaskell, 2000: 938)[51]. Once the supermarket chains realised the extent of public concern, they responded by removing GMOs from their store shelves and products[52]. Activists were also instrumental in playing the supermarket companies off against one another through the use of a public scorecard which rated them by their 'social responsibility' defined in terms of how responsive they were to public concerns about GMOs.

What happened in the supermarket sector was significant not only because it dramatically reduced European market demand for GM foods, but also because it provoked new demands for *GMO-free* agricultural products. In order to uphold their promises to consumers, European supermarkets began searching for production sites that could provide them with GMO-free corn, soy, canola, and cottonseed oil, leading to a new geography of production and the creation of new supply chains and relationships[53]. Moreover, as Europe's trade partners, particularly its former colonies in Africa, observed what was happening on the continent, they became reticent to plant GM crops, as we will see below. In the US, a new industry emerged around grain testing to determine whether export shipments were GM-

---

[50] The anti-GM movement in Europe has a much longer history than this, of course, and began in the early 1980s. Here we are referring to the second half of the 1990s, when the movement grew substantially and kicked into high gear with the delivery of the first GM crops into Europe (See Purdue, 2000; Schurman and Munro, personal communication; Schweiger, 2001; Thomas, 2001 for more on the history of the anti-GM movement in Europe).

[51] Between 1996 and 1999, the fraction of the publics in Greece, Luxembourg, Belgium, and Britain opposed to GM food rose over 20 percentage points, with changes in Portuguese, French, and Irish public opinion not far behind (Table 1). By 1999, only one-fifth of Western Europeans was strongly supportive of GM food, and only a third was supportive of GM crops (Gaskell, 2000, Table 2).

[52] Recent evidence suggests this consumer distrust of GMOs has continued unabated, and even gotten worse. 2005 Eurobarometer data suggest that among 25 EU countries, support for GM food averages only 27% of the population. In eighteen of the 25 EU countries, 30% or less of the population thought GM food technologies 'should be encouraged' (See Gaskell, 2006).

[53] Whether or not these locales are (or can remain) truly GM free is another question, given the mobility of transgenic plants and seeds.

free. Europe's rejection of GM food also helped stimulate an industry dedicated to producing 'identity preserved' crops for export to Europe. In short, the activists' actions reverberated widely throughout the global food system.

Second, in those European countries in which governments had initially shown strong enthusiasm for the technology (Germany, Britain and France), activists were very effective in pushing the 'state oversight' question and pressuring these governments to play a more serious regulatory role when it came to evaluating the risks of the technology (Gottweis, 1998; Jasanoff, 2005). In Germany, for instance, activists used their access to the Bundestag and their connections to the Green Party to force the German government to regulate biotechnology more stringently than it otherwise would have (Gottweis, 1998). In Britain and elsewhere, anti-GM groups seized every opportunity to point to the state's weak controls over the technology, which put the onus on governments to show that they really *were* regulating GM crops[54]. In many countries, activist groups mobilised discourses of public participation and transparency to force their governments to reveal what they were doing and to augment their oversight of the technology.

Activist pressure for increasing the regulation of GMOs also extended to the level of the European Union. In the late 1980s, anti-GM activists working with the Green Group in the European Parliament pushed hard to ensure that a 'process-based' rather than 'product-based' GMO regulatory system would be developed for the EU. Once a process-based system was successfully established[55], activists then worked with a few sympathetic EU member states to augment the particulars of the scientific risk data that was required of companies in order to assess their GM crop applications (Schurman and Munro, personal communication). Ultimately, the pressure that GM critics and sympathetic member states managed to exert through the EU system resulted in a five year moratorium on new GM crop approvals, which lasted from 1998 to 2003. In 2003, the EU created a new set of challenges for firms seeking to introduce GMOs into the region by establishing a stringent set of labeling and 'traceability' rules which require documentation of a food's ingredients from farm to fork. These new rules reflected the insistence by a politically mobilised European public that it had a 'right to know' what it was eating. Along with continued public disapproval of GMOs (Gaskell, 2006), these rules have continued to seriously impede the technology's deployment in Europe. Indeed, only a handful of imported GMOs have been permitted into the EU since the moratorium ended, and those that have come in have been limited to imports of GM food and animal feed (Bloomberg News/Reuters, 2007). Not a single new GM crop has been approved for cultivation in Europe over the last ten years (*ibid.*).

Through this combination of avenues, then, social activists dramatically altered the political reception and trajectory of agricultural biotechnology in Europe. The most immediate and concrete effect of their discourses and actions was to turn the European public against GMOs, and to stem the spread of GM crops and foods within the subcontinent. Yet just as important as this was in derailing the Biotechnology Project *within* Europe was the impact that European anti-GM activism had on *other*

---

[54] In one particularly noteworthy case, a political firestorm erupted in 2000 when Advanta Seeds UK admitted that GM rapeseed had inadvertently been mixed with conventional seed imported from Canada and sold in the UK and Europe over the previous two years. Activist organisations immediately called for the destruction of the entire contaminated crop, berating the Food Standards Agency for failing in its public watchdog role. The issue also ignited a rancorous debate in Parliament during which the government was assailed not only by opposition parties but by its own backbenchers for its 'frightening complacency' (Hickman and Roberts, 2000; Waugh, 2000).

[55] This took the form of European Council Directive 90/220, which specifies the process for getting a new GM crop approved in Europe.

countries and on the global debate about GMOs. As indicated below, Europe's effective market closure to GM food led many governments to worry that their agricultural exports would be rejected by Europe if they contained GMOs, and thus, to try to prevent GMOs from entering their national agricultures[56]. Europe also had very powerful 'demonstration' effects with respect to the regulation of the technology, and in terms of anti-GM activism. In the first case, it legitimated a model of regulation that was based on a robust precautionary principle and also established a strong precedent for GMO labeling[57]. In the second case, the successes of the European movement energised and galvanised anti-GM activists around the world. Activists from India to Kenya to the United States suddenly had the sense that 'another agriculture' *was* possible, and that their seemingly fruitless actions *could* make a difference[58].

4. Slowing the development and deployment of GM technologies in the global South

The dramatic success of anti-GM activism in Europe has not been repeated elsewhere. Nevertheless, that success has reverberated through the global agrofood system, helping to politicise biotechnology and to subject it to a level of political – if not public – scrutiny that is perhaps unprecedented in the modern history of technology transfer/diffusion. Even though there is much data that suggests that consuming publics in many developing countries have a limited awareness or knowledge about GMOs, activists have played a critical role in pushing often sympathetic governments to tighten their regulatory oversight and to carry out more stringent biosafety testing. Through their public watchdog role and other organising activities, they have drawn new constituencies into regulatory debates by reaching beyond the typical 'regulatory community' of scientists, agricultural departments, and science and technology policy makers to include public interest groups, farmers' organisations, and officials from other ministries. In so doing, they have increased the transparency of regulatory systems and decreased the ability of science professionals and/or biotechnology companies to write their own regulatory tickets. This has led to a significant slowing down of the technology's deployment – and, as we will argue below, an opening up of the political moment. In this section, we focus on the effects of anti-GM activism in Africa, where the potential for agricultural GMOs to solve social problems such as hunger and low farmer productivity is most profound. If a hegemonic system of bio-power comes to be, Africa is also where the 'distancing effect' of GM technologies will be most pronounced, given the region's weak research capacity, its virtually complete dependence on foreign entities (whether multinational companies or private foundations) to provide GM seeds, and the widespread disempowerment of farmers.

The work of activists outside of Africa, including their roles in shaping the construction of global regulatory regimes and the constriction of European markets, has had important impacts on the local

---

[56] Europe's market closure was not the only reason why some governments chose to reject GMOs, of course. Brazil, for instance, did not legalise GMOs until 2004 because critics of GMOs had convinced the Minister of the Environment that genetic engineering posed important threats to biosafety. It should also be noted that most governments' efforts to stop GMOs from coming into their countries have not been successful, both because GMOs are a highly mobile and for the most part, 'invisible' technology that is difficult to control, contain and police, and because farmers use GM seeds illegally.

[57] The EU Novel Foods Legislation, developed in 1997, established that products developed through the process of genetic engineering had to be labeled. See (Sand, 2006) for more detail.

[58] Although we do not have room to discuss it here, the anti-GM movement's success in shutting GMOs out of Europe also had an important economic effect on the largely-US based biotechnology industry. After 1998, many US based biotech companies experienced significant market turmoil, a loss of investor confidence, and declining profitability. Between 1999 and 2001, these disruptions became so intense that they provoked a dramatic downsising and restructuring in the industry. Although many of the companies that experienced this restructuring have since managed to bring their businesses back into the black (especially Monsanto, the industry leader), they have not experienced anything that approximates the smooth sailing that seemed so likely in the mid-1990s.

politics of agricultural biotechnology within African countries. In the realm of international regulation, as we have argued elsewhere (Schurman and Munro, 2003), activists made crucial contributions to the successful negotiation of the Cartagena Protocol on Biosafety, which regulates trade in transgenic organisms. An outgrowth of the 1992 Convention on Biodiversity, the Cartagena Protocol calls for governments to adopt a precautionary approach in designing their regulatory mechanisms, and to carry out stringent risk and impact assessments. It also allows parties to use socioeconomic considerations in reaching decisions on importing GMOs (see Segarra and Fletcher, 2001). In recent years, some analysts have argued that the Cartagena Protocol is inoperable and should be scrapped. But whatever its actual shortcomings, the Biosafety Protocol has become a key reference point in national political conflicts over transgenic technology. Governments in the global south signed on partly in order to assert their sovereignty against the neoliberal impulses of the United States and its allies. Subsequently, activists have used the Protocol to hold their governments' feet to the regulatory fire by demanding that they live up to their international obligations. In particular, activists have stressed the importance of taking a more precautionary approach to the acquisition and deployment of technologies that may threaten biodiversity and farmers' livelihoods in fragile agrarian systems. Activists have also appealed to the Protocol as a way of pressing national regulatory agencies to widen the scope of public participation and social criteria in their risk and impact assessments for GMOs (see Tjaronda (2006). Consequently, as Gupta and Falkner (2006: 12) have noted, 'the process of [the Protocol's] negotiation and implementation has created greater awareness of biosafety concerns and has strengthened domestic constituencies pushing for greater caution in testing and commercialisation of biotech products'. In Africa, where thirty-seven states have signed the Protocol, government officials have found it repeatedly necessary to explain publicly how their attempts to construct regulatory regimes are consistent with its tenets.

As suggested above, the ability of activists to close European markets also resonated powerfully in the local regulatory politics of African countries. This is well illustrated by the southern African food aid crisis in 2001-2002, when several African countries refused to accept emergency shipments of US food aid, even though many of their citizens faced possible starvation, on the grounds that it was genetically modified. As Zambian Vice President Enoch Kavindele explained to UN aid workers, 'Our decision to reject some of these foods is out of fear... We have been told that we will lose our European market if we start growing GM foods... Hungry we may be, but GM foods pose a serious threat to our agriculture sector... and [could] grind it to a halt'[59]. Moreover, as the leaders of Zambia, Zimbabwe, Mozambique and Malawi saw it, there was no proof or even scientific agreement that these commodity crops were safe for human consumption.

Not surprisingly, the decision to reject GM food aid set off a firestorm of overheated rhetoric – and diplomatic wrangling – on both sides of the GM debate. Proponents, including USAID, argued vehemently that the African governments were recklessly and baselessly endangering the lives of their poorest citizens. Critics countered that USAID and the World Food Program were recklessly and immorally using hunger to launch the seeds of their imperial project into the continent. These arguments played an important role in politicising the technology and highlighting the distancing effects of a power regime grounded on imported GM grains and seeds. But a more consequential effect of this crisis, perhaps, was that it focused attention on the absence of regulatory regimes in African countries. These governments were suddenly directly confronted with the fact that they lacked effective

---

[59] Quoted in (Michael, 2006).

mechanisms for managing either the process or the potential effects of disseminating GM maize to the hungry. None of the countries involved had biosafety policies in place. Moreover, governments realised that they not only lacked the institutional capacity to control the cross-border movement of GM seed, but had no established legal frameworks for adjudicating the rights and responsibilities of plant breeders, seed companies, farmers and consumers. Under these circumstances, the risks of losing potential international markets because of crop and farmland 'contamination' seemed extraordinarily high. In the minds of many activists and policymakers, these risks were compounded by the potential risks of biodiversity loss through GMO 'contamination' in farming systems and environments that were already extremely fragile. Given the established tradition of crop improvement in southern Africa through farmers' 'on-farm experimentation' in collaboration with public research institutions (what Carl Eicher represents as a kind of slow green revolution), even the smallest threat of losing that capacity was very frightening indeed to a poor government whose plant breeding R&D capacity was tied up in such techniques. This was particularly true since public funding and institutional capacity for agricultural R&D had been declining steadily for two decades (Pingali and Raney, 2005). In short, central and southern African governments saw in GM crops a profound and multi-faceted threat to their sovereign capacity to feed their populations.

This food aid crisis demonstrates one of the many ways in which closure of European markets reverberated politically through the global agrofood system. As Mnyulwa and Mugwaga (2003) have noted, the crisis constituted a wake-up call for African governments to focus more attention on devising their regulatory frameworks for biotechnology, including the protection of biodiversity resources. In effect, many states saw the need to re-assert their political agency in a disarticulated agrofood system based on an increasingly distant mode of bio-power[60]. The effect was somewhat paradoxical. On the one hand, under the scrutiny of activists and critical civil society organisations, African states recognised the erection of effective regulatory frameworks as a matter of considerable urgency. On the other hand, that same scrutiny forced them to move very slowly and deliberately to construct biosafety frameworks and biotechnology policies, rather than simply adopting the permissive regulatory frameworks for which proponents of the Biotechnology Project have pushed. Some countries have embarked on new initiatives to protect local genetic resources. In 2005, for example, the multilateral New Partnership for African Development (NEPAD) set out a strategy for developing a regional network of gene banks to protect biodiversity even as biotechnology research proceeds. Even in countries that are most eager to win the 'race to regulate,' so that they can play the leadership role of early adopters, such as Kenya, Burkina Faso, and South Africa, the construction of regulatory mechanisms has slowed down and become more contentious.

However, the stalling of biotechnology regulation in Africa was not simply a rebound effect of the closing of European markets. It also reflects the persistent efforts of *local* anti-GM activists to stop the deployment of the technology. From the late 1990s, activist organisations began to emerge and organise against the technology, working through increasingly extensive national and transnational networks. For instance, the first South African anti-GM organisation, BioWatch, was formed in 1997 'to publicise, monitor, and research issues of genetic engineering and [to] promote biological diversity and sustainable livelihoods'. Alarmed at what they perceived to be a wholesale, inchoate, and non-transparent process of approval for permit applications, these activists declared their aspiration 'to prevent biological diversity from being privatised for corporate gain'. In 2000, the South African Freeze Alliance against Genetic

---

[60] This pressure was exacerbated by the on-going negotiations over the Cartagena Protocol (finally achieved in 2003), which required governments to exercise stringent control over the movement of GMOs.

Engineering (SAFeAGE) was formed to press for a GM moratorium modeled on the European initiative. Over time these organisations built alliances and coalitions with other citizens' rights, environmental justice, and sustainable development organisations both nationally[61] and internationally, with such NGOs as the Biotechnology Trust of Zimbabwe and the Institute of Sustainable Development in Ethiopia (which is directed by Tewolde Egziabher, who played a prominent role in the Cartagena Protocol negotiations). As Ian Scoones (Scoones, 2005) has noted, these organisations were also able to consolidate networks and links globally by participating in international events such as the anti-WTO mobilisations in Seattle in 1999, the World Social Forum meetings (most recently in Nairobi in 2006), and the World Summit on Sustainable Development in Johannesburg in 2002[62]. In turn, these transnational networks provided local activists channels of sharing information, building solidarity, and designing tactics of contention in their confrontations with government officials, industry operatives, and pro-GM activists.

Given that new GM crops cannot be commercially deployed until they can be subjected to an effective regulatory process, activists have focused a considerable amount of energy on blocking or tightening the regulatory system. In South Africa, which is the only African country so far to legalise the production of GM crops, activist organisations have directly – and sometimes successfully – challenged the approvals of particular biotechnology events, including maize and sorghum varieties. As a result of this rising tide of challenges, the regulatory system in which the approval of GM applications was relatively easy ten years ago is almost paralysed today (South Africa, 2005). In addition, South African activists have used right-to-information laws to take regulators to court to force them to collect certain types of information, or to share the information they collect. In doing so, they have publicised the principle of consumers' right to know what they are consuming – a principle that has become entrenched in local discursive struggles over the release of GMOs into the food supply. Indeed, the high-end food retailer Woolworths responded to this argument by declaring in 2000 that it would go GM-free wherever possible and label its products accordingly where it was not possible. In July 2007, ten years after the passage of the Genetically Modified Organisms Act which regulates biotechnology, activists were able to convince the South African Department of Environment to call public Parliamentary hearings on the adequacy of labeling requirements for GMOs, thus keeping the controversy in the public eye. In Kenya, the Kenyan Biodiversity Coalition, comprised of some forty-three civil society organisations, was able to prevent the passage of an industry-friendly Biosafety Bill as it was being rushed through Parliament by organising public protests and demonstrations in Nairobi (Daily Nation (Kenya), 2007; Udongo.org, 2007). Thus, as David Dickson has suggested, the opposition from vocal critics has placed governments 'on the defensive, reluctant to move forward for fear of alienating voters' (Dickson, 2007). The construction of effective regulatory mechanisms in all but a few African countries seems a long way off.

There is no question that these efforts have had a significant impact on the development trajectory of this technology, at least for the time being. In the first place, by stalling the apparent juggernaut of the US-based, multinational-driven Biotechnology Project, they have opened new possible directions for agricultural technology development and deployment. Arguably, it was the convergence of an urgent

---

[61] These networks include such organisations as the Environmental Justice Networking Forum, Earthlife Africa, Ekogaia, etc. Many of these organisations are very small and fluid. But they are able to mobilise sizeable populations for particular local actions such as marches or supermarket drives.

[62] Freidberg and Horowitz (Freidberg and Horowitz, 2004) show how the WSSD meeting became a site for confrontation between pro-GM and anti-GM organisations from across the African continent.

food and agriculture crisis, on the one hand, and the challenges of constructing effective regulatory mechanisms in weak states, on the other, that pushed the newly formed Alliance for a Green Revolution in Africa (AGRA) to pull back from its argument that genetic engineering should spearhead efforts to resolve agricultural malaise in Africa and to place more emphasis on conventional plant technologies where African public research institutions have their greatest capacity, and have established effective partnerships with international institutions such as CYMMIT and CGIAR (see Ogodo, 2007; Pingali and Raney, 2005)[63]. One of AGRA's first initiatives has been to launch a program for training African crop breeders to work 'on local crops, in local environments [using] the power of applied plant breeding on African crops...to develop effective solutions to long-standing problems facing African farmers' (Alliance for a Green Revolution in Africa, 2007). Indeed, the Rockefeller Foundation, long a stalwart champion of the Biotechnology Project, has also substantially withdrawn its support for the argument that GMOs are the silver bullet of African agricultural rejuvenation (interview with South African scientist, July 2007). Today, the idea that agricultural biotechnology can only be effective if it is substantially indigenised, with greater public support and resources, carries a currency in official circles that was absent three years ago (see for instance Dhlamini, 2006; Dickson, 2007; NEPAD, 2005)[64]. Ironically, perhaps, the 'slow race' of conventional breeding procedures may turn out to be the 'faster race' to higher and more sustainable productivity[65].

In the second place, activists have not only opened the Biotechnology Project up to intense and critical political scrutiny, but have mobilised new players and new discourses that may be able to challenge the social-technical structure of this bio-power system. As efforts to construct regulatory frameworks across the continent have slowed down and become more contentious, there have been heightened, more wide-ranging, and more inclusive debates about the potential impacts and risks associated with this technology. As Ian Scoones (2005: 10) has noted in the South African case: 'A decade ago, GM crops were barely a concern in South Africa. The government, together with industry and a small cabal of scientists, set the terms. Today, this has all changed. A combination of high-profile court cases, ongoing demonstrations, a growing media profile and long-term engagement with legislators, bureaucrats and scientists has meant that the GM debate has been opened up to greater scrutiny'. As noted above, much of this increased surveillance has focused on whether the state has sought to regulate or to facilitate the deployment of GM technology into African environments. But activists have also generated high levels of scrutiny of the multinational companies promoting the technology, painting them as being driven by a predatory self-interest that would likely increase the dependence of resource-poor farmers, and perhaps destroy them. Across Africa, indeed, the most crucial element of farmers' distrust of GM technology is a fear that it will permit foreign multinationals to extend the stranglehold that they already hold over markets into the very farming practices of farmers themselves.

---

[63] In interviews in South Africa during August 2007, both 'pro' and 'anti' activists were perplexed at this development, and not able to easily interpret its meaning for the place of GM technology in African agricultural strategies.

[64] Most African states have a strong incentive to push for this shift towards a 'post-GM' agricultural R&D agenda because they lack the technological know-how and capacity to contribute substantively to the development of GM technology; they are 'technology takers' rather than 'technology makers'. A GM technology agenda places them almost entirely at the mercy of private interests located overwhelmingly in other countries. South Africa, Egypt, and to a lesser extent Kenya and Nigeria are the only real exceptions.

[65] In October 2007, the Maize Breeders' Network (MBNet), an organisation of maize breeders, seed producers, and development specialists in southern and eastern Africa, called for faster regulatory approvals of locally-bred conventional varieties that are adaptable to local conditions. This initiative was spearheaded by the Alliance for a Green Revolution in Africa (2007).

In response, those companies have had to elaborate and develop their claims about the *social* benefits of this technology (ie. the argument that this is a public good technology). To date, it has not been easy for companies to show these benefits, in part because no new 'pro-poor' (drought resistant, nutritionally enhanced, etc.) genetically engineered crops have yet emerged from the laboratory onto the market[66]. All the commercial approvals that have been sought are for the 'mainstream' crops of maize, soy, and cotton, where proprietary rights in the seed rest with a small number of northern multinationals (notably Monsanto – which recently acquired Delta and Pine Land, which had acquired Syngenta's global cotton operations shortly before that). Many of these applications are for gene 'stacking' which places a premium on the company's assertion of intellectual property rights (and potentially increases the genetic volatility of plants). Although there are some 'orphan crops' under development, such as cassava, pigeon pea, and sorghum, it is clear today that resources for such work are inadequate, that the multinational corporations have little interest in committing significant resources to such efforts, and that the problem will require extensive public funding and support[67].

This situation has created an interesting discursive political field in which GM technology is being negotiated. On the one hand, activists have not been able to demonstrate that the technology is inherently more risky – either to producers or to consumers – than conventional plant technologies. On the other hand, technology proponents have not been able to demonstrate that the technology has systematic advantages over conventional plant technologies, either for producers or for consumers[68]. As 'splitters' and 'spoilers' they have reached a kind of political impasse which might nevertheless turn out to be productive. Most notably, in this context both sides in the GM debate have set out to recruit 'resource poor farmers' to their position, and in so doing have brought them into the heart of political struggles over agricultural biotechnology. Indeed, activist organisations brokered the formation of a broad coalition of farmers' organisations in eastern and southern Africa to constitute an oppositional pressure group on states in the region to reject GMOs[69]. In a more direct initiative, the South African sustainable development organisation, BioWatch, has set out to develop alternatives to GM – and indeed industrial – farming through outreach programs with small-scale farmers to promote organic farming or 'low-cost' farming methods, supplying local markets. The organisation seeks to hook into the small but growing elite organic agriculture business[70]. Its aim is both to create an alternative viable farming system for poor farmers and to generate resistance to the deployment of GMOs because of

---

[66] Much-heralded trials of GM sweet potatoes in Kenya, conducted by KARI, ended in a failure that was widely publicised by activists. One significant exception, however, is a drought-resistant GM maize seed produced by Monsanto has recently been approved for field trials.

[67] As AGRA noted in launching its training programme for crop breeders: 'Most of the crops important to Africa – such as cassava, sorghum, millet, plantain, and cowpea – the so-called 'orphan crops,' are of little importance to researchers and educators in the developed world. As a result, there is a serious shortage of breeders of these crops. For example, there are under a dozen millet breeders in all of Africa. Yet millions of people in sub-Saharan Africa depend on millet as an important part of their diet. Conversely, most of the more than US$35 billion invested by private firms in agricultural research is concentrated in North America and Europe, on a handful of commercially important crops'.

[68] Even the claim that GM technology will solve problems of hunger has recently been muddied by the argument that GM crops will increase the production of agricultural products (e.g. cassava) that can be turned into biofuels.

[69] The same thing has happened in India, where activists and farmer-activists have sought to mobilise farmers around the issue of what GM seeds will mean for *them*.

[70] This effort coalesces with the effects of more distant activism against GMOs. When European supermarkets rejected GMOs, the upscale South African department store Woolworths, which modeled itself on Marks and Spencers, decided that if it was to establish itself as the standard for high quality food retail, it would also need to reject GM food. Consequently, it began to source some of its food products from poor African farmers who could easily become organic farmers because they were too poor to afford non-organic inputs such as pesticides, etc.

the dangers of 'contamination'[71]. In response, the pro-GM organisation, AfricaBio, has set out to demonstrate to poor African farmers the virtues of genetically engineered seeds by setting up GM maize demonstration plots in a variety of locations. Its aim is to prove to farmers the effectiveness and benefits of its Biotechnology Project, using the well-tried (colonial) method of teaching African farmers through the power of demonstration, rather than engagement: once people see what the seed can do, they will recognise it as good[72].

Each of these initiatives is small, but they indicate a significant respect in which the Biotechnology Project, as a system of bio-power, has opened up. These developments show that both the 'pro-splitters' and the 'anti-spoilers' recognise the need to take small farmers more seriously as agents within, or against, the system of bio-power. In the initial days of the GM debate, neither farmers nor local conditions were included in the dominant discourse of the Biotechnology Project, which focused very heavily on the scientific sanctity of the seed: the seed's performance was determined by its internal characteristics; evaluations in the global south were often desktop rather than field-based, and trials conducted in the US were frequently regarded as adequate for regulatory purposes elsewhere (see Gupta and Falkner, 2006); local field trials were carried out in secret locations; and the portability of the information contained in the seed was assumed. To the extent that farmers' interests and concerns have become part of the calculus of seed development and deployment, the 'distancing' character of the bio-power system is somewhat reduced, as is the degree of control expressed at its core.

5. Conclusion

As we have argued in this paper, the turn of the millennium was indeed a pivotal moment in the history of the corporate-led project of agricultural biotechnology. For two decades, anti-biotech activists and opponents of this project had developed and mobilised a trenchant critique of 'actually existing' biotechnology rather than buying into and accepting the local social constructions and global imaginaries the industry and its cheerleaders had conjured up. In Europe in particular, these alternative discourses took hold, with various conjunctural, structural, and cultural conditions facilitating a substantive challenge to the biotechnology industry's best laid plans. This was the first major disruption of the Biotechnology Project, and as such, to a system of bio-power that in part rests upon its foundations.

These developments in Europe bolstered and informed challenges to the Project in other parts of the world which have been spearheaded by a combination of activist movements and concerned governments in the global South. These challenges have focused mainly on environmental issues, biosafety, and biopiracy, but an equally important dimension has been a concern with public participation, transparency, and the extent to which these technologies advance human welfare. Today, at least in the global South, there is no possibility of going back to an 'apolitical' technology discussion which excludes questions of technology access and whether or not the technology benefits poor farmers, or its implications for fragile ecosystems and vulnerable biophysical environments. In other words, the notion of biotechnology as a good in itself is increasingly inadequate; today, the value of the technology is beginning to be measured against new yardsticks, including considerations of food sovereignty, farmer autonomy, and the preservation of agricultural communities as meaning-laden, socio-cultural units.

---

[71] This strategy resonates with the campaign by SAFeAGE to put in place GM-free zones, modeled on the European campaign.

[72] There is an extensive literature on the failures of this approach to agrarian development in colonial Africa. It also takes inadequate account of the fact that poor farmers are interested in multiple traits of a particular variety, of which yield is only one (see especially Bellon (2006)).

While it is not clear in which direction this technology will go and under what conditions it will be used, it *is* clear that many more people will be watching and participating in shaping its future.

Returning directly to Ruivenkamp's claims about the emergence of a hegemonic bio-power system, we contend that the future actually looks surprisingly open – certainly far more open than it did ten years ago. With their exclusive and unrelenting focus on the 'bads' of genetic engineering, Ruivenkamp's '*anti's*' may have helped to narrow debates about possible futures to focus on GM technologies, but their challenges have had important political and material effects that have destabilised the Biotechnology Project. Anti-GM activists helped stimulate the search for viable alternatives from organics to non-GM-based biotechnologies. They have also underscored the need to find technologies that do not wrest more autonomy from farmers, but that actually increase their power and control. The pressure GM critics have exerted on states to develop stringent and precautionary regulatory systems and to struggle against US government and industry pressures to enact US-style private property laws could also serve to reduce the 'distancing' inherent in a hegemonic bio-power system, if states use their power to constrain private capital. Of course, it is also very possible (and perhaps more politically probable) that states will choose not to play this 'anti-distancing' role, and will instead seek to foster the efforts of private biotechnology companies and seed producers to be the primary providers of GM technologies more or less under conditions of their own choosing.

As Gramsci (1971) pointed out, hegemony is never complete but is always fragile, an ongoing project of social, political and cultural struggle. As we have sought to show here, the struggle the anti-GM movement has waged against the Biotechnology Project has been counterhegemonic in two important senses. First, it has challenged the political-economic project of large multinational corporations to introduce GM seeds all over the world, and to exert control over agriculture from a distance. Second, and perhaps more importantly, it has challenged the cultural project of making a corporate-led, intellectual property-right protected technology seem normal, unquestionable (in all senses of the word), and as if it were the only possible future. Nevertheless, anti-GM activists have adopted an analytical framework and certain discursive positions that have tended to fetishize the technology. In so doing, they have made it clear that they themselves will not be in the vanguard of efforts to construct any agricultural future that involves genetic engineering. If agricultural biotechnology is to be reconstructed to meet different social ends, that task will have to be undertaken by others with a different social-technological imagination.

Acknowledgements

The authors would like to thank Teresa Gowan and Joost Jongerden for their comments on an earlier version of this paper; and the participants at the Tailoring Biotechnologies Kyoto Conference, Kyoto, Japan, Nov. 2-3, 2007, for a stimulating conversation.

References

Alliance for a Green Revolution in Africa. (2007). As Schools Begins, Unique African Partnership Announces Launch of Critical PhD Program for Crop Breeding in Africa. Nairobi.
Bellon, M. (2006). Crop Research to benefit poor farmers in marginal areas of the developing world: a review of technical challenges and tools. *CAB Reviews: Perspectives in Agriculture, Veterinary Science, Nutrition and Natural Resources* 1.

Bloomberg News/Reuters. (2007). EU blocks approval of genetically modified crops. *International Herald Tribune*: 10 October 2007. Available at: http://www.iht.com/articles/2007/10/10/business/gmo.php.

Carr, S. (2000). EU Safety Regulation of Genetically-Modified Crops: Summary of a Ten-year Country Study. Milton Keynes, UK, The Open University.

Daily Nation (Kenya). (2007). Biosafety Bill too Rushed (Editorial). 5 October 2007.

Dhlamini, Z. (2006). The role of non-GM biotechnology in developing world agriculture. SciDevNet. 7 February 2006. Policy Brief.

Dickson, D. (2007). Africa must create its own biotechnology agenda. SciDevNet, 12 June 2007.

Freidberg, S. and Horowitz, L. (2004). Converging Networks and Clashing Stories: South Africa's Agricultural Biotechnology Debate. *Africa Today* 51: 3-26.

Gaskell, G., Allum, N., Bauer, M., Durant, J., Allansdottir, A., Bonfadelli, H., Boy, D., de Cheveigne, S., Fjaestad, B., Gutteling, J. M., Hampel, J., Jelsoe, E., Jesuino, J. C., Kohring, M., Kronberger, N., Midden, C., Nielsen, T. H., Przestalski, A., Rusanen, T., Sakellaris, G., Torgersen, H., Twardowski, T. and Wagner, W. (2000). Biotechnology and the European Public. *Nature Biotechnology* 18: 935-938.

Gaskell, G., Stares, S., Allansdottir, A., Allum, N., Corchero, C., Fischler, C., Hampel, J., Jackson, J., Kronberger, N., Mejlgaard, N., Revuelta, G., Schreiner, C., Torgersen, H. and Wagner, W. (2006). Europeans and Biotechnology in 2005. Patterns and Trends: Final Report on Eurobarometer 64.3. *A Report to the European Commission's Directorate General for Research*, pp. 87. Available at: http://ec.europa.eu/research/biosociety/pdf/eb_64_3_final_report_second_edition_july_06.pdf

Gottweis, H. (1998). *Governing molecules: the discursive politics of genetic engineering in Europe and the United States.* Cambridge, Mass., MIT Press.

Gramsci, A. (1971). *Selections from the Prison Notebooks.* New York, International Publishers.

Gupta, A. and Falkner, R. (2006). The Cartagena Protocol on Biosafety and Domestic Implementation: Comparing Mexico, China and South Africa. In: *EEDP Briefing Paper 06/01.* Chatham House.

Hickman, M. and Roberts, B. (2000). Commons Statement over GM Seed Row. *PA News (wire service).*

Jasanoff, S. (1995). Product, Process or Programme: Three Cultures and the Regulation of Biotechnology. In: *Resistance to New Technology,* (Ed. M. Bauer), Cambridge, England, Cambridge University Press, pp. 311-331.

Jasanoff, S. (2005). *Designs on nature: science and democracy in Europe and the United States.* Princeton, N.J., Princeton University Press.

Kloppenburg Jr., J.R. (1988). *First the Seed: The Political Economy of Plant Biotechnology.* Cambridge, Cambridge University Press.

Levidow, L. (2000). Pollution Metaphors in the UK Biotechnology Controversy. *Science as Culture* 9: 325-351.

Michael, M.T. (2006). Africa Bites the Bullet on Genetically Modified Food Aid. Worldpress.org.

Mnyulwa, D. and Mugwaga, J. (2003). Agricultural Biotechnology in Southern Africa: a Regional Synthesis. Working Paper no.1, prepared for FANRPAN/IFPRI Regional Policy Dialogue on Biotechnology, Agriculture, and Food Security in Southern Africa. Johannesburg, April 23-25.

NEPAD. (2005). Africa's Science and Technology Consolidated Plan of Action.

Ogodo, O. (2007). Kenyan maize variety resistant to boring pest. *SciDevNet*: October 18 2007. Available at: http://www.scidev.net/news/index.cfm. retrieved on October 24, 2007.

Pingali, P. and Raney, T. (2005). From the Green Revolution to the Gene Revolution: How Will the Poor Fare? ESA Working Paper No. 05-09.

Purdue, D.A. (2000). *Anti-genetiX: the emergence of the anti-GM movement.* Aldershot, Ashgate.

Ruivenkamp, G. (2005). Tailor-made Biotechnologies: Between Bio-power and Sub-politics. *Tailoring Biotechnologies* 1: 11-33.

Sand, P.H. (2006). Labelling Genetically Modified Food: The Right to Know. *Review of European Community & International Environmental Law* 15: 185-192.

Schurman, R. (2004). 'Fighting Frankenfoods': Industry Structures and the Efficacy of the Anti-Biotech Movement in Western Europe. *Social Problems* 51: 243-268.

Schurman, R. and Munro, W. (2003). Making Biotech History: Social Opposition to Agricultural Biotechnology and the Future of the Biotechnology Industry. In: *Engineering Trouble: Biotechnology and its Discontents*, (Eds. R. Schurman and D.T. Kelso), Berkeley, University of California, pp. 111-129.

Schurman, R. and Munro, W. (2006). Ideas, Thinkers, and Social Networks: The Process of Grievance Construction in the Anti-Genetic Engineering Movement. *Theory and Society* 35: 1-38.

Schweiger, T.G. (2001). Europe: Hostile Lands for GMOs. In: *Redesigning life?: the worldwide challenge to genetic engineering*, (Ed. B. Tokar), London, Zed books, pp. 361-372.

Scoones, I. (2005). Contentious Politics, Contentious Knowledges: Mobilising against GM Crops in India, South Africa, and Brazil. Institute for Development Studies Working Paper 256.

Segarra, A. and Fletcher, S.R. (2001). Biosafety Protocol for Genetically Modified Organisms: Overview. Library of Congress. Washington DC: Congressional Research Service. The Library of Congress (US).

Smith, E. and Rajan, K.S. (2007). Therapeutic Values: Food, Drugs and the Promise of Health. Presentation at the conference Tailoring Biotechnologies: Reconstructing Agro-Biotechnologies for Development?, Kyoto, November 3-5, 2007. Available at: http://www.tailoringbiotechnologies.com/Kyoto2007/Elta_Smith_&_Kaushik_SRajan_ppt.pdf

South Africa, Department of Agriculture. (2005). Genetically Modified Organisms Act, 1997.

Thomas, J. (2001). Princes, Aliens, Superheroes and Snowballs: The Playful World of the UK Genetic Resistance. In: *Redesigning life?: the worldwide challenge to genetic engineering*, (Ed. B. Tokar), London, New York, Zed Books, pp. 337-350.

Tjaronda, W. (2006). What are the Dangers of Biotechnology? Namibia, New Era.

Udongo.org. (2007). Biosafety Bill Kenya - Next?

Waugh, P. (2000). MPs demand destruction of GM rape crop. *The Independent (London)*: 19 May 2000.

Wright, S. (1998). Molecular politics in a global economy. In: *Private Science: Biotechnology and the Rise of the Molecular Sciences*, (Ed. A. Thackray), Philadelphia, University of Pennsylvania Press, pp. 80-104.

Part II. Opposition and participation

Tidy back yards or global justice? Types of rural GMO opposition in Austria and France and their wider implications

Franz Seifert

1. Introduction

In line with the thrust of this book this article proceeds on the assumption that social conflict constitutes a potentially productive force: by elaborating a critique of a given socio-political order believed to be based on natural necessity, social conflict reveals the possibility of alternatives to this order and creates leeway for political maneuver (see Ruivenkamp, this book). Social conflicts, however, vary widely in shape and salience as well as in their impact on public discourse and political decision-making. In fact, only a small portion of movements actually musters the critical mass in terms of public support to stimulate significant debates and provoke policy change. It is therefore worthwhile to explore the factors shaping those social movements that come to mobilise mass publics and examine their wider consequences. In our case, we will ask for the impact of public controversy against the backdrop of this book's guiding vision - the reconstruction of biotechnologies to meet the needs of the poor in the global South.

The conflict we will deal with is one the most effectual controversies in recent decades – the mobilisation against agro-food biotechnology in the European Union. The movement, to cite only its most outstanding features, resulted in a moratorium on approvals of genetically modified organisms (GMOs) in the EU, in protracted tensions between these Member States and the European Commission, and in a considerable tightening of the EU's biotechnology policy. To the U.S., as the world's largest developer and producer of commercial GMOs, these developments came as a challenge to vital trade interests. Therefore the U.S. along with Canada and Argentina filed a complaint with the World Trade Organisation (WTO) against the EU for its halting biotechnology policy. These European and transatlantic tensions amount to a significant perturbation of a (bio-) technological trajectory that has drawn criticism not only for the environmental risk agro-biotechnology is alleged to entail, but also for a socio-politic regime going along with it – a regime marked by exclusionary property rights, global trade liberalisation, corporate power concentration and agricultural intensification – all developments particularly aggravating the current crisis of farmers around the world. Furthermore, the consequences of these transatlantic developments are were not restricted to the industrialised nations directly involved but also have significant repercussions on the global governance of agro-food biotechnology and on biotech-policies of developing countries (Falkner and Gupta, 2006; Seifert, 2006a; Schurman and Munro, this book).

As a detailed enquiry into these consequences in their entirety goes beyond the scope of this article, we concentrate on particular aspects of the European anti-GMO mobilisation, namely on the way farmers in certain EU Member Countries – here we focus on Austria and France – resist to agricultural biotechnology and, further, the role these farmers come to play in these countries' anti-GMO policies respectively. After describing these forms of resistance and analysing their causes – including the reasons for national variations – we will reflect on the impact of these contentious processes on the European and global reconstruction of biotechnology.

Several reasons account for the choice of this approach. First, the focus on farmers (or peasants) draws on the insight that the fate of agricultural producers is at the very heart of the debate on reconstructing biotechnologies (Jongerden, this book). As this contribution will demonstrate, this does not only hold for farmers, peasants and rural workers in the global South, but also for those in the industrialised West. In fact, as these countries still dominate global politics due to their political and economic weight, critical discourses on the rural world prevalent in Western societies and the impact of these discourses on national or regional agro-food policies are more likely to impinge on global developments than comparable discourses and policies in the global South. Furthermore, to understand rural resistance to agro-food biotechnology in Northern countries can inform reflection on the possible linkages between arenas of rural biotechnology-reconstruction in the industrialised Northern countries and their counterparts in the global South.

Second, choosing a comparative approach and exploring how rural resistance to agro-biotechnology plays out in different locales sheds light on the significance of national contexts. Austria and France lend themselves to such a comparison as in both countries rural GMO-opposition significantly impinges on national policies while, at the same time, both countries extremely differ in this respect: both countries passed through heated GMO-controversies, are marked by a public opinion hostile to agro-food-biotechnology and defend a controversial anti-GMO policy in the EU; Yet, variation going along with national contexts is considerable: while Austria runs a consensual, State supported, purely inward looking 'NIMBY' (Not-In-My-Back-Yard) policy, designed to protect the interests of organic farmers, in France, public discourse on agro-biotechnology is highly contentious, with a vocal radical farmers association driving biotechnology opposition, employing tactics of civil disobedience and radical activism, and pursuing a global vision of an alternative agriculture in a 'NIABY' (Not In Anybody's Back Yard) world.

This introduction is followed by four parts. The next part will give a sketch of the European anti-GMO movement. It is followed by the two accounts of farmers' resistance in Austria and France. The fifth, and concluding part will draw an analytical comparison of the national cases, outlines the reasons for these differences and eventually discuss the contribution these two observed models – State protected NIMBYism versus contentious NIABYism – can deliver for the potential development of alternative, locally adapted forms of agro-biotechnology.

2. The European anti-GMO movement

The power of the European anti-GMO movement owes itself to a number of factors; first, the coincidence of two sensitising public events; second and probably most important, perturbations of the EU's GMO approval system due to certain Member States' recalcitrant biotechnology-policies, and third, targeted activism of a number of influential, mostly international NGOs. In most cases, rural resistance to GMOs had a minor impact on national and European policy change, however, rose in importance in more recent phases.

In 1996, the institutional crisis over the mad cow disease and the arrival in Europe of first shipments of non-labeled GM products sensitised the – until then largely indifferent – European public to the emergence of new foods believed to carry hitherto unknown health risks. The ensuing movement against the food-technology took two to three years to gain momentum, and spread unevenly across national publics: The first controversy occurred in Austria, followed by public debates in Greece and Ireland in

1996 and 1997; in 1999 public clashes followed in France, the UK, Belgium and Italy (Seifert, 2006b). Under the impression of popular anti-GMO sentiments governments adopted precautionary policies. A growing number of States – among them Austria and France – issued safeguard bans on GMOs already approved for marketing in the Common Market und thereby undermined EU harmonisation goals[73]. The EU approval system's eventually came to a standstill in summer 1999, when an alliance of five countries – France, Greece, Denmark, Italy and Luxembourg – declared to block further approvals until completion of the then ongoing amendment of the EU regulatory framework[74].

This Community-wide ban on GMO approvals is the most important episode in the European movement. It had two principal – an somewhat contradictory – effects. First, the moratorium heightened pressure on the Commission to design a strict regulatory regime able to appease Member States and consumer concerns. Member States' demands coupled with dysfunctional Community procedures led to a 'trading up' of EU regulatory standards, moving them closer to the position of the most critical Member States (Vogel, 1997). In order to restore consumer confidence, the new regulatory framework that went into force in 2003 imposed severe restrictions on agricultural biotechnology by enhancing the status of the precautionary principle in the approval procedure and setting up a strict labeling and traceability regime. Even though a regulatory cleavage had appeared in early 1990s, the revised framework sealed the divergence between EU and US regulatory standards which maintain a far more permissive approval policy and do not require labeling (Bernauer and Meins, 2003; Vogel, 2001).

As a second and countervailing effect, the moratorium heightened pressure on the European Commission to comply with WTO standards. The WTO regards product bans that cannot be scientifically proven as obstacles to international trade. In 2003, the U.S. together with Canada and Argentina, raised this charge against the European moratorium with the WTO, which a WTO jury later confirmed, declaring that national bans and the Community-wide blockage of GMO approvals violated free trade principles in causing undue delays to importers (WTO, 2006). The WTO case reflects the authority of international free trade disciplines weighing heavily the European Commission, forcing the EU's executive body to unblock the approval process, to lift illegal national safeguard bans and, in spite of the EU's 'different way', to ensure compatibility with the US regulatory system (Skogstad, 2006).

What these transatlantic and European developments meant for the global uptake of modern biotechnology or the tailoring of biotechnologies suited to the needs of the poor in the global South is yet to be assessed. It is, however, beyond doubt that the European anti-GMO movement and its inner-European consequences *have* such effects. For example, in international negotiations on agreements and institutions governing the global regulation of biotechnology (like the Biosafety Protocol or the Codex Alimentarius) the EU systematically promotes its precautionary and consumer oriented approach in order to bring the global governance of biotechnology in line with its own regulatory concept (Von Homeyer, 2006; Seifert, 2006a; Skogstad, 2001). Other types of consequences, like direct influence on national or regional regulatory frameworks in the global South, processes of linkage and adaptation in markets or civil society or concerted policy responses and alliance building on the part of developing countries are plausible but still need further empirical examination.

---

[73] In 1997, Austria and Luxemburg issued such bans, later other Member States followed. Between 1997 and 2000, such were decreed on 13 occasions by Austria (3), France (2), Germany (1), Italy (4), Luxembourg (1), Greece (1) and the United Kingdom (1) (who later withdrew its ban).

[74] Later on, in 2000 and 2001 respectively, Austria and Belgium joined the blockade group.

Examining the driving forces of the European movement, as first and decisive factor we identify recalcitrant Member States who adopted national bans, blocked the approval process and thus pushed the Commission to tighten biotech-regulations (Seifert, 2006b). Beyond States, however, NGOs were pivotal in organising public protest which, in turn, forced these States to revise biotechnology policies. The ambit of involved groups ranges from environmentalist NGOs to consumer associations to critics of neo-liberal globalisation. Activists employ a wide mix of tactics and forms of direct action: destroying GM crop test sites, lobbying governments to support GMO-bans or major food retailers to go GM-free, monitoring nations and companies for compliance with declared GM-free or precautionary policies, staging sensational protests against cases of GM contamination or bio-patenting, or challenging scientific claims over the safety of GM product by confronting them with differing scientific expertise. NGOs' geographical reach varies, too: While many groups are rooted in their local constituencies, international NGOs are both organising national campaigns and masterminding trans-national campaigns. It is therefore international NGOs like Greenpeace International and Friends of the Earth Europe (FoEE) who play the decisive role in the coordination of the pan-European anti-GMO movement, and whose 'activity overshadows that of any other group' (Ansell *et al.*, 2006: 103).

Compared to international NGOs, groups representing agricultural producers in general take a low profile, even though in most countries peasant associations are part of national anti-GMO advocacy coalitions. Among agricultural interest groups there tends to be a divergence between the predominant, corporatist organisations lobbying for middle-sized and big farmers, who embrace a technology which is potentially beneficial to them and thus adopt wait-and-see attitudes or openly advocate agro-biotechnology, and such groups standing up for economically vulnerable or quality food producers (see Levidow, this book), who tend to oppose the genetic technology.

In Europe, organic farmers make up a large portion of quality food producers. Organic farmers are among the most ardent adversaries of agro-biotechnology because the definition of what is an organic product is stipulated by European directives and these, since the early 1990s, categorically rule out the use of modern biotechnology in organic production. The adventitious presence of GMOs in organic produce therefore seriously undermines the marketing potential of organic products and poses a threat to their producers. This is the reason why farmers', in particular organic farmers' involvement in the anti-GMO movement has intensified in the wake of the EU's 'coexistence policy': the policy recognises the right of farmers to either opt for or against agro-biotechnology and seeks a way to ensure the orderly side-by-side of both production types. As the specification of this policy is left up to Member States, whose attitudes vary widely, and its terms still are unsettled, in a number of countries GM-free producers, organic farmers in particular, emerged as vocal actors in the policy field (Seifert 2006c).

Following these general remarks on the European anti-GMO movement, a closer examination of rural resistance in two countries will sharpen our eyes for the manifold and contrasting ways of how these conflicts play out in different locales, societies and national contexts. Our next case, Austria, will highlight the role of organic farmers and protectionist agricultural policies in political conflict management. The French case, contrastingly, illustrates how political conflict management fails in the face of radical peasant resistance that challenges law, science and the State and promotes a strong global justice agenda.

3. Austria

Among EU member countries Austria is one of the most vigorous agro-biotechnology opponents, domestically, the country does whatever possible prevent the introduction of GM crops. The origin of this stance goes back to the mid-1990s, when a first, illegally conducted field trial caused an intense anti-GMO mobilisation resulting in a popular initiative that brought a resounding victory to GMO opponents. The mobilisation had been supported by the highly influential, conservative tabloid *Neue Kronen Zeitung* (NKZ), which maintained its anti-GMO stance ever since. Public opinion, as evinced by countless opinion surveys, is staunchly anti-GM to a degree such that nowadays virtually no public decision-maker dares to speak out in favor of the agricultural technology.

In essence, the policy adopted against the backdrop of hostile public opinion is a State-run NIMBY policy: Any government responsible for a first GMO release 'In Our Back Yard' risks sensitive losses in popular support, and therefore any effort is being made to prevent GMO-releases on Austrian ground. Up to now, this policy was successful: to date not even GMO releases for scientific purposes have been conducted. Against the commercial growing of GM crops, which is under EU jurisdiction, Austria sought 'protection' through its three safeguard bans. For eight to ten years – depending on the time of issue – Austria managed to maintain these bans, riding out two Commission initiatives to overturn them. A third attempt, at the Council of Ministers on 30 October 2007, finally succeeded, when environment ministers could no reach a qualified majority to rebuke the Commission proposal.

But things are never as bad as they seem. The decision does not seriously destabilise Austria's NIMBY policy. First, the motion only applies to a ban on importing and processing into food and feed of controversial corns but does not question the cultivation ban. The NIMBY's major concern – No GMO-cultivation in My Back Yard – thus remains unaffected.

Second, in recent years Austrian agricultural policy makers adopted a new pre-emptive strategy that promises to keep Austria 'GMO-free' even after an eventual abolition of the ban on GMO cultivation. The policy takes advantage of the EU's crystallising coexistence policy and setting up legal regimes at the regional level which, in theory, are in line with the EU's liberal demand but prohibitive in practice (Seifert, 2006c).

The Austrian NIMBY policy is based on an all-embracing political consensus: Government and political opposition concur in their denouncing of agro-biotechnology; Among NGOs, Greenpeace and Global 2000 who, in the campaigning phase of the late 1990s, had pressurised government on GM food labeling, bio-patenting and bans – have become part and parcel of a State-run NIMBY strategy. Certain prominent NGO-activists, for example, have become important State partners: Some have embarked on successful party careers in government, others have specialised on the furnishing of scientific expertise used to justify anti-GMO policies (*ibid.*).

Politically organised farmers, organic farmers in particular, also play a crucial – albeit quite different – role in Austria's NIMBY policy. Their significance does not lie in the public arena, where the international NGOs Greenpeace and Global 2000, allied with the NKZ, figure as most visible and effectual movement actors, discontented farmers, however, never became main voices in the public and hardly ever staged public protest. Yet, the role of *organic farming* in Austria's agro-biotechnology policy is key. Organic farmers, are the main beneficiaries of a national policy that ultimately aims at making the whole country a 'GMO-free zone' and, indeed, government commonly justifies its NIMBY stance as a measure for the protection of organic farmers.

The prominence of organic farmers shows most clearly with the Austrian co-existence policy (*ibid.*). Intended by the Commission to institute an orderly side-by-side of GM- and non-GM-producers, Austria turns co-existence into an instrument to pre-empt GMO cultivation: With roughly 10 percent of farmers embracing this production type, the density of organic farms is high. Separation distances set to prevent the intermingling of crops therefore enormously expand the area ineligible for GMO cultivation. Combined with an overly bureaucratic hurdles and local, often State sponsored 'GM-free' campaigns this makes for a sufficiently prohibitive environment for agro-biotechnology.

While the genetic technology's unpopularity is the immediate political cause for Austria's NIMBY-policy, this policy also serves the interests of organic farmers, despite the fact that organic farmers are a small minority among farmers. Until the late 1990s, other influential agricultural stakeholders representing the majority of conventional farmers opposed an outright GMO-free policy and warned that agro-biotechnology's potential uses should not be forfeited. In later years, however, Austrian agricultural policy underwent a shift and succeeded to persuade majority stakeholders to accept the NIMBY-strategy: In the wake of the European anti-GMO movement GMOs became unpopular among European consumers, thus the argument proved convincing that a rigorous GMO-free policy would give a competitive edge to *all* Austrian producers.

Nevertheless, the fact that organic farmers are particularly privileged by this policy warrants explanation. The reason is that Austria's GMO-free approach fits well into the more general strategic outlook of the country's agricultural policy. Since the late 1980s Austria embarked on a 'greening' of its agricultural policies, seeking to reconcile agricultural production, environmental protection, rural development and European integration. Organic farming came to play a crucial role in this reform, as this production type appeared as model solution for these demands: organic farming delivers popular niche products, enjoys a positive image in the public and allows for the allocation of State subsidies in the context of liberal integration of European and global agro-food markets. Organic farms receive payments within the agro-environmental program (AEP), accounting for a third of the entire agricultural budget. With about 80% of farms participating in the AEP (in 2005), Austria is among the European forerunners in environmental policy making (BMLFUW, 2006: 10). And with roughly 9 percent of agricultural area under organic cultivation and a 10 percent share of organic farms Austria is among the European leaders in this respect too (Groier and Gleirscher 2005: 15).

This large share is owed to heavy State subsidies, which are a result of a political deal struck in the public debates around Austria's accession to the EU: as both the quality of food and the insecure fate of Austrian farmers – many of them uncompetitive mountain farmers – became issues in these debates, the promise to support organic farming reconciled various policy ends: it created public acceptance of Austria's accession to the EU, introduced an additional subsidy for producers in disadvantaged mountain regions, and provided protection to an important political clientele. In addition, Austrian retailing chains pioneered the marketing of 'bio'- products and thus confirmed the commercial viability of organic agriculture. In the longer term, organic farming became the 'sacred cow' of Austrian agricultural policy so that anything which might corrupt this sector provokes aversive reactions by policy-makers.

Yet, all this is not to say that, in Austria, farmer associations do not carry the *potential* to challenge mainstream agricultural policies or to pursue internationalist agendas. The (albeit very small) Austrian Mountain Farmers Association (OEBV), for example, was among the organisers of the public initiative. In fact, it was the OEBV who put forward the idea for a plebiscite already in 1995, months before its most visible protagonists took it up; Only later, when the NKZ stepped in and created the opportunity for a successful mobilisation, the other organisations jumped on the bandwagon to employ their well-tried PR strategies. Furthermore, in contrast to the leading farmers' representatives, the OEBV is internationalist in outlook and member of progressive, international farmers' organisations such as the *Coordination Paysanne Europeenne* (CPE) and the international peasant organisation *Vía Campesina*[75]. The tiny Austrian NGO with its 300 odd members holds a much broader perspective than is entailed by the national NIMBY policy.

In brief, even though certain Austrian farmer organisations have the potential to disturb the ostensible harmony in Austria's agricultural policy and to think in dimensions beyond the narrow national or clientelistic interest, the influence such groups have on the Austrian anti-GMO movement comes discreetly, for example, through conceptual background work. In general, however, a receptive agricultural policy prioritising ecological functions and organic farming creates a situation obfuscating tensions within the agricultural policy field. To this adds the unrivalled hegemony of the governing Austrian People's Party and agro-lobbies. We hence conclude that Austria's NIMBY policy is not the result of power struggles but of negotiation processes within an receptive policy context against the backdrop of agro-biotechnology's extreme unpopularity with the general public.

4. France

In the European context, France is, like Austria, a strong opponent of agro-food-biotechnology. Like Austria, France supported the European moratorium on GMO-approvals, and in fact was – together with Greece – one of its instigators. France also maintained national safeguard bans contrary to the opinions of the European Commission and the WTO. And like in Austria, this recalcitrant position reflects a largely hostile public opinion. Differences between the two recalcitrant countries, however, begin with the way this opinion played out in the course of a public controversy: in France the controversy first climaxed in summer 1999 – two years later than Austria; and – unlike Austria, where local chapters of international environmental NGOs like Greenpeace and FoEE steered the framing process – in France the farmers association CP and its spokesman José Bové took centre stage. Through these, the GMO discourse took on a dimension that goes far beyond a concern of keeping one's backyard clear of GMOs. Rather, GMOs were made a key element in a wider reaching, radical critique of modern State and global capitalism linked to the emerging global justice movement. To understand why critical French discourse took this NIABY-turn we will first give a sketch of the CP and José Bové and, in a subsequent step, illustrate their role in the French anti-GMO movement.

4.1. The Confédération Paysanne

The *Confédération Paysanne* (CP) was founded in 1987 as a fusion of two peripheral farmer associations at the radical left, which both had emerged from the *Paysans-Travailleurs* founded in 1972. Both groups were split-offs from the hegemonic, conservative FNSEA (*Fédération nationale des syndicats d'exploitants*

---

[75] In fact, the OEBV is a founding member of the CPE in 1974. (homepage CPE: http://www.cpefarmers.org/, homepage *Vía Campesina*: http://viacampesina.org/)

agricoles), who, internally, muted criticism from discontented farmers of Western and Southern France. The two groups thus merged to augment their capacity to challenge the dominant FNSEA who, in turn, marginalised the new contenders (Martin, 2005).

The key episode in the emergence of the new left current, later to converge into the CP, was the struggle for the *Larzac* – a limestone plateau in the Southern *Massif Central*, where, in 1971, farmers had begun non-violent protests to block the expansion of a military base. The conflict lasted until 1981, when François Mitterrand finally dropped the plan. The episode with its enormous waves of solidarity among farmers rallying in the *Larzac* was the formative period of the CP, which developed a peculiar set of principles inspired by critical, anarchist and ecological thinking: a defense of small scale agriculture respectful of the environment; the refusal of the predatory capitalist competition prevalent in the farming sector; the denunciation of the cultural homogenisation that go along with liberal globalisation. What is more, the *Confédération*, in contrast to other agricultural lobbies, entertains an *internationalist* vision and advocates anti-colonialism and anti-imperialism, global solidarity among farmers, workers and consumers, and it condemns the liberalisation of global agro-markets as well as compensatory regimes of agro-subsidisation in the EU and U.S.. Correspondingly, and in contrast to the typical farmers association, the CP, being integrated in a global network of movement actors and activists of a multitude of shades, resembles an NGO or social movement actor rather than a corporatist farmers union. Furthermore, similar to the Austrian OEBV, the CP entertains institutionalised links to the European CPE, and at global scale, to the international small-holders' organisation *Via Campesina*. But in contrast to the tiny and rather specialist Austrian NGO the CP is a mass organisation. One factor that, in recent years, contributed to the CP's growing popularity was the charismatic activist and CP spokesman José Bové.

4.2. José Bové

While José Bové is a founding member of the CP, he is by no means a typical farmer. Born in 1953 to parents who are both agricultural scientists, Bové spent most of his childhood in the U.S. and most of his teen age in Paris, then vibrating with revolutionary ideas. There he developed a pacifistic and anarchistic posture that led him to refuse draft and go underground in the countryside. Anti-militaristic commitment brought him and his wife Alice to the *Larzac*, where the couple took part in the huge rallies organised by the *Paysans-Travailleurs*, and took possession of a deserted farm, initially to prevent the army from seizing the land, later to set up home and, in following years, establish a profitable production of Roquefort cheese. Yet, José Bové never gave up to engage in non-violent ecological, anti-nuclear and anti-colonial activism, exercising tactics of civil disobedience that sometimes resulted in court trials and manifestations of solidarity; but it was not before summer 1999 he gained prominence as a key figure of the CP.

4.3. Role in the French GMO-debate

We now turn to the question of how the CP and Bové got involved in the French anti-GMO struggle. In its initial phases the CP kept a low profile. In fact, it was not until early 1997 that it adopted an outspoken anti-GMO stance (CP, 1997). Greenpeace France, by contrast, had built up critical expertise on agro-biotechnology since the early 1990s. It was therefore well prepared when, in 1996, the NGO's headquarters decided to launch a first European anti-GMO campaign[76]. But while attempts made by Greenpeace France to sensitise the French public achieved only moderate press coverage, it were opposing government decisions that gradually raised public awareness in 1997.

In 1998, the CP embarked on a mobilisation strategy – well established in preceding protest struggles – that, in the longer run, made it the opinion leader on GMOs: in January 1998, a group of CP activists entered a Novartis warehouse in *Nérac* to destroy a reserve of GM maize seeds. When the three responsible activists – René Riesel, José Bové and Francis Roux – appeared before court and were convicted to a suspended sentence, the CP brought together a selection of vocal biotechnology critics to turn the court hearing into a public trial against GMOs. In summer, CP activists increasingly resorted to acts of field vandalism, and in summer 1999 the CP scored its next spectacular coup: on 2 June in the village *Gaudiès*, CP members and a group of South Indian peasant farmers, touring Europe to protest the G8 meeting, first destroyed test fields planted with GM colza in the public research facility Cetiom[77]; and later, led by René Riesel and José Bové, destroyed transgenic rice plantlets in a greenhouse of the public research centre CIRAD[78], which deals with innovations for rural use in developing countries. With this series of direct actions in company of Indian activists, the CP extended its critique of GMOs beyond a purely ecological and national frame to denounce the complicity of corporate interest and public research and bestow their action with internationalist meaning; a NIABY critique of GMOs took shape.

The breakthrough came a month later in the wake of the '*Millau* McDonald's incident': on 12 August 1999, José Bové and about a hundred CP activists dismantled a McDonald's store under construction in the town of *Millau*. Rather than the deed itself it was the ensuing repression by State authorities that turned the balance. What had begun as a routine protest against WTO-backed U.S. punitive tariffs on French Roquefort aroused the public ire when Bové was arrested and kept in custody for several weeks. Suddenly, Bové was propelled into nation-wide celebrity – and turned out a charismatic communicator. He seized the opportunity to stage the CP's radical critique of industrial agriculture, liberal globalisation and cultural homogenisation. One motive created by Bové that particularly stroke a chord in the French public was that of the *malbouffe* – a synonym for junk food, unhealthy, industrial, and GM food products. In view of the notorious French preoccupation with taste, *cuisine* and dining, it was an easy task to politicise the ongoing French debate on alimentation and to make *malbouffe* a symbol for the power of multi-nationals and the destruction of dear traditions, identities, communities by the forces of globalisation (see Nicolosi, this book, for the cultural dimensions of food).

Not only in France Bové proved to be 'the right man at the right time'. He also managed to bring the CP's critical discourse to a global audience: a few months after the *Millau* incident, Bové established himself as figurehead of the rising international anti-globalisation movement by staging his criticism amid the widely televised protest storm in Seattle (Heller, 2002: 29-33). In the following years, Bové had

---

[76] Within the Greenpeace organisation, the French expert Arnaud Apoteker was among the first to recognise the pertinence of agro-food biotechnology to the environmental NGO's agenda and to advocate a pan-European approach. First attempts to organise a common strategy, however, failed in 1994. Only later in the following year, as in became clear that, in 1996, a first, massive wave of GM products was to enter the European market, the decision to prepare a European campaign was taken. (Interview, Greenpeace, 6.7.2005, Zurich).

[77] *Centre d'études technique interprofessionnel des oléagineux métropolitains*. The following year, on 13 April 2000, the same Cetiom test site in *Gaudiès*, newly planted with transgenic colza, was destroyed by CP activists, supported by and officials of the French Greens and led by José Bové.

[78] *Centre de coopération internationale en recherche agronomique pour le développement.*

appearances in most WTO-meetings and World Social Forums held in places like Doha, Porto Alegre and Hong Kong. Bové's sudden international prominence allowed him to establish a new framing: from hazardous life forms, he turned GMOs into symbols of global capitalism and cultural homogenisation forced upon citizens around the world just like MacDonald's fast-food (Heller, 2002).

In France, Bové never lost hold of his popularity, even augmented it over the years to push on the GMO-debate. The principal tactic employed to stay in public focus were GMO field destructions hitting test sites and commercial plantations alike[79]. What aroused most public discussion was the series of spectacular court trials and, at times, prison sentences these events entailed. Since, usually, only instigators were held accountable, José Bové stood trial time and again to be convicted to prison sentences and monetary penalties[80]. While Bové insisted on the political and symbolic nature of field destructions, prosecution denounced them simply as malicious injury of property. Most fervent debates turned around the disputed legitimacy of jail sentences imposed on a syndicate leader, who was thus turned into a political prisoner. The wave of sympathy and solidarity prompted by the trials reached out far beyond France's ecologist and radical left into the socialist and conservative camp and made it extremely difficult for the State to impose strict sanctions. A tacit game of provocative civil disobedience and mitigated juridical counteraction unfolded: On the one hand, Bové's perpetrations are too evident to go unpunished, on the other, the prosecution of a folk hero is likely to backfire on authorities. Bové, in turn, exploits this ambiguity in a tacit game of provocation, applause, mitigated State repression and blame.

Over the years, thousands of activists followed suit: At the meeting, '*Larzac 2003*' in August 2003 – a 30 years jubilee of the protest episode in the 1970s that became a milestone of the *altermondialisme* or French global justice movement (Agrikoliansky *et al.*, 2005) – Bové called on followers to join the *faucheurs volontairs*, an activist network dedicated to ransacking GMO plots. Immediately, hundreds of sympathisers signed up growing to 6.000 members in 2006. In summer 2003 and subsequent years, the *faucheurs* 'mowed' around half of all of authorised field trials. Field destructions entailed a range of reactions: in rare cases direct confrontations with police forces, counter-demonstrations by farmers defending their right to grow GMOs and, in one instance, an encounter with a farmer armed with a gun. Most important, again, is the series of judicial proceedings prompted by the actions; since *Larzac 2003*, at least 17 field operations conducted by the *faucheurs* led to court trials some of which received considerable media attention and provided a stage to spread the GMO-adversaries' issue framing and trigger general debates on GMOs, civil disobedience and the (il-)legitimacy of State repression.

Beyond challenging State authority, peasant opposition to GMOs, driven by the CP and José Bové, is marked by a deliberate alliance policy: José Bové soon became the figurehead of the nascent *altermondialist* movement, which made him appealing as an ally for political parties on the left and extreme left and, in particular, the French Greens. Yet, Bové stressed the CP's distance to party politics which he found too constraining and too ready to compromise. Instead, both the CP and Bové established themselves as *movement* actors, appealing to civil society *because* of their detachment from party politics. They nevertheless created alliances with political parties, where these were opportune.

---

[79] The latter are justified as symbolic attacks on the intermingling of research and corporate interest in the life sciences. As commercial GMO cultivation resumed not before 2005, only recently commercial GMO plantation became targets for activists.

[80] From 1998 to early 2007, in a series of Court trials followed by objections and their rebuttal, Bové was found guilty of malicious injury of property in at least five instances.

Most successful was the alliance backing the campaign of the referendum on the EU Constitution in spring 2005, which united the radical left and parts of the Green Party and Socialists[81].

Overall, the *faucheurs'* Luddite tactic of GMO test-site sabotage is at once categorical and symbolic[82]. It is categorical in that it attacks *any* GMO release into the environment regardless of whether it serves commercial or alternative ends. Thus, anti-GMO activism targeted GMO test sites with plants designed to meet developing countries needs, as in the case the CIRAD plantlets of transgenic rice, and scientific test sites, as was the case with the Cetiom tests for studying the environmental risks of transgenic colza. In both cases the CP transformed the ensuing trials on malicious injury of property into trials on the legitimacy of public GMO research intended for developing countries or risk research. The CP thus successfully transformed the legal dispute into a public debate questioning the very possibility of scientific neutrality on the subject of GMOs.

But anti-GMO activism is symbolic, too, as it does not merely attack GMOs as physical objects but also as *pars pro toto*: by attacking GMOs, opponents denounce the socio-economic order that goes along with the technology: cultural homogenisation, corporate power, neo-liberal globalisation and the State's complicity therein. Linked to this broader ideological agenda is – atypical for a farmers lobby – the CP's internationalist outlook: advocacy of North-South solidarity, rejection of global market integration, and support of small farmers and peasants around the world[83]. Seen in this context, French GMO opposition is diametrically opposed to the Austrian NIMBY approach, shutting out the unwanted intruder from the – national – backyard; rather it envisions a NIABY world, banning GMOs – and the social ills that come with them – from *anybody's* precincts, particularly those in the Developing World.

5. Conclusions

The two national cases highlight the crucial influence of national contexts on the way this rural resistance to GMOs is being played out: in France, oppositional farmers challenge State authority and are key actors in a global justice movement that promotes a demanding global agenda; In Austria, oppositional farmers, organic farmers for the most part, hardly display any opposition since they are protégées of a consensual, State-run policy of national seclusion. Apparently, forms of resistance to GMOs on the part of those who are affected by technological globalisation do not keep to a simple, universal stimulus-response model but highly depend on national circumstances[84]. In the following we will discuss two questions: First, we will point out the causes of these differences. Second, we will turn to the question

---

[81] The fact that Bové emerged even more popular out of the campaign finally made him give up his resistance against the plan to run for Presidency in 2007 – with the known, disappointing results: rigid party politics fragmented the radical left, and Bové succeeded to attract only a small fraction of its potential voters. Later in 2007, Bové turned back on anti-GMO activism.

[82] On the historical predecessor of the *faucheurs*, the 19th century English Luddites, see Jongerden in this book.

[83] The three key points of the CP's program defying neo-liberal globalisation are *global* visions: food sovereignty (the right of peoples and countries to freely produce their food and protect their agriculture of world market competition); food safety (the right to guard against food risks for human health); and the conservation of biological diversity.

[84] This finding should also add to our caution with general interpretations of farmers resistance as 'struggles to regain control over the means of production'. (Jongerden, this book). Regaining or defending economic autonomy might be a stated goal of French peasants, but hardly of Austrian organic farmers, whose large share in Austrian agriculture is largely due to State subsidisation. Given the tight framework of conditional rules to follow in order to qualify for organic subsidies, this does not necessarily mean a higher degree of autonomy.

as to whether and how these forms of rural resistance in two European countries contribute to the potential creation of biotechnologies tailored to meet the needs of the poor in the global South.

First, what can be learned about the role of national circumstances from our cases? Social movement research typically provides the notion of opportunity structures – the institutional as well as cultural-symbolic qualities of a State or political culture that predispose a social movement to take place and succeed. (Kriesi, 2004; Berclaz and Guigni, 2005) We distinguish idiosyncratic – locally specific – and general opportunity structures. Idiosyncrasies that partly explain the Austrian GMO-policy are, for example, the high public influence of the NKZ. For France, we have to take into account the legacy of the Paris spring, giving rise to the *Larzac* and, later, the ecologist movement and, eventually, even the *altermondialist* renaissance. Another French peculiarity: Greenpeace – major agent in the European anti-GMO movement – is less popular in France than elsewhere and, finally, José Bové – 'the right man at the right time' – is no doubt quite a unique personality.

With general opportunity structures we refer to *political cultures* and policies: As to political cultures, violent action – from road blocks to demolition of offices – is habitual in French farmer protest. The field destructions of the *faucheurs* thus fit into a general pattern of rural conflict behavior. Furthermore, *agricultural policy* is a field giving rise to dissimilar patterns of resistance: Austria's agricultural policy – 'greened' over the years and privileging organic farming – facilitated the adoption of a non-GMO stance; French agricultural policy, by contrast, has only hesitantly adopted ecological elements, not least because France is one of the world's largest agro-exporting countries which is why productivity rewarding policies are deeply entrenched. Finally, *power struggles* in agricultural politics – or the lack thereof – are a factor influencing GMO Luddism: in France, the ascend of GMOs as a public issue corresponds to the political rise of the CP, asserting itself against the dominant FNSEA. The – at least temporarily – rising popularity of the CP owes itself to a major part to José Bové and to his spectacular handling of the GMO issue. In Austria, no such dynamic ever occurred; today, agricultural politics is the same closed field, overseen by the People's Party and agricultural lobbies, it used to be. Impulses by the environmental movement have been incorporated into agricultural policies, and potential contenders never made inroads into power circles. In sum, national variations in rural GM activism cannot be reduced to a simple explanation but rather results from a number of different circumstantial factors – national idiosyncrasies, political cultures and recent agricultural policies.

In the following we examine the impact these two types of farmers resistance have on the potential reconstruction of biotechnologies in the developing world. This can only be done in a very tentative way; While we agree that this impact is substantial and has 'forced open the future of agriculture, rather than ensuring [...] a hegemonic system of bio-power' (Schurman and Munro, this book), it is yet unclear *how* the agro-food battle on modern biotechnology in the OECD world actually affects this future in developing countries. Nor is it clear how Northern and Southern farmers concerns are linked; if farmers interests differ within and across national policies, they certainly do so across the global North South divide.

We will thus approach the question in two distinct ways general enough to allow for relatively reliable claims: first, we reflect on the impact of Austria, France and the EU respectively had on the global governance of biotechnology; second we look into discursive effects the two models might have on countries and actors around the world. The first demarche abstracts from the variant forms of rural resistance in the two countries to look into their behavior as *States* in a multi-level governance environment. (A behavior which is presumably influenced by rural – but also a great number of other – GM resisters in their respective public arenas.) It proceeds on the assumption that States – or State supported supra-national entities – are still the major actors shaping global rule systems which, for example, set the terms for the global spread of agro-food-biotechnology. Now, in this – European and global – context, Austria and France play quite similar roles, notwithstanding their dissimilar policy motives. Both countries maintained national safeguard measures and supported the European de facto moratorium thus put pressure on the European Commission and helped in the 'trading up' of EU regulatory standards. There are nuances though: France took the lead in the moratorium with the stated goal to significantly change the EU regulatory framework, while Austria later jumped on the bandwagon mostly concerned with preserving its NIMBY status; in later years, Austria had to face several direct confrontations with the Commission who tried to lift its safeguard ban, while the French bans had become irrelevant as their market approvals had expired. However, basically both countries shared the same goal: blocking agro-biotechnology in Europe as long as possible and pushing the EU regulatory revision towards a framework as restrictive as possible. This undertaking was successful to an extent, as the EU regulatory regime is certainly among the most rigorous and best policed in the world, even though it is equally designed to render possible agro-food-biotechnology in order to conform to the global free trade disciplines of the WTO.

But the EU's policy amounts to more than a response to internal pressures externally constrained by free trade disciplines. As the U.S. and the EU follow two essentially different regulatory approaches – process versus product based regulation, precautionary versus risk based risk assessment, labeling and traceability versus non-labeling – it is to be expected that both great powers will compete to establish their own regulatory models with as many allies as possible (Drezner, 2005). In fact, it can be demonstrated that the EU (like the U.S.) acts in such a manner – i.e. as global policy entrepreneur seeking to align an emergent system of global governance with the principles and norms of its own regulatory system. This is particularly the case with the Biosafety Protocol which, in contrast to the set of rules inherent WTO law, recognises precautionary thinking about modern biotechnology, but it can also be shown for other international regulatory arenas such as the Codex Alimentarius Commission[85] (Seifert, 2006a). Thus we can conceive of the influence of single States on the global governance of agro-food-biotechnology: GMO aversive governments like Austria or France push the European Commission to adopt an ever tighter regulatory framework for agro-food biotechnology who in turn, seeks to influence processes of global rule making in such a way as to universalise the EU's regulatory approach[86].

The future consequences of this regulatory competition for the global South have yet to be assessed. Up to now, the EU framework with its traceability and labeling requirements caused many developing countries to abstain from growing GM crops for fear of contaminating non-GM crops for EU markets. Future scenarios on behalf of the developing world vary. Some authors see developing countries as caught between the lines in a 'GM cold war', but concede them the chance to gain greater autonomy by establishing bio-political alliances to face the dominance of Northern powers (Meijer and Stewart, 2004). Others highlight the pivotal role of environmental risk assessment in the global regulation of agro-food biotechnology and suggest developing countries to exploit the scientific ambiguity inherent in risk assessment in order to create greater leeway for bio-political choices and rules tailored to their

---

[85] This kind of EU 'regulatory foreign policy' is more than the empirically substantiated outcome of theoretical predictions. In fact, it is openly declared in recent policy statements by the European Commission. See: Tobias Buck: 'EU wants rest of the world to adopt its rules' The Financial Times, 19 February 2007.

[86] Apparently, even NIMBYs – so obsessed with their clean backyards – can have external effects.

particular circumstances (Millstone and Van Zwanenberg, 2003). At any rate, we may hold – again in line with Schurman and Munro – that the emergence of a rivaling centers of influence in the global regulation of ago-food biotechnology has disrupted the project of establishing a uniform set of global regulatory procedures and standards under the aegis of the WTO and thus opened new possible directions for agricultural technology development in the global South.

We now return to the specific modes of farmers' resistance in Austria and France to examine their *discursive* implications for the development of alternative biotechnologies in the global South. There are commonalities and differences. An important discursive feature both types of opposition have in common is the *categorical* rejection of agro-biotechnology. While this attitude forgoes potentially productive options in the debate on reconstructing biotechnologies to meet the needs of the poor, it also adds to justified criticism of the technology's corporate underpinnings.

The question whether modern biotechnology carries the potential to deliver crops meeting the needs of the poor is one of the most controversial topics in the current controversy. Promoters of modern biotechnology point to the many possibilities of creating locally adapted transgenic plants and argue that only increases in agricultural productivity can feed a growing world population. (FAO, 2004) Adversaries retort that malnutrition is a consequence of poverty rather than of low productivity, and point to the fact that, over twenty years of GM crop development, locally and socially adapted crops never saw the light of day as multinationals focus on the development of products promising profits to be reaped through a tight system of intellectual property rights[87] (Altieri, 2000).

Both Austrian and French oppositional discourses persist in the entrenched pro-anti-dichotomy. Whether State subsidised organic farmers or radical GMO-Luddites, the message they convey is that sustainable agriculture cannot walk along the path of genetic engineering. Both discourses are unequivocal on this point: Organic farming rules out modern biotechnology by its very statutes, radical French anti-GMO activism deliberately turned against non-commercial test releases linked to projects for the benefit of developing countries. To the extent biotechnology is equated with genetic engineering, this position rules out any further discussion or developmental work on locally adapted *modern* biotechnologies. Still less, it gives any clue how genetic engineering could contribute to adapted and socially equitable biotechnology applications.

The categorical rejection of GMOs creates blind spots, too. In fact, E.E. Schattenscheider's dictum 'The definition of alternatives is the supreme instrument of power', seems to apply here - of counter-hegemonic, discursive power, one might add. As an alternative to GM crops, opponents suggest that agriculture should bank on more humble forms of farming deliberately renouncing the pull of higher profits and technological efficiency enhancement. While this discourse resonates with the European public, it remains to be asked whether the juxtaposition of modern biotechnology and sustainable agriculture is *universally* warranted. Rather than based on objective, everlasting facts the categorical rejection of GMOs being part and parcel of both national NIMBY and internationalist NIABY framings can be viewed as the outcome of a discursive strategy fit to provoke public resonance in Northern societies. It is questionable whether there is a 'natural' incompatibility between agro-biotechnology and sustainable, socially benign modes of rural development, and an open question whether GM plants

---

[87] This leads to the fact that, after more than ten years of GM crop cultivation, global market for GM seeds covers only four cash crops: GM soybean, occupying 57 percent of global biotech area, followed by maize (25 percent), cotton (13 percent) and canola (five percent of global biotech crop area) (James, 2006).

Reconstructing biotechnologies: critical social analyses

wont ever provide solutions to resource poor, small scale agriculture. On the other hand, there is no doubt that agro-biotechnology development, since its beginnings in the 1980's, closely conforms to corporate productivity and profitability imperatives, while locally adapted GM crops meeting the needs of the poor have been advertised over and over again but never materialised. Against this background, a discursive strategy that creates a catchy albeit not necessarily true opposition, but helps denouncing the logic of profit maximisation underlying the biotechnology project, has its merits. We can expect this criticism to last as long as the marginalisation of public research in industrial and developing countries goes on and sustainable applications of agro-biotechnology remain mere promises.

At last, we look at the *different* parts insular NIMBYs and internationalist NIABYs have in the European and global GMO discourse. First, rural GMO protest in France promotes a far more categorical and principled critique of biotechnology than NIMBY-Austria. NIABY activists feed the public discourse with explanations for why they do not only want to put an end to GM crops but to the entire exploitative socio-political order that goes along with them, and this not only in France but everywhere in the world. Contrastingly, NIMBYs – eager to protect their national farming system – display a rather pragmatic readiness to come to arrangements with promoters of GM crops: As long as GM free areas (which might add up to an entire country) are kept safe, NIMBYs do not care much about harm the object of their dislike might cause elsewhere.

Second, the French critique is more widely visible than Austria's NIMBYism. While few people know about the latter, it is probably safe to say that, today, the whole world knows about José Bové's crusade against corporate biotechnology. Bové has given a face to peasant resistance in the age of neo-liberalism, and he has used the precious advantage of global reputation to reformulate GMO opposition as a radical critique to neo-liberal globalisation. Rural GMO-opposition in France has excelled in reframing the global discourse on agro-food biotechnology to incorporate discursive links between issues such as the crisis of the countryside; the power of multinational corporations; the unjust North-South-divide; and the annihilation of long-established ways of life by the forces of modernisation. Its radical critique of supranational power centers have helped to bring to public scrutiny the arcane politics of supra- and international decision making. Moreover, it has invested public discourse with new ideas such as food-sovereignty which, while they possibly look utopian today, might usher in the emergence of localised alternatives to global food markets[88]. Actors like Bové and the CP have raised awareness of the political nature of food consumption. Given the 'orthorectic' culture of modern affluent society obsessed with the sanitary, dietetic and moral right and wrong of consumption (Nicolosi, this volume), this discursive stimulus invigorates the rise of a 'moral food economy' – at least in the affluent world – spawning niche markets like 'fair trade' and other kinds of 'quality food'. (see Levidow, this volume) These, in turn, have the potential to stimulate the development of alternative agricultural models in both the global North and South.

Third, by contrast to Austria's NIMBYism, the French CP with its close ties to the internationalist *Via Campesina* and the networks of the global justice movement, participates in a process which attracts global mass attention, generates a constant 'brainstorming' for alternative ideas, and breeds

---

[88] If food-sovereign regions ever become reality, it remains open under which circumstances and at what point in time this is going to happen. We recall that, while the conceptual foundations for organic farming have been laid in the 1920s, it was not until the 1990s organic farming became a serious alternative to conventional agriculture - not least for agro-political reasons, as the Austrian illustrates. During the peak years of agricultural modernisation from the 1950s to the 1970s this career must have appeared utopian, too.

a multitude of local-global social networks and attempts of social experimentation that provide the social ground for real world alternative forms of living. All this can impact on the creation of locally adapted biotechnologies in developing countries and contributes to the fact that today the future of agro-systems in the South looks more open than ten years ago.

Then again, programmatic stringency, public visibility and worldwide interconnectedness do not necessarily mean political effectiveness. The immediate political effect of the global justice movement is difficult to discern. Its critical ideas feed into popular discourse, but how far these feed into real world politics is an open question. When it comes to bar agro-biotechnology from a given territory, however, Austrian NIMBYism turned out as working strategy, particularly in the context of the EU's embracing of organic farming, its emerging coexistence-policy and CAP reforms towards multi-functional agriculture. A similar approach might therefore be tempting to other European governments facing similar problems: public rejection of GMOs and a fragile agricultural structure. Austria's combination of State subsidised organic farming and GMO NIMBYism could thus become an object of *institutional* export article.

This, however, pertains to the European context only. In the EU organic agriculture has further potential to expand, as a feasible alternative to industrial agriculture and biotechnology alike and as a practical way to save agricultural subsidies in the context of liberal CAP reforms. But it is far less likely that State subsidised organic farming provides a viable model for subsistence farmers in developing countries. In sum, seen from a NIMBY perspective, the Austrian approach appears more successful as against the lingering French controversy. But the universal critique springing from this controversy draws attention to the socio-political conditions of development and the need for international solidarity among farmers and consumers. While it has been argued that the transatlantic regulatory dispute had the most powerful effect on developing countries, only future developments will allow a closer assessment of the impact the described discursive changes have on the emergence of alternative agro-systems in the global South.

Acknowledgements

I am grateful for encouragement and comments on a first draft of this paper presented at the conference 'Critical Construction of Agrarian Biotechnologies' November 3-4, 2007 in Kyoto/Japan. My special thanks to Professor Shuji Hisano from Kyoto University for his exceptional dedication and hospitality.

References

Agrikoliansky, E., Fillieule, O. and Mayer, N. (2005). *L'altermondialisme en France. La longue histoire d'une nouvelle cause*. Paris, Flammarion.

Altieri, M.A. (2000). No: Poor Farmers Won't Reap the Benefits. *Foreign Policy* 119: 123-131.

Ansell, C., Maxwell, R. and Sicurelli, D. (2006). Protesting food: NGOs and political mobilization in Europe. In: *What's the Beef? The Contested Governance of European Food Safety*, (Eds. C. Ansell, D. Vogel), MIT Press, Berkeley, pp. 97-122.

Berclaz, J. and Guigni, M. (2005). Specifying the Concept of Political Opportunity Structures. In: *Economic and Political Contention in Comparative Perspective*, (Eds. M. Kousis and C. Tilly), Boulder, Paradigm, 15-32.

Bernauer, T. and Meins, E. (2003). Technological Revolution Meets Policy and the Market. Explaining Cross-National Differences in Agricultural Biotechnology Regulation. *European Journal of Political Research* 42 (5): 643-683.

BMLFUW (Federal Ministry of Agriculture, Forestry, Environment and Water Management). (2006). *47. Gruener Bericht*. Vienna, BMLFUW.

CP (Confédération Paysanne). (1997). Technologies génétiques: pour un moratoire sur la mise en culture et la commercialisation, pour l'application du principe de précaution. Communiqué de la Confédération paysanne, 17 mars 1997, CP: Siège de Bagnolet.

Drezner, D. (2005). Globalization, harmonization, and competition: the different pathways to policy convergence. *Journal of European Public Policy* 12 (5): 841-859.

FAO (Food and Agricultural Organization). (2004). Agricultural Biotechnology: Meeting the Needs of the Poor? The State of Food and Agriculture 2002-2004. FAO: Rome.

Falkner, R. and Gupta, A. (2006). Real Impact of GM decision will be felt in developing countries. *Financial Times*: Monday, February 13, 2006.

Groier, M. and Gleirscher, N. (2005). *Bio-Landbau in Österreich im internationalen Kontext. Bd1: Strukturentwicklung, Förderung und Markt*. Vienna, Federal Institute for Less-favoured Mountainous Areas.

Heller, C. (2002). From Scientific Risk to paysan savoir-faire: Peasant expertise in the French and global debate over GM Crops. *Science as Culture* 11 (1): 5-37.

James, C. (2006). *Global Status of Commercialized Biotech/GM Crops: 1996-2006*. ISAAA: Ithaca, NY.

Kriesi, H.-P. (2004). Political Context and Opportunity. In: *The Blackwell Companion to Social Movements*, (Eds. Snow, D.A., S. Soule and H.P. Kriesi), Oxford, Blackwell, pp. 67-90.

Martin, J.-P. (2005). La Confédération paysanne et José Bové, des actions médiatiques au service d'un projet? *Ruralia* 15 June 2005. Available at: http://ruralia.revues.org/document142.html

Meijer, E. and Stewart, R. (2004). The GM Cold War: How Developing Countries Can Go from Being Dominos to Being Players. *RECIEL* 13 (3): 247-262.

Millstone, E. and Van Zwanenberg, P. (2003). Food and Agricultural Biotechnology Policy: How Much Autonomy Can Developing Countries Exercise? *Development Policy Review* 21: 655-667.

Seifert, F. (2006a). Divided we stand: The EU as dissonant player in the global governance of agro-food biotechnology. UNU-IAS Working Paper. UNU-IAS: Yokohama, Japan. Available at: http://www.ias.unu.edu/resource_centre/Seifert.pdf

Seifert, F. (2006b). Synchronised national publics as functional equivalent of an integrated European public. The case of biotechnology. European Integration Online Papers, 10 (8), Available at: http://eiop.or.at/eiop/index.php/eiop/article/view/2006_008a/26

Seifert, F. (2006c). Regional GM opposition as multilevel challenge? The case of Upper Austria. *Tailoring Biotechnologies* 2 (3): 11-36.

Skogstad, G. (2001). The WTO and Food Safety Regulatory Policy Innovation in the European Union. *Journal of Common Market Studies* 39 (3): 485-505.

Skogstad, G. (2006). Regulating Food Safety Risks in the European Union: A Comparative Perspective. In: *What's the Beef? The Contested Governance of European Food Safety* (Eds. Ansell, C. and D. Vogel), MIT Press, Berkeley, pp. 213-236.

Vogel, D. (1997). Trading up and governing across: transnational governance and environmental protection. *Journal of European Public Policy* 4 (4): 556-571.

Vogel, D. (2001). *Ships passing in the Night: GMOs and the Politics of Risk Regulation in Europe and the United States. European and American Perspectives on Regulating Genetically Engineered Food*. INSEAD, Fontainbleau.

Von Homeyer, I. (2006). The EU Deliberate Release Directive: Environmental Precaution versus Trade and Product Regulation. In: *Institutional Interaction in Global Environmental Governance*. (Eds. S. Oberthür and T. Gehring), Cambridge, MA, MIT-Press, pp. 259-283.

WTO (World Trade Organization). (2006). European Communities – Measures Affecting the Approval and Marketing of Biotech Products. Reports of the Panel. Available at: http://www.wto.org/english/news_e/ news06_e/291r_e.htm

Democratising agri-biotechnology? European public participation in agbiotech assessment

Les Levidow

1. Introduction

Public participation in technoscientific issues has recently gained mainstream support in Europe, in response to greater conflict around innovation and regulation of controversial technologies. STS scholars have played key roles in stimulating or organising such participation. The exercises have attracted diverse views regarding their appropriate design, roles and consequences. And they have attracted various criticisms – e.g. that participants were not representative of the public, or that the government did not make a prior commitment to follow views expressed there, or that technical aspects were separated from other issues.

Those criticisms may be descriptively accurate but imply particular benchmarks, even simplistic models of direct democracy. Together they imply that participants truly representing the public could guide government decisions – as if the government had no agenda of its own, nor a wider accountability to representative democracy. Amidst proposals for participation, there are diverse models of what would count as a democratic assessment of technology (Joss, 1998: 4). According to a survey of participatory TA exercises, these rarely have a demonstrable impact on political decision-making (Bütschi and Nentwich, 2002). Perhaps such exercises matter in more subtle ways, which therefore need different analytical questions about democratic accountability.

For some analysts of participatory TA, at issue is 'how to make those in charge accountable' and thus 'how to organise effective accountability' for government decisions (Hagendijk and Irwin, 2006: 56-57). Some have echoed concerns that participatory methods may 'subvert broader democratic political processes' or that they may not be 'fit for purpose', (Burgess and Chilvers, 2006). Participatory TA has been seen as supplementing older political forms of accountability with broader social forms. But this aim leaves open some difficult questions: 'who is holding whom accountable, and by which means?' (Abels, 2007: 111).

As a basis to evaluate various participatory methods, criteria for success include the following: the quality of deliberative processes, consensual proposals from the process, influence on policy, etc. (Rowe and Frewer, 2004). Although those concerns and criteria are valid, they imply that state-sponsored TA exercises could have a clear purpose in promoting technological democracy and citizenship. Why should this be the case?

As a case study for such analysis, this paper focuses on agricultural biotechnology, a sector which has faced extraordinary public protest in Europe. Agbiotech has attracted diverse forms of public participation, e.g. open mass meetings, protest, boycotts, mass-media stunts and sabotage. Through these means, an emerging citizenry has demanded government accountability for innovation choices. Among the various responses, many state bodies across Europe have sponsored formal participatory exercises, beyond simply access to regulatory procedures. So agbiotech provides a rich, multi-country case study.

This paper discusses the following questions:
- How and why did state bodies sponsor participatory TA of agbiotech?
- What aims arose in designing, managing and using those exercises?
- What was their relevance to democratising agbiotech?

2. Democratising technology – or managing conflict?

Participatory technology assessment has been promoted as a means to democratise technology, especially by enhancing the public accountability of innovation trajectories. To do so, participatory design should acknowledge that science and innovation are social, cultural and institutional activities.

> 'As such, public engagement offers a way to be more accountable for the particular values and interests, which underpin both the governance of science and the general use of science in governance... Public engagement holds greatest value when it occurs 'upstream' – at the earliest stages in the process of research or science-informed policy making... In practice, the relationship between representative democracy and participatory methods becomes most clear and complementary, when engagement is approached as a means to 'open up' the range of possible decisions, rather than as a way to close this down. Choice among the options thereby identified then becomes a clearer matter of democratic accountability' (Stirling, 2006: 5; cf. Stirling, 2005).

Achievement of such accountability depends upon the aims, design and management of the process.

However, public participation in technological issues has had diverse agendas. According to Lars Klüver (2006, cf. 1995), a long-time advocate at the Danish Board of Technology, public participation has recently become mainstreamed, along with changes in its policy role. Originally it was promoted as a vehicle for democratisation and citizen empowerment, so that people could challenge policy assumptions and influence decisions. Now public participation goes hand-in-hand with liberalism: politics is seen as a market of opinions, so citizens should be invited into the open market (cf. Popper, 1962).

Participation now becomes yet another governance tool among others, e.g. for adjusting, supplementing or enhancing the policy process. Aware that they often lack public confidence, policymakers seek methods of upstream conflict-management. These professional reasons have recently driven interest by mainstream institutions in public participation and will continue to do so, he argues (Klüver, 2006).

Upstream conflict-management restricts the role of participants. In the UK, for example, there have been various proposals for 'upstream public engagement' between the public and scientists at an early stage (e.g. HM Treasury/DTI/DfES, 2004: 105). Such engagement has been advocated as means to deliberate possible innovation choices and to make them more accountable (e.g. Wilsdon and Willis, 2004). By contrast to those ambitious aims:

> '[public engagement] is sometimes portrayed as a way of addressing the *impacts* of technology – be they health, social, environmental or ethical – rather than helping to shape the trajectory of technological development. The hope is that engagement can be used to head off controversy...' (Wilsdon *et al.*, 2005: 33).

Indeed, conflict-avoidance or conflict-management may be built into the design of public engagement. Rather than evaluate participatory TA according to an ideal model, each case should be seen as an arena for diverse strategies for how to represent agbiotech, the public and the relevant expertise.

3. Denmark 1987: sustainable agriculture?

The Danish consensus conference has been advocated as a 'counter-technocracy' – a means to challenge expert claims through a deliberative process. The lay panel has no vested interest different than the general public, and its report helps to promote technology assessment (TA) as a broad societal process. It extends a Danish tradition of *folkeoplysnig* – people's enlightenment through an adult education network which builds a reflective, informed citizenry (Joss, 1998: 20).

As its guiding principle, 'a well-functioning democracy requires a well-educated and engaged population'. Successful participation is understood in those terms: as a participant commented, for example, 'We initiated a really good assessment process among the public' (cited in Klüver, 1995: 41, 43). In the Danish consensus conference, then, 'interested citizens' personify a political culture in which technological decisions are held accountable to public debate, mediated by Parliament.

Denmark's debate on agricultural biotechnology was initiated in the mid-1980s by environmental NGOs. A series of 'debate booklets' were issued by NOAH, the Danish affiliate of FoE, proposing new legislation to regulate GMO releases. In response to public concerns, a Parliamentary 'green' majority imposed a statutory ban in the 1986 Gene Technology Act; GMO releases would not be permitted unless there was sufficient knowledge about the ecological consequences (Toft, 1996). With this wording, the government could be held accountable to demonstrate such knowledge for risk assessment; this burden of evidence meant a *de facto* ban for several years.

Parliament also mandated funds for an information campaign on biotechnology. Some funds were specially earmarked for NGOs, especially NOAH and some trade unions, in order to stimulate further debate on advantages and disadvantages of biotechnology. In these ways, environmental NGOs gained extra resources and political opportunities to frame the issues for further public debate. NOAH organised ten public conferences on the wider environmental consequences, on sustainable agriculture including organic agriculture, on food labelling, on animal welfare and ethics, on the Third World, on seed diversity (including patents), and on biological warfare. These debates were reported through a series of publications and statements from NOAH[89].

In that context the Danish Board of Technology held its first consensus conference in 1987 on 'Gene Technology in Industry and Agriculture', timed to coincide with Parliamentary debate on the issue (Hansen *et al.*, 1992; Klüver, 1995: 44). In its report the lay panel took up risk issues as well as ethical ones (Teknologinævnet, 1987). Accepting a key recommendation, Parliament voted to exclude animals from the 1987-90 national R&D program for gene technology. The conference eventually had more profound effects on the Danish regulatory regime through wider public debate.

A further information campaign was coordinated by the Board of Technology and Danish Adult Education Association. During 1987-1990 they supported more than 500 local meetings all over

---

[89] Much information here, supplied by Jesper Toft, is not available in English-language documents.

the country in order to stimulate debate on human and non-human uses of biotechnology, including concerns about risk and ethics. Environmental NGOs were often invited to speak, as the most visible critical actors on the scene.

The government also funded a subsequent program, organised by trade unions, to stimulate further debate on advantages and disadvantages of agbiotech. Their educational materials posed questions about sustainable agriculture: for example, would genetically modified crops alleviate or aggravate the existing problems of crop monocultures? (Elert, 1991: 12). Through that wider debate, the consensus conference indirectly influenced Parliament and thus regulatory policy.

In the EU-wide regulatory procedure, dominant member states implicitly took for granted eco-efficiency benefits of herbicide-tolerant crops, while disregarding the herbicide implications or assuming them to be benign (Levidow *et al.*, 1996, 2000). By contrast to those EU-level assumptions, Danish regulators were held publicly accountable for assessing the broad implications of GM crops for agricultural strategy, herbicide usage and the environment. Such judgments were scrutinised by the Parliament's Environment Committee, often by drawing upon specific questions from NGOs. Under such domestic pressures, Danish representatives in turn proposed that risk assessments evaluate those implications at the EU level (Toft, 1996, 2000).

Thus citizen participation enhanced government accountability for regulatory criteria, going beyond optimistic assumptions about environmental benefits. GM crops were subjected to criteria of sustainable agriculture, which in turn were opened up to the lay expertise of agbiotech critics. Environmental NGOs found greater scope to influence regulatory procedures and expertise.

Agri-innovation choices became more contentious in the late 1990s, however; NGOs demanded alternatives to agbiotech and to intensive agricultural methods. In a 1999 consensus conference, the lay panel asserted the need for extra measures – not only for product safety, but also to prevent GM products 'becoming controlled by monopolistic companies', as well as measures to evaluate ethical aspects (Einsiedel, 2001). As the conference organisers emphasised, those proposals were expressing citizens' viewpoints, thus providing a basis for dialogue with decision-makers (Teknologinævnet, 1999). The panel's proposals challenged the assumptions and limits of the EU legislative framework. Yet public demands for accountability were being channeled into more stringent measures to regulate biophysical risks. This pervasive tension has parallels in later TA exercises.

4. Germany 1991-92: participation trap

Since the time that the German government promoted agbiotech in the 1980s, this policy provoked widespread protest – e.g. from the Green Party, environmentalist groups and local campaigns. Although critics gained high-profile attention in the mass media and civil society, their views remained marginal to official procedures, unlike German corporatist arrangements for labour issues. Opposition to agbiotech split civil society and the major political parties (Gill, 1996).

4.1. TA exercise

German public controversy focused on herbicide-tolerant crops, given their potential for spreading that trait and for changing patterns of herbicide usage. To address such conflicts, the government sponsored a TA exercise on GM herbicide-resistant crops in the early 1990s. Funding came from the Ministry of Industry and Research, which was strongly promoting biotechnology. It was initiated and coordinated by the Berlin Wissenschaftszentrum (Science Centre) as an experiment in environmental conflict management. The 50-odd participants had quasi-expert roles; they included overt proponents and opponents of HR crops, as well as representatives of regulatory authorities, agricultural associations, consumer organisations, etc. From the start, conflict erupted over how to define the relevant scientific issues and the expertise needed to evaluate them.

A broad participation was needed to deliberate the arguments arising in the polarised public debate on agbiotech, according to the organisers. The TA was designed to evaluate those arguments for and against herbicide-resistance GM technology, especially its possible consequences – but not alternative options for weed control in agriculture. Thus the procedure was 'a technology-induced TA, not a problem-induced TA' (van den Daele, 1995: 74).

Environmental NGOs counterposed the latter approach. They wanted the TA to compare biotechnology products with other potential weed-control methods, as alternative solutions to agricultural problems. However, the NGOs' proposal was rejected by the organisers (Gill, 1993). Consequently, the narrow remit set difficult terms for participation by the broadly representative individuals from NGOs – indeed, terms for their expert status.

As the organisers acknowledged, 'The TA implicitly accepted the matter-of-course development of technology as the starting point', as well as possible risks as the main grounds for state restrictions: 'If critics fail to provide evidence of relevant risks, the technology cannot be banned'. So critics held the burden of evidence for any risks. Advocates held the burden to demonstrate benefits, though failure to do so would have no bearing upon regulatory decisions (Van den Daele, 1995: 75). This framework marginalised alternative agronomic solutions, while reinforcing the dominant system: 'intensive farming as the reference system'. Within that framework, participants themselves defined their controversies as debates about empirical evidence, e.g. regarding the possibility of environmental damage – not about values and goals (*ibid.*: 76, 77).

The organisers aimed to include and deliberate all viewpoints on the risk-benefit issues. By subjecting expert views to scrutiny, the TA could reach conclusions about empirical claims, rather than political or ethical ones. 'This procedure placed participants under massive pressure either to admit consensus or justify dissent', especially through detailed empirical evidence (*ibid.*: 80).

From NGOs' standpoint, the technology-induced TA framework effectively favored experts in specialised technical areas, e.g. gene flow and herbicide effects. In practice, the TA exercise set a lower burden of evidence for demonstrating benefits than for demonstrating risks, in a period before much empirical research had been done on risk scenarios. Consequently, the discussion emphasised environmental benefits, especially the prospects for farmers to use less harmful herbicides and/or lower quantities of them (Gill, 1993).

On the basis of the expert reports, the TA symbolically normalised any risks. According to agbiotech proponents, echoing the government's advisory body, any risks from GM herbicide-tolerant crops were similar to those from conventional crop plants and herbicide usage. 'In many areas it was argued that there was no need for political action because the identifiable problems could be dealt with in the established registration procedures ... if one agreed to the 'normalisation' of the risks' (Van den Daele, 1995: 82). In this way, the exercise undermined NGO claims about novel or unknown risks; once normalised, any risks would be manageable through regulatory procedures, even contemporary ones.

4.2. Science court or parliament?

The technology-induced TA framework posed a dilemma for participation by agbiotech critics. Once inside such an exercise, 'They have to criticise a technology which promises to satisfy some needs which may even be produced by the technology itself...' (Gill, 1993: 74). That is, putative benefits satisfy 'needs' which are predefined by biotechnological solutions for intensive monoculture. Thus a technology-induced TA tends to accept and reproduce the social vision built into the technology.

Environmental NGOs and their associated research institutes faced a difficult choice: either play a quasi-expert role within that framework and thus help legitimise it, or else abandon that role and be treated as merely lay voices. After much conflict, they withdrew before the TA exercise could report its conclusions. They gave several reasons for withdrawal, e.g. that their voluntary participation was occupying too much time, especially the task of commenting on long expert reports (van den Daele, 1995: 81). According to an NGO expert, 'I had not imagined that you could destroy participation by throwing paper on top of people' (cited in Charles, 2001: 107). By withdrawing from the TA, they could devote greater resources to public protest and preserve their credibility with NGO members and activists (Gill, 1993: 81-82).

After this withdrawal decision, they were criticised by the WZB coordinator:
> 'One cannot present one's position in public as scientifically substantiated and then cast fundamental doubt on science as neutral... Participation in the procedure implies the readiness to submit oneself on the empirical issues to the judgment of science' (Van den Daele, 1995: 84; also 1994).

As the WZB coordinator told the story many years later, he had been skeptical of claims that herbicide-tolerant crops had special risks or special benefits, so he saw NGO arguments about risks as a proxy for political ones:
> '...the idea of special risks is not a good argument. We should turn to the issues of democracy and who's going to decide how society develops... Apparently it would have been difficult for them [NGOs] to declare explicitly that the conflict was not about risks, but about social goals and political reforms ...' (Van den Daele, cited in Charles, 2001: 107).

However, that distinction was not so clearly drawn by the organisers beforehand; it became more explicit in later retelling the story. According to a social scientist who attended the TA exercise, some NGO participants saw it as analogous to a parliament which could evaluate agbiotech in terms of societal goals. However, Van den Daele retrospectively portrayed it as a science court, whose remit the NGOs did not understand or accept; this portrayal offers a *post hoc* legitimation for the failure to integrate them (personal communication, Bernhard Gill, 2006).

Moreover, the distinction between a science court and parliament is not so straightforward; neither is the distinction between risk assessment and socio-political goals. At issue was the range of questions to be answered by science, their normative assumptions, and the alternative technological options to be considered as comparators for agri-environmental assessments. Some questions from participants were pre-empted or marginalised by the TA exercise, especially by constructing particular boundaries between expert and lay voices.

Societal futures were reduced to scientific issues, readily assessable by experts in 'the state of the art'. Civil society representatives found themselves in a 'participation trap'; they could either participate within the government's risk-benefit framework for GM crops per se, or else be marginalised. Overall the exercise reinforced the government's policy framework and its public unaccountability. In a similar way, societal conflict over agri-innovation issues was channelled into risk assessment through regulatory procedures. Together these practices extended and reinforced the *Rechtstaat*, at least until government policy began to change in 2002.

5. UK 1994: risk-benefit framework

Before the UK had any significant public debate on agbiotech, a National Consensus Conference on Plant Biotechnology was held in 1994. Proposed by staff at London's Science Museum, it was funded by the Biotechnology and Biological Science Research Council (BBSRC). Initially reluctant to sponsor the event, the BBSRC was persuaded by the focus on GM crops as 'the least contentious' area of biotechnology, especially as compared to animal biotech. Yet civil servants criticised that focus because agbiotech was not being considered in policy debate at that time (Joss, 2005a: 211).

The exercise was coordinated by the Science Museum, whose staff implicitly diagnosed the problem as public misundstanding or anxiety. The coordinators had previously obtained funds in the name of diagnosing and overcoming public unease about biotechnology. At the beginning and end of the Consensus Conference, the funders made clear their aim to enhance 'public understanding' of biotechnology and thus support for it. Underlying the exercise was a presumed cognitive deficit of the public.

The Consensus Conference centred upon a lay panel of relative newcomers to the biotechnology debate; they would question and learn from designated experts – whose selection was contested within the Steering Committee. Two members attempted to exclude representatives of 'extreme' anti-biotech groups from expert status – and thus from a list prepared by the organisers – though this effort did not prevail (Joss, 2005b: 211). The organisers portrayed themselves as neutrally mediating between experts and the public. However, the exercise demarcated a boundary between 'expert' and 'public-interest' views, thus demoting the latter (Purdue, 1995, 1996).

A particular lay/expert boundary was performed by expert witnesses, in the process of being questioned by the lay panel. The panel expressed views about economic, political, legal and ethical issues of agbiotech.

> 'Yet the key questions – and the experts' responses – were largely framed within the technocratic discourses of specialist expert knowledge... It was largely taken for granted that the task of technology assessment depended primarily upon the technical and professional skills of research scientists' (Barns, 1995: 203).

The structure implied that experts are needed to help overcome the deficient understanding of the public, though the lay panel often challenged the supposed neutrality of official expertise (*ibid.*).

'[This] set up a functional division of labour: "lay" people ask questions, while "experts" provide the answers. Indeed to play out their "lay" role properly, the "lay" panel was obliged... to show appropriate deference to the "experts" and the organisers. The "lay" panel was thus encouraged to take on the challenge of investigating biotechnology, but from an exaggerated position of innocence and ignorance' (Purdue, 1995).

'The whole construction of their layness induced an undue deference to the experts, irrespective of the expert's actual level and area of competence' (Purdue, 1996: 533).

The lay/expert boundary was reinforced in the final, public stage of the process. There the chairman tended to give pro-biotechnology speakers the status of 'mobile experts', knowledgable on diverse aspects. By contrast, NGO activists were put on the defensive to demonstrate their expertise (*ibid.*).

The process raised wide-ranging questions and disagreements, even within the Panel. Nevertheless, the organisers instructed the panelists to present a single report, permitting no minority views (Purdue, 1996: 537). Consequently, some critical views were marginalised in the panel's report, as if there were consensus on how to define risks and benefits.

Particularly marginalised were concerns about who would legitimately direct biotechnological innovation. Among themselves, panel members raised issues about who was 'in control' – e.g. concerns about R&D priorities, environmental monitoring and accountability (Joss and Durant, 1995: 82). In the panel's report, these issues were largely reduced to safety controls and patent issues.

Having listed potential benefits and risks, the report concluded: 'Biotechnology could change the world, but in order for it to be used effectively – maximising benefits and minimising risks – we also need to adapt economic and social structures to take account of the changes it might produce'. By contrast to government policy, the panel opposed any extension of patent rights; it also advocated mandatory labelling of GM food for the public right to choose. In particular: 'Regulatory control in the UK is among the most stringent; however, there is still room for improvement' (Science Museum/BBSRC, 1994: 7, 14). Although questioning some pro-biotech arguments, the report reinforced a common societal problem of product safety, while adding the principle of consumer choice.

After the panel presented its final report, the document was interpreted in divergent ways. According to the organisers, 'the lay panel has given the field of plant biotechnology its qualified support' (Science Museum, 1994: 2). However, the report could just as well be read as skeptical; it emphasised not only risks, but also predictable disadvantages of agbiotech. It also criticised inadequacies of government regulation, along lines similar to criticisms by NGOs. One excerpted the report as campaign material, entitled 'Whose consensus?', emphasising differences between the panel's report and government policy (Genetics Forum, 1994).

The UK exercise sought mainly to explore 'the public understanding of science' in Britain (Joss and Durant, 1995: 76, 96, 104: note 14), according to the conference organisers. They claimed 'to adopt the Danish model of the consensus conference', yet this aims to generate a wider societal debate that could influence the Parliament and government. The UK exercise anyway had little potential for such influence: Parliament had no relevant policy decision at that time (*ibid.*: 99), and there was little public debate on agbiotech.

In any case, the lay panel had little means to challenge the UK risk-benefit framework, even if it had presented minority views. A more significant policy challenge was coming from the opposite direction. UK regulatory procedures then were facing deregulatory pressure from the agbiotech industry and other Ministries, amidst a Europe-wide campaign against 'over-regulation' (Levidow, 1994). Environment Ministry officials saw the lay panel's report as helpful for protecting their regulatory procedures and expertise from such pressure.

In all those ways, the UK Consensus Conference reinforced an expert/lay boundary within the UK's risk-benefit policy framework. The Panel recommended regulatory adaptations to ensure that agbiotech would be kept beneficial and safe. Although individual panel members raised issues about corporate-biotechnological control over the agri-food chain, these were reduced to regulatory control measures, e.g. safety regulation and product labelling. This framework implied little scope for public participation in definitions of risk or benefit, much less in innovation priorities. Policy issues could be implicitly delegated to expert bodies through normative assumptions in their advice.

6. France 1998: the benign technocratic state

By 1997 French regulatory policy faced a legitimacy crisis. France had led efforts to gain EU-wide approval for GM crops, yet these were now opposed by a broad range of organisations. The Confederation Paysanne, representing farmers who elaborated a peasant identity, opposed agbiotech while counterposing 'quality' alternatives to industrialised agriculture (Heller, 2002). An oppositional petition was signed by many prominent scientists, not necessarily anti-agbiotech, but all of them concerned about regulatory failures to develop appropriate ecological expertise and risk research (Marris, 2001).

In February 1997 the Prime Minister decided not to authorise commercial cultivation of Ciba-Geigy's Bt176 GM maize in France, even though French regulators had led EU authorisation of the same product. This unstable policy indicated a crisis of official expertise within an elite-technocratic political culture. According to some critics, an official 'objectivity' too narrowly defined the relevant expertise. As an alternative approach, expert procedures would open up a scientific critique of possible options; this space would provide the expertise necessary for decisions (Roqueplo, 1996: 67, my paraphrase). By incorporating counter-expertise, regulatory procedures would develop an *expertise contradictoire* (contradictory expertise), which would enhance democratic debate and state accountability for decisions.

In November 1997 the government announced a set of measures, including a plan to sponsor a consensus conference on GMOs, by reference to the Danish Model. This event was later officially called a Citizens' Conference. As an official rationale, this event would provide 'a new way of elaborating decisions' and a means to implement 'participatory democracy', according to the Ministry of Agriculture. Yet the government never clarified the relation between the citizens' conference and its own decision-making procedure (Marris and Joly, 1999). This relation was subtly played out within the conference process, especially by defining expert roles.

From the start, the conference was designed to re-assert the benign expertise of the state, especially the Parliament, which saw itself as the only legitimate representative of the Nation. Organisation of the citizens' conference was delegated to a Parliamentary unit, Office Parlementaire d'Évaluation des Choix Scientifiques et Technologiques (OPECST), which symbolised a political neutrality separate from the government. OPECST appointed the steering committee, which in turn decided that the panel membership should represent diverse views of ordinary citizens – rather than stakeholders in the debate. It also decided which 'experts' – all of them scientists – would give briefings or testimony to the panel, thus framing the issues in advance (Marris and Joly, 1999). The organisers saw those arrangements as necessary 'to prepare a public debate which is not taken over by one side or the other', i.e. to correct or avoid biases in the existing public debate (OPECST, 1998a). Implicitly, such biases included anti-agbiotech NGOs on one side and Monsanto on the other side, especially from the perspective of the Left-Green Parliamentary majority.

Held in 1998, the conference included different framings of the policy problem. At the public hearings, the citizens' panel often challenged claims by experts about risks and benefits of GM crops. According to the panel's report, control by multinational companies could threaten farmers' independence. Genetically altered species pose a risk of standardisation. And GM rapeseed poses known risks of uncontrolled proliferation, both through pollen and seeds. Nevertheless GM crops could bring economic benefits to European agriculture (OPECST, 1998b; Boy *et al.*, 1998). Together these arguments implied the need for national public-sector expertise in agbiotech innovation.

The panel's recommendations focused on institutional arrangements for better managing agricultural biotechnology. Such measures included the following: greater social participation in scientific advice; public-sector research on ecological risks and agbiotech innovation; a system to ensure traceability of food derived from GM crops; and adequate labeling to inform consumer choice. 'Until these conditions are satisfied, part of the panel believes that a moratorium would be advisable' (*ibid.*). By advocating state funds for agbiotech innovation, the panel accepted the government's problem-definition of a national technological gap whose solution requires public-funded science, presumed to be benign. The panel's concerns about rapeseed complemented the French government's decision to oppose approval of GM herbicide-tolerant rape, on grounds that gene flow could complicate weed control (Marris and Joly, 1999).

The panel's conclusions were translated into policy advice by the Parliamentary organisers, as if they were neutral experts in the public good. Moreover, having attended the proceedings, the OPECST President presumed to speak for the panel:

> 'Taking all these views into account he then himself adopted a position on a number
> of topics... He has identified the issues and looked into peoples' fears and concerns'
> (OPECST, 1998b).

This translation can be illustrated by the strategic issue of how to structure expert advice. The panel had proposed that a citizens' commission should be part of the scientific advisory committee. Yet OPECST recommended instead that it be kept separate; this proposal could better perpetuate a neutral image of scientific advice, thus reinforcing a boundary between expert/lay roles.

The panel's advice anticipated the general direction of government policy: more stringent regulatory criteria, risk assessment by a broader scientific expertise, and 'independent' risk research, which was equated with public-sector institutes. It helped to legitimise and reinforce such initiatives, which had not been universally accepted within the government beforehand. In June 1998 the government announced measures along those lines (Marris and Joly, 1999). Institutional reforms emphasised expert procedures to minimise the risks and enhance the benefits of a controversial technology.

Despite its limitations, the citizens' conference initiated a new form of active public representation and knowledge-production. Panel members explored techno-scientific and social aspects together from the perspective of ordinary citizens. They sought to inform decision-makers about the views of those who do not normally speak out – and who do not feel represented by political parties, trade unions, or environmental and consumer NGOs. This potential for participatory evaluation, especially for considering alternative options, was limited by the overall structure, especially the small opportunity to interact with designated experts (Joly *et al.*, 2003).

Overall the citizens' conference was used to legitimise state claims to represent the public good, especially through expert roles. OPECST selectively promoted some accounts of agbiotech and its regulation as the expert ones, while explicitly speaking on behalf of citizens. The Agriculture Ministry had claimed to implement 'participatory democracy', yet the exercise extended the French tradition of technocratic governance (Marris and Joly, 1999).

Within this framework, expert roles remained the exclusive realm of the state authorities and their officially designated advisors. Ordinary people could question experts and recommend institutional reforms, but Parliamentary experts would officially speak for them. Thus the process reinforced lay/expert boundaries, in the face of public challenges to the official expertise for agbiotech.

7. UK 2003 public dialogue: policing boundaries

From the late 1990s onwards the UK had a widespread public controversy over agbiotech. Protest actions and attacks on field trials gained public support by linking GM crops with various issues – BSE, other food scares, globalisation, 'pollution', etc. (Levidow, 2000). The government faced an impasse over regulatory decisions, especially the criteria for permitting a GM herbicide-tolerant maize which the EU had approved in 1998. As a key issue, conservation agencies had warned that changes in herbicide usage could harm farmland biodiversity, so the government funded farm-scale trials to monitor such effects.

To address wider issues beyond risk regulation, the government had created the Agricultural and Environment Biotechnology Commission in 2000. Its report, *Crops on Trial*, advised the government to initiate an 'open and inclusive process of decision-making' within a framework that extends to broader questions than herbicide effects. It proposed a 'wider public debate involving a series of regional discussion meetings' (AEBC, 2001: 19, 25). The government was persuaded to sponsor this – alongside the intense, sporadic debate which was occurring anyway.

Called 'GM Nation?', the official public debate was carried out in summer 2003. Beforehand the government vaguely promised 'to take public opinion into account as far as possible'. The exercise was intended for the organisers to gauge public opinion, rather than for participants to deliberate a collective view on expert matters (Horlick-Jones *et al.*, 2006). 'GM Nation?' also aimed to elicit views of the ordinary public, rather than organisational representatives – an artificial distinction, given that most civil society organisations and wider social networks had discussed agbiotech in previous years.

An overall Public Dialogue had a tripartite structure which explicitly distinguished between lay and expert issues. 'GM Nation?' was designed mainly for the lay public. An expert panel carried out a Science Review of literature relevant to risk assessment. And a government department carried out a Costs and Benefits Review of GM crop cultivation in the UK.

The Public Dialogue was designed in those three separate parts, with an explicit aim that they would work closely together. The three procedures were kept formally separate, yet the supposedly lay and expert issues became intermingled in practice. The official boundaries were both challenged and policed, thus constructing the participants in contradictory ways.

7.1. Representing public views?

'GM Nation?' featured several hundred public meetings open to anyone interested, drawing over 20,000 participants (DTI, 2003). When participants in 'GM Nation?' largely expressed critical or skeptical views towards agbiotech, arguments ensued over whether they were 'representative' of the public. According to a pro-agbiotech coalition, the Agriculture and Biotechnology Council, the exercise was hijacked by anti-biotech activists, so the format was not conducive to a balanced deliberation of the issues.

According to academic analyses, however, that criticism frames the public as atomised individuals who have no prior opinion. The exercise predictably drew a specialised public which was largely suspicious or hostile to agbiotech. Participants represented both themselves as individuals and wider epistemic networks. The debates were filling an institutional void, in the absence of any other formal opportunity to deliberate the wider issues (Reynolds and Szerszynski, 2006).

The government sponsors had asked the contractors to involve 'people at the grass-roots level whose voice has not been heard'. As the official evaluators noted afterwards, however, it was problematic to distinguish clearly between 'an activist minority' and a 'disengaged, grass-roots minority'. Many participants in 'GM Nation?' were politically engaged in the sense that their beliefs on GM issues formed part of their wider worldview. Yet policymakers tend to construct 'the public' as an even-handed majority – and therefore legitimately entitled to participate in engagement exercises (Horlick-Jones *et al.*, 2004: 135; 2006). Indeed, 'grass-roots' conventionally means local organised activists, yet this term was strangely inverted to mean a passive, uninformed public.

As envisaged by the sponsors, separate focus groups would allow the public to frame the issues according to their own concerns, yet special measures were needed to realise the policymakers' model of the public. They saw the open meetings as dominated by anti-biotech activists, unrepresentative of the general public. Politically inactive citizens were seen as truly representative and thus as valid sources of public opinion, by contrast to 'activists'. To exclude the latter individuals from focus groups, candidates underwent surveillance and screening. 'Perhaps paradoxically, the desire to allow the public to frame the discussion in their own terms led the organisers to rely on private and closely monitored forms of social interaction'. According to this ideal model of the focus groups, the organisers would be listening to the *idiotis*, by analogy to ancient Greek citizens too ignorant to fulfill their responsibilities (Lezaun and Soneryd, 2006: 22-23). In this way, the more informed, expert citizens would be excluded from representing the public.

'GM Nation?' was intended to canvass all views and concerns about agbiotech, yet there were boundary disputes over issue-framings, admissible arguments and participants' roles. Some used the opportunity as politically engaged actors in their own right, not just as indicators of public opinion. Attending shortly after the US-UK attack on Iraq, some participants drew analogies between government claims about agbiotech and about Weapons of Mass Destruction. They suspected that the government was concealing or distorting information in both cases; they wondered whether it would ignore public opinion towards agbiotech, as in the attack on Iraq. Initially the chair tried to steer the discussion back to agbiotech, on grounds that 'GM Nation?' was not about the Iraq war, though participants still elaborated the analogy. Thus the public consultation had a disjuncture between public politics and government policy as understood by the sponsors of the exercise (Joss, 2005b: 181).

7.2. Expert/lay roles

For the carefully selected focus groups, the organisers commissioned 'stimulus material', so that participants would have a common knowledge-basis for discussion. The Steering Group asked the contractors to supply 'objective' information. Yet there were grounds to include 'opposing views' because this is often how people encounter information in real life', according to the official evaluators of 'GM Nation?' The ultimate material did include divergent views, but their sources were removed from the workbook for focus groups. Afterwards the official evaluators questioned 'the extent to which information is meaningful if it is decontextualised by stripping it from its source' (Horlick-Jones *et al.*, 2004: 93-94; Walls *et al.*, 2005).

Indeed, people often make judgments on the institutional source of expert views, but they had little basis to do so in the 'GM Nation?' focus groups. Omission of the sources was not simply a design deficiency in the exercise. By default, the issue of expert credibility was diverted and reduced to scientific information about biophysical risk. Participants had little basis to evaluate such information, so the exercise constructed a lay/expert boundary, constraining public roles even more narrowly than in the wider public debate.

Separate from 'GM Nation?', the GM Science Review was officially limited to a panel of experts evaluating scientific information. At the same time, relevant NGOs were consulted about experts who could represent their views on the panel. In this way, panel members were selected along relatively inclusive lines, encompassing a wide range of views about GM crops. As these selection criteria recognised, the public did not regard scientific expertise as a neutral resource (Hansen, 2006: 580), so the Panel's public credibility would depend upon a diverse composition. Although the Panel's report identified no specific risks, it emphasised uncertainties and knowledge-gaps important for future risk assessment of GM products (GM Science Review, 2003). These uncertainties implied scope for a wider public role in expert judgments.

As a high-profile part of the GM Science Review, the Royal Society announced a meeting to 'examine the scientific basis' of various positions. Opening the event, the chair announced the laudable aim 'to clarify what we know and do not know' about potential effects of GM crops. In the morning, agro-ecological issues were analysed in a rigorous way, especially for their relevance to the prospect that broad-spectrum herbicides may be widely used in the future. But those complexities were ignored when considering GM herbicide-tolerant crops in the afternoon (Levidow, 2003). By downplaying expert ignorance, the overall structure did not facilitate a debate about knowledge versus ignorance, nor provide much basis for public involvement.

Moreover, the boundaries of 'science' were policed along pro-biotech lines. Inconvenient issues, findings or views were deemed non-scientific. For example, speakers freely advocated the need for agbiotech to solve global problems, e.g. environmental degradation, the food supply, etc, but the chair cut off anyone who questioned these claims – for going beyond science (*ibid.*). Thus biotechnological framing assumptions were reinforced as 'science', along with the expert status of their proponents – while skeptics were marginalised as merely expressing lay views on extra-scientific issues.

In sum, the UK Public Dialogue involved a struggle over how to construct the public, especially in relation to expertise. The structure and management imposed boundaries between apolitical grassroots versus activist, as well as between lay versus expert status. Nevertheless participants challenged those boundaries, performed different models of the public and questioned dominant expert assumptions.

8. Conclusions: democratising agbiotech?

The Introduction posed the following questions:
- How and why did state bodies sponsor participatory TA of agbiotech?
- What aims arose in designing, managing and using those exercises?
- What was their relevance to democratising agbiotech?

Since the 1980s various state bodies in Europe have sponsored a participatory technology assessment (TA) of agbiotech; this can be analysed as a specific type of arena with diverse forms and contexts. In most national contexts, agbiotech was being officially promoted as an essential source of eco-efficient GM products, whose safety would be validated by experts as the only necessary scrutiny. These neoliberal policy frameworks were increasingly challenged by autonomous citizen initiatives.

In responding to or anticipating public concerns about agbiotech, participatory TA exercises were sponsored with diverse, overlapping, even contradictory aims. From various deficit models of public unease, sponsors and other advocates sought to democratise technology, to educate the public, to counter 'extreme' views, to gauge public attitudes, to guide institutional reforms, and/or to manage societal conflicts. Such aims had a bearing upon the design, management, staging and process of each exercise. Each process manifest diverse accounts of technology, the public, expertise and democracy (cf. Joss, 2005a).

In these TA exercises, individuals were pre-selected to participate in a group process, questioning expert claims in order to reach a group view. Participants deliberated the normative, value-laden basis of expert claims, thus developing a lay expertise; they went beyond simply questioning experts (cf. Kerr, 1998; Wakeford, 1998). By contrast to a negotiation among interest-groups, participants addressed the public good by appealing to common societal interests and problems (cf. Hamlett, 2003).

However, dominant problem-definitions incorporated or marginalised critical voices. Some problems were treated as common ones for group deliberation, while others were ignored or marginalised as uncommon ones, inconvenient for a group consensus or for a thinkable government policy. Some participants questioned whether agbiotech would provide a means for sustainable agriculture and a benign control over the agri-food chain; some suggested the need for alternatives. These questions were generally channeled into regulatory criteria and were reduced to control measures.

In such ways, participatory TA exercises biotechnologised democracy. The innovation trajectory was protected as societal progress, partly by diverting any challenge into managerial problems. Discussion generally focused on appropriate regulatory arrangements for agbiotech, represented as a series of potentially beneficial products; at issue was how to minimise risks and maximise benefits. Citizens' roles were modeled according to the biophysical 'risk' frameworks of EU and/or national legislation, thus generating conflict over such roles. In analogous ways, public participation in regulatory procedures manifest tensions between the broad comments submitted and the official 'scientific' criteria for relevant evidence (Bora and Hauseldorf, 2006; Ferretti, 2007). Conflicts over societal futures can be managed 'by re-absorbing discourses of polarity into a system of 'legitimate differences' and by defining the locations where differences can be articulated' (Gottweis, 1998).

Regardless of other views held by TA participants, any wider deliberation was constrained – by a search for consensus, by the design of each exercise, and by the government policy framework. This overall context limited what could be said with influence on the process, and thus what roles could be credibly performed by participants (cf. Hajer, 2005). The process internalised and reinforced policy assumptions about agbiotech as essential progress – albeit perhaps warranting more rigorous, publicly accountable regulation. Through a discursive depoliticisation, contentious issues were displaced onto the management problems of an inevitable future, as in neoliberal governance more generally (cf. Goven, 2006; Pestre, 2008). Consequently, tensions arose between discussing a 'common' problem – how to make agbiotech safe or acceptable – versus encompassing problems of political-economic control, innovation choices and societal futures.

Those tensions took the form of various boundary conflicts, which erupted more starkly in some cases. In the German TA exercise, as an extreme case, the NGO representatives could maintain their official expert status only by accepting a risk-benefit framework. Instead they rejected these terms for participation, demanded a broadly comparative assessment, and thus were relegated to the lay public or irrational objectors. In the 2003 UK Public Dialogue, the official structure nominally separated all relevant issues into three components – public concerns, scientific risk assessment, and economic benefits; accordingly, expert matters were formally separated from other issues for discussion by lay participants. Despite that official tripartite structure, all the issues became mixed in practice; their boundaries were both contested and policed.

In the design and management of the TA exercises, then, boundaries were imposed – between biotechnological imperatives versus alternative options, between scientific versus policy issues, as well as between expert versus lay roles – thus closing down issues. By contesting those boundaries, some participants opened up policy issues and performed different models of the public, implying broader roles for citizens. These performative interactions produced different understandings of the policy problem (cf. Hajer, 2005). If analysed in this way, then public engagement can 'clarify what conflict is really about' (de Marchi, 2003).

In each TA exercise, issues were opened up beyond the government policy framework, though they remained more narrow than in the parallel public debate. Ultimately the process reinforced official boundaries between scientific and extra-scientific issues, as well as between expert and lay roles (though these boundaries took different forms across the national cases). Agbiotech was being co-produced along with particular models of expertise, citizenship and their relationship. Institutions were created or adapted in designing and managing each TA process accordingly.

To some extent, state-sponsored participatory TA exercises anticipated, stimulated or reinforced policy changes which enhance the state's accountability for regulatory frameworks. Such outcomes depended upon a longer-term socio-political agency beyond the TA exercise and its panel. However, the TA exercises did not help publics to hold the state accountable for its commitment to agbiotech as an objective imperative.

What does this mean for efforts to democratise technology?
'... appraisal conducted in 'opening up' mode might be seen as substantively more coherent and normatively more consistent with the prevailing institutions and procedures of representative democracy' (Stirling, 2005: 229).

'In practice, the relationship between representative democracy and participatory methods becomes most clear and complementary, when engagement is approached as a means to open up the range of possible decisions, rather than as a way to close this down. Choice among the options thereby identified then becomes a clearer matter of democratic accountability' (Stirling, 2006: 5).

In the state-sponsored cases analysed here, participatory methods and representative democracy do not seem complementary. Or perhaps they are perversely so. In performing publics, participation symbolically set boundaries on citizen roles and closed down innovation (non)choices. By default, if not by design, such issues were channeled into regulatory arenas, which thereby carried the burden of conflicts over societal futures.

Thus state-sponsored participatory TA readily complements neoliberal representative democracy and its unaccountability, while reproducing its contradictions through contested boundaries. Democratic accountability remains a task for a wider societal contest over normative policy commitments and pre-empted futures. The prospects will depend upon wider, autonomous forms of participation — neither sponsored nor welcomed by state bodies.

Acknowledgements

Helpful editorial comments on previous versions were received from the following colleagues: Gabi Abels, Bernhard Gill, Dominic Glover, Joanna Goven, Christine Hauskeller, Maria Paola Ferretti, Pierre-Benoit Joly, Mercy Kamara, Huei-Chih Niu, Dominique Pestre, Stefan Sperling, Andy Stirling, Zoe Sujon, Jesper Toft, John Walls, Sue Weldon, Bernhard Wieser and Richard Worthington. Also helpful were comments on related talks at these events: Participatory Approaches to Science and Technology (PATH) conference, June 2006, Edinburgh; Critical Social Science for/on Contentious Technologies, workshop in November 2006, Egenis Centre, Exeter, UK; Critical Issues in Science and Technology Studies, 6th Annual IAS-STS Conference, May 2007, Graz, Austria; Science and Democracy Network annual meeting, June 2007, Cambridge, UK; Reconstruction Agro-Biotechnologies for Development', 3-4 November 2007, Kyoto, Japan. A longer version has more systematic reference to analytical perspectives (Levidow, 2007).

References

Abels, G. (2007). Citizen involvement in public policy-making: does it improve democratic legitimacy and accountability? The case of pTA. *Interdisciplinary Information Sciences* 13(1): 103-116.
AEBC (2001). *Crops on Trial*. Available at: www.aebc.gov.uk

Barns, I. (1995). Manufacturing consensus? Reflections on the UK National Consensus Conference on Plant Biotechnology. *Science as Culture* 5(2): 200-216.

Boy, D., Donnet-Kamel, D. and Roqueplo, P. (1998). A Report on the Citizens Conference on Genetically Modified Foods (France, June 21-22, 1998), including the report prepared by the French Lay Panel. Available at: http://www.loka.org/pages/Frenchgenefood.htm

Bora, A. and Hauseldorf, H. (2006). Participatory science governance revisited: normative expectations versus empirical evidence. *Science and Public Policy* 33 (2): 479-488.

Burgess, J. and Chilvers, J. (2006). Upping the *ante*: a conceptual framework for designing and evaluating participatory technology assessment. *Science and Public Policy* 33(10): 713-728.

Bütschi, D. and Nentwich, M. (2002). The role of participatory technology assessment in the policy-making process. In: *Participatory Technology Assessment: European Perspectives.* (Eds. S. Joss and S. Bellucci), London, Univ. of Westminster Press, pp. 233-256.

Charles, D. (2001). *Lords of the Harvest: Biotech, Big Money and the Future of Food.* Cambridge, MA, Perseus.

De Marchi, B. (2003). Public participation and risk governance. *Science & Public Policy* 30 (3): 171-176.

DTI. (2003). Reports on GM Nation? London: Dept of Trade & Industry. Available at: www.gmnation.org.uk/

EEC. (1990). Council Directive 90/220 on the Deliberate Release to the Environment of Genetically Modified Organizms. *Official Journal of the European Communities,* L 117 (8 May): 15-27.

EC. (2001). European Parliament and Council Directive 2001/18/EC of 12 March on the deliberate release into the environment of genetically modified organizms and repealing Council Directive 90/220/EEC. *Official Journal of the European Communities,* L 106: 1-38.

Einsiedel, E.F., Jelsøe, E. and Breck, T. (2001). Publics at the technology table: The consensus conference in Denmark, Canada, and Australia. *Public Understanding of Science* (10): 83-98.

Elert, C. *et al.* (1991). *Biotechnology at Work in Denmark.* Copenhagen: Danish Board of Technology (Teknologinævnet).

Ferretti, M.P. (2007). What do we expect from public participation? The case of authorising GMO products in the European Union. *Science as Culture* 16(4): 377-396.

Genetics Forum. (1994). Whose consensus? *The Splice of Life:* December.

Gill, B. (1993). Technology assessment in Germany's biotechnology debate. *Science as Culture* 4 (1): 69-84.

Gill, B. (1996). Germany: splicing genes, splitting society. *Science & Public Policy* 23 (3): 175-179.

GM Science Review. (2003). First Report: An open review of the science relevant to GM crops and food based on the interests and concerns of the public. London: GM Science Review Panel. Avialable at: www.gmsciencedebate. org.uk/report/pdf/gmsci-report1-pt1.pdf.

Gottweis, H. (1998). *Governing Molecules: The Discursive Politics of Genetic Engineering in Europe and the United States.* Cambridge, MA, MIT Press.

Goven, J. (2006). Dialogue, governance and biotechnology: acknowledging the context of the conversation. *Integrated Assessment Journal* 6 (2): 99-116.

Hagendijk, R. and Irwin, A. (2006). Public deliberation and governance: engaging with science and technology in contemporary Europe. *Minerva* 44 (2): 167-184.

Hajer, M. (2005). Setting the stage: a dramaturgy of policy deliberation. *Administration & Society* 36 (6): 624-647.

Hamlett, P. (2003). Technology theory and deliberative theory. *Science, Technology and Human Values* 28 (1): 122-140.

Hansen, L. *et al.* (1992). Consensus Conferences. Copenhagen: Danish Board of Technology.

Hansen, J. (2006). Operationalising the public in participatory technology assessment. *Science and Public Policy* 33 (8): 571-584.

Heller, C. (2002). From scientific risk to *paysan savoir-faire*: peasant expertise in the French and global debate over GM crops. *Science as Culture* 11 (1): 5-37.

HM Treasury/DTI/DfES. (2004). *Science and Innovation Investment Framework 2004–2014*. London, HM Treasury.

Horlick-Jones, T., Walls, J., Rowe, G., Pidgeon, N., Poortinga, W. and O'Riordan, T. (2004). *A Deliberative Future? An Independent Evaluation of the GM Nation? Public Debate about the Possible Commercialization of Transgenic Crops in Britain, 2003*, Understanding Risk Working Paper 04-02, Norwich, University of East Anglia.

Horlick-Jones, T., Walls, J., Rowe, G., Pidgeon, N., Poortinga, W., O'Riordan, T. (2006). On evaluating the GM Nation? Public debate about the commercialization of transgenic crops in Britain. *New Genetics and Society* 25 (3): 265-288.

Joly, P.-B., Marris, C. and Hermitte, M.A. (2003). A la recherche d'une 'démocratie technique. Enseignements de la conférence citoyenne sur les OGM en France. *Natures, Sciences et Sociétés* 11 (1): 3-15.

Joss, S. (1998). The Danish consensus conferences as model of participatory technology assessment. *Science & Public Policy* 25 (1): 2-22.

Joss, S. (2005a.) Lost in translation? Challenges for participatory governance of science and technology. In: *Wozu Experten/Why Experts?* (Eds. H. Torgersen and A. Bogner), Wiesbaden, Verlag für Sozialwissenschaften, pp. 197-219.

Joss, S. (2005b). Between policy and politics. In: *Democratization of Expertise? Exploring Novel Forms of Scientific Advice in Political Decision-Making* (Eds. S. Maasen and P. Weingart), Dordrecht, Springer, Sociology of the Sciences Yearbook, pp.171-188.

Joss, S. and Durant, J. (Eds.) (1995). *Public Participation in Science: the Role of Consensus Conferences in Europe*. London, Science Museum.

Kerr, A., Cunningham-Burley, S. and Amos, A. (1998). The new genetics and health: mobilizing lay expertise. *Public Understanding of Science* 7: 41-60.

Klüver, L. (1995). Consensus conferences at the Danish Board of Technology. In: *Public Participation in Science: the Role of Consensus Conferences in Europe* (Eds. S. Joss and J. Durant), London, Science Museum, pp. 41-49.

Klüver, L. (2006). New trends in public participation. PATH conference, http://www.macaulay.ac.uk/pathconference

Levidow, L. (1994). Biotechnology regulation as symbolic normalization. *Technology Analysis and Strategic Management* 6 (3): 273-88.

Levidow, L. (1998). Democratizing technology – or technologizing democracy? Regulating agricultural biotechnology in Europe. *Technology in Society* 20 (2): 211-226.

Levidow, L. (2000). Pollution metaphors in the UK biotechnology controversy. *Science as Culture* 9 (3): 325-351.

Levidow, L. (2003). Policing the Scientific Debate on GM Crops: the Royal Society meeting of 11th Feb 2003. Available at: http://www.gmsciencedebate.org.uk/topics/forum/0070.htm

Levidow, L. (2007). European public participation as risk governance: enhancing democratic accountability for agbiotech policy? *East Asian Science, Technology and Society (EASTS): an International Journal* 1(1): 19-50. http://www.springerlink.com/content/7wv97641w42r7867/fulltext.pdf.

Levidow, L., Carr, S., Von Schomberg, R. and Wield, D. (1996). Regulating agricultural biotechnology in Europe: harmonization difficulties, opportunities, dilemmas. *Science & Public Policy* 23 (3): 135-157.

Levidow, L., Carr, S. and Wield, D. (2000). Genetically modified crops in the European Union: regulatory conflicts as precautionary opportunities. *Journal of Risk Research* 3 (3): 189-208.

Lezaun, J. and Soneryd, L. (2006). *Government by Elicitation: Engaging Stakeholders or Listening to the Idiots?* London, LSE CARR Discussion Paper no.34. Available at: www.lse.ac.uk/Depts/carr

Marris, C. (2001). Swings and roundabouts: French public policy on agricultural GMOs since 1996. *Politeia* 60: 22-37.

Marris, C. and Joly, P-B. (1999). Between consensus and citizens: public participation in technology assessment in France. *Science Studies* 12 (2): 3-32.

OPECST (1998a). Letter from Steering Committee to participants in the preparatory weekends, 16 April.

OPECST (1998b). Conférence de Citoyens Sur l'Utilization des Organizmes Génétiquement Modifiés en Agriculture et dans l'Alimentation. Paris: L'Office Parlementaire d'Évaluation des Choix Scientifiques et Technologiques (OPECST), Assemblée Nationale. Available at: www.senat.fr/opecst/

Pestre, D. (2008). Challenges for the democratic management of technoscience: governance, participation and the political today. *Science as Culture* 17 (2).

Popper, K.R. (1962). *The Open Society and Its Enemies*, vol. 1. London, Routledge.

Purdue, D. (1995). Whose knowledge counts? 'Experts', 'counter-experts' and the 'lay' public. *The Ecologist* 25 (5): 170-172.

Purdue, D. (1996). Contested expertise: plant biotechnology and social movements. *Science as Culture* 5 (4): 526-545.

Reynolds, L. and Szerszynski, B. (2006). Representing GM Nation, PATH conference, proceedings at http://www.macaulay.ac.uk/pathconference

Roqueplo, P. (1996). *Entre savoir et decision, l'expertise scientifique*. Paris, INRA.

Rowe, G. and Frewer, L. (2004). Evaluating public-participation exercises: a research agenda. *Science, Technology and Human Values* 29(4): 512-557.

Rowe, G., Marsh, R. and Frewer, L. (2004). Evaluation of a deliberative conference. *Science, Technology and Human Values* 29(1): 88-121.

Roy, A. and Joly, P-B. (2000). France: broadening precautionary expertise? *Journal of Risk Research* 3 (3): 247-254.

Science Museum. (1994). UK National Consensus Conference on Plant Biotechnology: Final Report. Available at: http://www.ncbe.reading.ac.uk/NCBE/GMFOOD/conference.html

Stirling, A. (2005). Opening up or closing down? Analysis, participation and power in the social appraisal of technology. In: *Science and Citizens: Globalization and the Challenge of Engagement* (Eds. M. Leach, I. Scoones and B. Wynne), Zed, London, pp. 218-231.

Stirling, A. (2006). From science and society to science in society: Towards a framework for 'co-operative research', Report of a European Commission workshop, 24-25 November 2005. Available at: eurosfaire.prd.fr/7pc/bibliotheque/consulter.php?id=308

Teknologinævnet. (1987). *Genteknologi i industri og landbrug*. Teknologinævnets Rapporter 1987/2 og 1987/4.

Teknologinævnet. (1999). Final document of the consensus conference on genetically-modified foods, Copenhagen: Danish Board of Technology, www.tekno.dk

Toft, J. (1996). Denmark: seeking a broad-based consensus on gene technology. *Science & Public Policy* 23 (3): 171-174.

Toft, J. (2000). Denmark, potential polarization or consensus? *Journal of Risk Research* 3 (3): 227-236.

Van den Daele, W. (1994). *Technology Assessment as a Political Experiment*. Berlin, Wissenschaftszentrum.

Van den Daele, W. (1995). Technology assessment as a political experiment. In: *Contested Technology: Ethics, Risk and Public Debate*, (Ed. R. von Schomberg), Tilburg, International Centre for Human and Public Affairs, pp. 63-89.

Wakeford, T. (Ed.) (1998). *Citizen Foresight: A Tool to Enhance Democratic Policy-Making, Part 1: The Future of Food and Agriculture*. London Centre for Governance Innovation and Science and the Genetics Forum. Available at: http://www.ncl.ac.uk/peals/assets/publications/Citizenforesight.pdf

Walls, J., Horlick-Jones, T., Niewöhner, J., O'Riordan, T. (2005). The meta-governance of risk and new technologies: GM crops and mobile telephones. *Journal of Risk Research* 8: 635-661.

Wilsdon, J. and Willis, R. (2004). *See-through Science: Why public engagement needs to move upstream*. London, Demos, www.demos.co.uk.

Wilsdon, J., Wynne, B. and Stilgoe, J. (2005). *The Public Value of Science: or how to ensure that science really matters*. London, Demos, www.demos.co.uk.

Part III. Potentialities of reconstruction: critical reflections

First the peasant? Some reflections on modernity, technology and reconstruction

Joost Jongerden

1. Introduction

Among the first books I read on the social shaping of (and by) agro-biotechnology were Henk Hobbelink's *New Hope or False Promise: Biotechnology and Third World Agriculture* (1987), Jack Kloppenburg's *First the Seed* (1988) and Guido Ruivenkamp's *The Introduction of Biotechnology in the Agro-Industrial Production Chain: The Transition to a New Organisation of Labour* (1989). The books are all critical of biotechnology, but take different perspectives. While Hobbelink and Ruivenkamp concentrate on the social shaping of society by biotechnology, Kloppenburg emphasises a social shaping of technology by capital. In spite of these differences, all three authors deconstruct biotechnology in terms of social relations. As a young student, reading these books was a transformative experience. I realised that technology was not to be considered a mere tool, to be used in a good or bad way, but a socially construct, configuring social relations. The focus of the authors was on three types of social relations: (1) the commodification of genetic resources and the transformation of the agricultural producers into entrepreneurs, (2) a new international division of labour between 'the North' and 'the South' emerging from the substitution and interchangeability of products, producers and markets, and (3) a new division of labour between the private sector and the public sector.

The process of commodification of germplasm referred to the transformation of genetic resources and seeds from potentialities into commodities. In this transformation process, breeding was turned into a mechanism for the accumulation and reproduction of capital, and the process and product of plant breeding became defined in terms of market or commercial value. In *First the Seed*, a social history of plant breeding in the United States, Kloppenburg reconstructed the pursuit of two routes by corporate plant breeding in order to achieve the commodification of germplasm: the legislative route (IPR, legislation) and the technological route (hybridisation). The legal route introduced the idea of intellectual property granted to the breeder of new varieties which complied with the DUS requirements (distinctness, uniformity and stability). Concealed in the concept of intellectual property is the idea of 'unauthorised propagation' of plant material. This brings attention to the notion that it is not the awareness of what belongs to oneself that is the most important feature of the concept of property, but rather, that of what does not, what belongs to someone else (Deibel, this volume). The concept of 'unauthorised propagation' tries to impress on agricultural producers the idea that the propagation of plant material is an inherently illicit deed, requiring that one pays for the seeds one uses. A result of the application of this concept is a disconnection between seed multiplication and the agricultural producer. The disconnection between the multiplication of productive seed and the agricultural producer is materialised in the technological route of hybridisation. This route eliminates the possibility of any reproduction of original varieties, and forces agricultural producers into the commercial seed market, every year.

Henk Hobbelink's *New Hope or False Promise, Biotechnology and Third World Agriculture* predicted a growing dependency of the agrarian South on the industrial North, mainly as a result of the increasing biotechnological possibilities of substitution, and the interchangeability of producers, products and markets. Biotechnology, it was argued, would break primary products down into components, and blur boundaries between product-groups and sectors, an issue also raised by Ruivenkamp. At that time, the 1980s, the phantom was the example of the production of corn-based sweeteners in the United States, substituting sugar, and making producers of sugar-cane/beet and producers of corn interchangeable. Basically, the production of sweeteners from corn contributed to the reduction of excess corn stocks that had resulted from a US embargo against the USSR, a main buyer of US corn. – but it also contributed to unemployment in the Philippines. In general, the substitution and interchange of producers, products and markets was said to be profitable for corporate industry, and causative of poverty and dependency in third world countries.

In *The Introduction of Biotechnology in the Agro-Industrial Production Chain*, Ruivenkamp emphasised the importance of the concept of the *agro-industrial production chain* and *remote control* in describing how control over agricultural production systems became increasingly transferred from agricultural producers to corporate suppliers of inputs and processors of agricultural 'raw materials'. Ruivenkamp distinguished three disconnection processes. First, there was a disconnection of agricultural production from the natural environment, initially by means of chemical interventions (fertilisers, pesticides) and which had been widely described and criticised for its perceived unsustainability. Second was a growing disconnection between the production of agricultural products and end-products, as a consequence of a devaluation of agricultural products into raw-materials and components for industry, and for the chemical industry as well as the food processing industry. Third, there was a disconnection between production and consumption, since agricultural producers increasingly produced for very distant and anonymous markets. Referring to these disconnection processes and the increasing possibilities for substitution and interchangeability by means of particular biotechnologies (enzyme technology, among others), Ruivenkamp introduced the concept of biotechnologies as politicising products.

These authors were not only among the first to criticise biotechnology development, but also proposed the idea of a reconstruction of biotechnology. Since they considered technology, and thus biotechnology too, as a social construct, it seemed possible (at least theoretically) to develop a biotechnology that would not produce profits in one place and poverty elsewhere, but one that would support and develop the livelihoods of peasants and the rural poor. A critical analysis of current developments in agriculture and industry and contemporary biotechnologies was followed by a commitment to the development of alternatives: both alternatives in development and alternatives in technology. Such alternatives, however, tended to be discussed at a macro-level, and appeared to be actor-less. It was not unusual to refer to the categories 'North' and 'South'. Yet categories such as North and South do not function as proper compass points to direct concerns regarding the actors and beneficiaries of alternative biotechnologies – and not just because these categories are of a too general kind, but because they ascribe human values and agency to abstract (quasi-geographical) entities. The question of who the subjects were of this alternative development and reconstruction of biotechnologies remained not only unanswered, it was not posed. Even now, the issue of the actual subjects of any alternative biotechnology is hardly raised.

Today, when we turn again to the work of the three authors mentioned, we may observe an orientation towards peasants and peasant production systems. Hobbelink[90], Kloppenburg[91], and Ruivenkamp[92] are all inclined to put their hopes for a future agriculture on the food sovereignty movement, which is basically a movement centered upon a peasant mode of production. Indeed, looking at grass-roots initiatives for the reconstruction of technologies, we may observe that reconstruction is, in fact, often taking place in a context in which local, smallscale agricultural producers try to strengthen, obtain or regain control over the means of production, in other words, in a context of the development of a peasant production system. In a district of the southern Indian state of Andhra-Pradesh, for example, we have seen the development and decentralised production of location-specific Bt. sprays against insects harmful to the oil-crop Castor. The main component of the spray – a powder made from Bt. – is stored under local conditions by farmers, to be used as and when needed. Production of the Bt.-based powder also takes place at a local level, close to the farmers, who are thus no longer dependent on middle-men and chemical substances of dubious origin (see Puente, 2007 for an extensive discussion). In the Netherlands during the 1980s and 1990s, we have witnessed several attempts by farmers to regain control over labour and the products of their labour. This took place mainly in the form of a withdrawl from bulk production for global markets and the establishment instead of closer linkages to local markets and consumers – which implied in most cases that the processing to food-products was also reorganised, with more control over the processing process by producers and consumers. A high profile initiative was the Zeeuwse Vlegel (Zeeuws refers to the Zeeland region in the southwest of the Netherlands, *vlegel* means both rascal and is an instrument to flail grains), which started with the production of a speciality bread, but diversified with a range of products including beer, pancake flour and biscuits. The attempt to produce food products of good quality also resulted in collective attempts to organise variety and seed-availability (Jongerden, 1996; Wiskerke, 1997). We see that technology reconstruction is a reflexive undertaking, it questions social relations in production and through this questioning itself changes.

This paper is not a systematic mapping or comprehensive analysis of reconstruction, but an associative undertaking, taking us from a peasant issue, through modernity and rationality and back again to reconstruction. The paper consists of three main parts. The first, *A great transformation?*, considers what is considered a main characteristic of the modernisation process: the transformation from a peasant to a farmer mode of production. The meaning and scope of this transformation is discussed through two

---

[90] Hobbelink co-founded the organisation Grain, one of the initiators of the People's Food Sovereignty Statement, which states: 'In order to guarantee the independence and food sovereignty of all of the world's peoples, it is essential that food is produced though diversified, community based production systems. Food sovereignty is the right of peoples to define their own food and agriculture; to protect and regulate domestic agricultural production and trade in order to achieve sustainable development objectives; to determine the extent to which they want to be self reliant; to restrict the dumping of products in their markets, and; to provide local fisheries-based communities the priority in managing the use of and the rights to aquatic resources. Food sovereignty does not negate trade, but rather, it promotes the formulation of trade policies and practices that serve the rights of peoples to safe, healthy and ecologically sustainable production.' See: http://www.peoplesfoodsovereignty.org/content/view/32/26/ (last accessed, 26-07-2007)

[91] In an interview with Seedling in 2005, Kloppenburg states: 'An essential part of the resistance is the emergence of food sovereignty movements in the South and the local food movements in the North. People around the world increasingly understand that they are not locked into a single, capital- and energy-intensive trajectory of agricultural development and that one can eat well, pleasurably and sustainably by improving the technologies we already have and looking towards agro-ecology and organic agriculture. What people need is not simply something to oppose but also something to replace what you are opposing, and to find a new paradigm for agriculture and for eating. I think that the food sovereignty and local food movements are providing that kind of concrete alternative.' See: http://www.grain.org/seedling/?id=414 (last accessed, 26-07-2007)

[92] Ruivenkamp is one of the initiators of a research program on technology-development based on the food-sovereignty concept. See: http://www.telfun.info/

novels by John Steinbeck and Yaşar Kemal, among the most important and socially engaged writers of the 20th century. The second part, *A clash of modernisations?*, looks at the existence of two modernities, arguing that those who have resisted so-called modern technologies should not be dismissed as mere rebels against the future, but ought rather to be considered as committed to another kind of modernity, as, to some extent, reconstructionists. In the third part the rationality of modern technology is discussed. It is argued that a reconstructionist strategy cannot be founded upon an instrumental rationality and should be based on a substantial rationality.

Before continuing, I would like to say a few words on the concept of technology. In this work, different and sometimes contradictory views on what technology is will arise (technology as an instrument, technology as 'shaker and mover', technology as social product). To clarify my own position: I consider technologies as processes through which people (human agents) change the material (production forces) and social conditions (production relations) of existence and the knowledge underlying these changes. This definition ascribes three main features to technology. First, since technology is about processes through which people transform the conditions of their existence, technology is considered to be social. Second, technology deals with both material conditions of existence (transformation of nature) and social conditions of existence (transformation of society and of culture). Three, technology is about systems of knowledge. Yet, technology is not only a social construct, but also shapes social relations. Social systems and technological systems co-create, they configure each other.

2. A great transformation?

In 'The Age of Extremes' the historian Eric Hobsbawm states that 'the most dramatic change in the second half of the [twentieth] century, and the one which cuts us forever from the world of the past, is the death of the peasantry' (Hobsbawm, 1995: 289). He concludes that from the peasant world of the past only three regions on the globe remain dominated by villages and fields, namely, sub-Saharan Africa, south and continental south-east Asia, and China. In Europe and the Middle East only one peasant stronghold remained at the end of the 20th century, Turkey (*ibid.*: 291), although even there the rural population has dropped from 62 percent of the total population in 1970 to 35 percent in 2000 (Jongerden, 2007: 129). The announced death of the peasantry is inextricable linked to the realisation of modern society and (wo)man. The modern world, as it increasingly became imagined, was no place for the peasant, and his (or her) destruction was simply considered part of the process of becoming modern. This destruction of a class (Hobsbawm defines the peasantry in terms of its relation to the means of production and market) had dramatic consequences for the lives of those involved. To understand the tremendous impact on people's lives of this process of 'becoming modern' in the 20th century, I would like to look at two novels by two of its best known and most committed story-tellers, John Steinbeck and Yaşar Kemal.

2.1. 'The Grapes of Wrath'

John Steinbeck's 1939 masterpiece, 'The Grapes of Wrath', describes in moving detail and great compassion the lives of ordinary people in a period of agricultural modernisation and economic crisis. The setting of the story is 1930s Oklahoma and California, during the time of the Dust Bowl (the extreme drought and dust storms in the North America prairies) and Great Depression. The story is told from the perspective of Tom Joad, a young man on parole after serving four years for manslaughter. The novel focuses on his family, a poor peasant family of sharecroppers, who are forced to leave their land and home as a result of economic hardship brought on by the destruction of their crops in the Dust Bowl drought, along with the mechanisation of agriculture. When Tom returns home from jail he sees the small unpainted house mashed at one corner, pushed out from its foundations, the fences gone and cotton growing all over. 'Hell musta popped here', says Tom Joad (Steinbeck, 2000: 42). The dispossessed Joad family invests the little money they have left in a journey to California in the hope of finding a better life. They sort out their belongings and sell what is of no use anymore or they cannot carry. Their tools, such as the hand plow, only have no value beyond the weight of the metal, since everything has become mechanised. The Joad family takes the great cross-country highway to the West, and meet others on their way; they discover that thousands, hundreds thousands of other families, sharing a similar fate of dispossession, have made the same decision, following the same dreams. These people on the move are families which had once lived on a little piece of land, but had been pushed out by banks and machines, and were now looking for work (Steinbeck, 2000: 295). On the road, the Joad family hear stories from disenchanted people returning from California, but they try to ignore these and stick to their hopes, since they are all they have. Upon arrival in California, the Joads discover that for each job there are many applicants and the likelihood of getting a decent wage is low. Farms have grown larger and owners fewer, fruit trees have taken the place of grains and tractors replaced human labour. The farms employ the dispossessed peasants, the homeless and the hungry, giving them food on credit for their work; but when the work is done, the labourers are likely to find out that instead of having earned money they are in debt to the farm. For some, the New Deal Resettlement Administration provides an opportunity for warmth and shelter and food. However, the program is frustrated by landlords, and the government does not supply it with sufficient space and money to enroll all the workers together in the working camps. The Joads accept a job in an orchard, not knowing that it is involved in a strike, and when things turn violent a friend of the family, the preacher Jim Casey, is killed, and Tom Joad is forced to kill again, which turns him into a fugitive.

'The Grapes of Wrath' is a passionate novel, its 400-odd pages written in about a hundred days, and its structure characterised by the interspersing of long, intimate narrative chapters detailing the Joad family drama as they flee to California with short panoramic chapters on the migrants as a group. Steinbeck attempted to involve the reader in the actuality, by writing about people's lives as really lived. This actuality was, in fact, the death of the peasantry, and not only as a manner of speaking. Many peasant migrants were literally dying, as illustrated by the words of John Steinbeck in a letter to a friend sent in the winter of 1938, just months before the publication of his book (Steinbeck, 2000):

> 'I must go over in the interior valley. There are about five thousand families starving to death over there, not just hungry but actually starving. The government is trying to feed them and get medical attention to them with the fascist group of utilities and banks and huge growers sabotaging the thing all long the line. [...] They think that if these people are allowed to live in camps with proper sanitary facilities, they will organise and that is the bugbear of the large landowner and the corporation farmer. [...] I've tied into the thing from the first and I must get down there and see it and see if I can't do something to help knock these murderers on the head. [...] I am pretty mad about it.'

2.2. 'The Legend of the Thousand Bulls'

'The Legend of the Thousand Bulls', by Yaşar Kemal, was published in 1971, almost 30 years after Steinbeck's Grapes, but the themes of the two novels are rather close. The story is set in Turkey in the 1950s, the period when Marshall help and mechanisation had started to carry modernity to the countryside. The main concern of the story is the Aydınlı nomads, a branch of the Yörük. The story explains how the Aydınlı, once a mighty nomadic tribe with no less than two thousand black tents, have now become a dwindling and marginalised group. In the summer they bring their flocks to the summer pastures in the Aladağ mountains near the city of Adana in the south of Turkey, and in the autumn they move and pitch up their tents down in the Çukurova plain. However, the spaces needed to sustain their economic and social life are under assault. Mechanised agriculture and the colonisation of land leads common pastureland to become private property and turned into cotton and wheat fields, ever expanding since tractors have moved in. Year by year, there is less space for the Aydınlı to herd their sheep, and the tolls levied on their camp sites are increased. Modernity is squeezing them out. Notwithstanding the fate of many of the other nomadic Yörük tribes, who surrender and settle in villages to become agricultural workers for rich landowners, or shopkeepers, the Aydınlı persist in their way of life, hoping for things to change. But their hope is in vain, and the Aydınlı slowly realise that nothing can save them, that they will be dispersed and shrunk, and will end up in a apartment in the city, 'sitting in a tiny room, pondering like barn-owls.'

The Aydınlı were not settled by force in the Çukurova plain in the last quarter of the 19th century, as many others were. In 1876, the military were guarding the road into the Taurus mountains from the Çukurova plain, and not a soul was allowed to pass from either the plains to the mountains or the mountains to the plain. New villages and towns were to be built, and officials were busy measuring and calculating squares, avenues and streets. However, malaria was rampant and epidemics ravaging the countryside. The Çukurova plain was strewn with the skeletons of men and animals. The Aydınlı had managed to avoid this fate by evading the military blockade, but by 1950 there was no strip of land untilled. Quarrels over land ownership became rife, and the Aydınlı could hardly find a place to pitch their tents, let alone graze their stock. They had been suffering for several years, but never dreamed it would become like this. For a while they managed to camp on village pasture land, paying small fortunes to the villagers for the privilege, but then the pasture land too began to be cultivated. In the mountains, meanwhile, they were harassed by forest guards. The Aydınlı could not count on much sympathy from their former fellowmen:

'What haven't we seen, what!' Beardless Agha said. 'What haven't we gone through! When did we ever live like human beings in this Çukurova plain? You see that railway that comes piercing through the mountains? That pair of iron rails stretching on and on, which the Germans laid? Well, every foot of it holds a human life... What haven't we seen till we reached this day, what! What haven't we suffered, what haven't we gone through!'

Beardless Agha was launched now. 'I don't have the heart to look back on the past. Not ever! Not one of us old Turcomans has the heart to remember.

'Before we came to live this sedentary life, the whole world was a paradise for us, everywhere, the Taurus Mountains that we call the Thousand Bulls, with their many peaks, Aladag, Düldül, Kayranli, Bent, and also the Gavur Mountains, and all the region of Payas and Dumlukalé, and the whole of the Çukurova plain... But all this came to an end with the settlement. And now these Aydinli nomads come to us and complain! After living like princes for a hundred years, they complain! Well, let them. While we were dying like flies down here on this bloody plain with the heat, the mosquitoes, the fever, the pestilences, the wars, the taxes, they were taking their ease up in the high-lands beside the cool springs, among the blue hyacinths and pennyroyal, in the dappled shade of the pines. Did they ever trouble about us? Did they once ask how we were faring?

No, they've parted company with us, they've become a separate people. It's they who've broken the old tradition, who've caused this division among the tribes.'[93]

Both of these novels tell the story of the death of the peasantry. Etymologically, the word 'peasant' is derived from the French *païsant*, meaning someone from the *pays*, the countryside. What really characterises the peasant, however, is the relationship to the land, worked upon by his/her own labour, the direct access to the means of production. In other words, the concept of the peasantry denotes a specific position within the relations of production and exchange: peasants labour on the land they own or have access to, and are only partially integrated and subordinated to larger econonmic systems (Hardt and Negri, 2004: 116)[94]. During the 20th century, an increasing number of people who had been involved in agriculture, those who had had direct access to the means of production and provided for their own subsistence, were dispossessed, 'depeasantised', and became rapidly and massively concentrated in urban centers (Araghi, 1995: 338). In this process, technology played an important role, especially the mechanisation of agriculture, an issue relatively under-studied compared to the biochemical aspects of the green-revolution (seed development, synthetic fertilisers, etc), but maybe of greater importance in the overall transformation of agriculture.

Crucially, both 'The Grapes of Wrath' and 'The Legend of the Thousand Bulls' tell the story of modernity as one of a 'productive destruction' (see also Harvey, 1989: 16). The destruction of the peasantry was productive in the sense that it was contingent on the creation of a new class of farmers. In other words, the death of the peasantry is associated to the birth of the farmer, a modern entrepreneur, integrated into agro-industrial chains and producing primarily for market consumption. The image of productive destruction was introduced in 1942 by the economist Joseph Schumpeter to depict the process of capitalist development (Schumpeter, 1975 [1942])[95]. This issue had already been touched upon, however, by Walter Benjamin in his famous 'Work of Art in the Age of Reproducibility', a first version of which was written in 1936, and in which modernity is described as 'inconceivable without its destructive, cathartic side: the liquidation of the value of tradition'[96] (Benjamin, 2003: 254). The idea of productive destruction is itself derived from the modernist project, or more specifically, from the

---

[93] (Kemal, 1976: 79-80).

[94] I would make two points here. First, although peasants produce for their own consumption, they are not subsistence producers. Second, although they have direct control over the means of production, they are not, and have never been, free individuals, but operate within a set of institutions (that of the landlords, for example) that provide services and protection, and for which they have to provide services (such as forced labour) that place severe constraints on their behavior (Migdal, 1974: 198-199).

[95] In Capitalism, Socialism and Democracy, Schumpeter writes (1975: 83-84): 'The opening up of new markets, foreign or domestic, and the organisational development from the craft shop and factory to such concerns as U.S. Steel illustrate the same process of industrial mutation – if I may use that biological term – that incessantly revolutionises the economic structure *from within*, incessantly destroying the old one, incessantly creating a new one. This process of Creative Destruction is the essential fact about capitalism. It is what capitalism consists in and what every capitalist concern has got to live in ... Every piece of business strategy acquires its true significance only against the background of that process and within the situation created by it. It must be seen in its role in the perennial gale of creative destruction; it cannot be understood irrespective of it or, in fact, on the hypothesis that there is a perennial lull.'

[96] Of course, one may argue that the very idea of productive destruction has been put forward by Karl Marx and Friedrich Engels already in 1848 in the Communist Manifesto. 'The bourgeoisie cannot exist without constantly revolutionising the instruments of production, and thereby the relations of production, and with them the whole relations of society. Conservation of the old modes of production in unaltered form, was, on the contrary, the first condition of existence for all earlier industrial classes. Constant revolutionising of production, uninterrupted disturbance of all social conditions, everlasting uncertainty and agitation distinguish the bourgeois epoch from all earlier ones. All fixed, fast frozen relations, with their train of ancient and venerable prejudices and opinions, are swept away, all new-formed ones become antiquated before they can ossify. All that is solid melts into air, all that is holy is profaned, and man is at last compelled to face with sober senses his real condition of life and his relations with his kind'. http://www.marxists.org/archive/marx/works/1848/communist-manifesto/ch01.htm

dilemma of how to create a new world that transcends the old one. The literary archetype is Goethe's Faust, who forces everybody to create a new (social) landscape and eventually deploys Mephistopheles to dispossess an old couple living in a small cottage by the sea for no other reason than that they do not fit his master plan of development. Mephistopheles, however, kills the couple and burns down their house (Harvey, 1989: 16)[97]. The couple, one may say, stands for the fate of the Joads and the Aydınlı in the 20th century.

3. A clash of modernisations?

The death of the peasantry was considered a logical, inevitable and irreversible consequence of modernity. The modernity project of the 19th and 20th centuries – as it was perceived by modernists like Wilbert Moore, Ernest Gellner, Talcott Parsons, Daniel Lerner, and Bernard Lewis – was typically not considered an open-ended process, excited by the liberation of the subject, but a preconceived picture of development. Rather than the agents of modernisation and subjects of their own history, people were the objects of a prescribed process of social and societal transformation. Modernity emerged in the 19th and 20th centuries as a narrow, sterile path directed by the deeds and discourses of political elites. It simply became an imposed condition[98].

This idea of modernity was mainly centered on the development of strong states (nation-states) and liberal economies. As a political and academic project, it became a major theme in the post-Second World War economic, social and political sciences; and as a dominant current in American and European scholarship during the 1950s and 1960s, it aimed at providing an alternative explanation of development to Marxist scholarship. More than just a research object, modernisation was treated as a guiding principle for social transformation under the guidance of the state. In the political domain it was equated with the creation of powerful authoritarian polity (with a particular role for military institutions, since the perceived impersonal social relations of the military were considered exemplary of what was universal in modern society), which managed two basic transformation processes. First, was the development of industry and capitalist relations of production in the economy ('free' labour, the 'free' market), including the transformation from a peasant to a farmer mode of production. Small-holdings, the dominant structure of peasant production were considered to be simply economically backward and inefficient, and not only because of technological and mechanical limitations, but also because of the relations of exchange the integration into wider markets, which implies loss of control, makes small-scale producers particularly vulnerable to the institutions dominating access and determining prices in these markets). The image and aspirational end-state of modern agricultural production was that of large industrial and high-tech holdings competing in world markets (Van der Ploeg, 2003; Hardt and Negri, 2004: 120). Second, in the cultural domain, modernisation and state-managed social transformation was held responsible for the production of a nation. Ernest Gellner's work on nations and nationalism, for example, argues that nations are the inevitable consequence of industrial society (Gellner, 1983)[99].

---

[97] For a synopsis of Faust see: http://www.ucalgary.ca/~esleben/faust/goethe/synopsis.html

[98] For a systematic critique of modernisation theory and its thinkers, see 'The Sociology of Development and Underdevelopment' by Andre Gunder Frank (1967).

[99] Gellner's argument runs thus: 'Mankind is irreversibly committed to industrial society, and therefore to a society whose productive system is based on cumulative science and technology. (...) We do not properly understand the range of options available to industrial society, and perhaps we never shall, but we understand some of its essential concomitants. The kind of cultural homogeneity demanded by nationalism is one of them, and we had better make our peace with it.' (1983: 39). For a critique of this, see Jongerden (2007).

Modern society had its imagined spatial form. Industrialisation was twinned to urbanisation, and the farmer-to-peasant production transformation would be accompanied by rural-to-urban migration as traditional countryside practices became redundant. The land and its people were industrialised. A flow of people from a retarded countryside (in conventional modernisation theory the rural is associated with negative social images) to cultured cities (the urban being associated with positive social images) was thought to be a part of the unfolding of history. Only few thinkers questioned the supposedly necessary and inevitable link between industrialisation and urbanisation (e.g. Kropotkin, 1912; Bookchin, 1987, 1990, 1992; Friedmann, 1996).

Even though few criticised modernity's premises – there seemed to have been a broad consensus on the inevitability of the death of the peasantry, and the idea that urbanisation and industrialisation should coincide – modernity has never been an unambiguous project. In very general terms, I would like to draw a distinction between early and late (recent) modernisation, with the caveat that such a historicising may be misleading, since different ideas about modernity do not only follow each other in time, but also compete with each other. We may distinguish between a libertarian and authoritarian modernity, linked to the early and late forms, and percieve a clash between the two.

Modernity in the 18th and 19th century has been associated with the Enlightenment, with its flow of thoughts that contested and condemned established ways of thought (religion, metaphysics, and speculative philosophy) and life (the imperial form of states, absolutism and monarchy) in Europe. It was critical to the thoughts and powers that were, being concerned with the newly discovered dynamics of the development of nature (e.g. evolutionary theories) and society (e.g. historical materialism), and rethinking and redefining the relationship between man and nature and man and society. Looking at the early conceptions of modernity, we should appreciate their radical understanding of democracy, comprising the idea that every citizen is a part of the sovereignty and can acknowledge no personal subjection (Paine, 1791-1792). The move from democracy as the rule from the few or the many to the that of everyone may seem a relatively small semantic shift, but it is one which has radical consequences. 'We can only all rule when we do so with equal powers, free to act and choose as each of us pleases' (Hardt and Negri, 2004: 240-241). This conception of democracy is to be traced back to the idea of the sovereignty of the subject, a sovereignty not only in the political domain, but also in other domains, and an idea which may be considered an important aspect of the food-sovereignty movement.

The many peasant rebellions in the last couple of centuries have been sparked by assaults on the means of existence and the subjugation to bosses and markets. This resistance has been a global phenomenon. '[T]he peasant rebellions of the twentieth century', Eric Wolf concludes in his 'Peasant Wars of the Twentieth Century', 'are no [...] simple responses to local problems [...]. They are but the parochial reactions to major social dislocations, set in motion by overwhelming societal change' (Wolf, 1969: 295). These peasant rebels may have looked backward to old customs they preferred to preserve, but also tried to reinvent old rights in a new time and place, looking forward as much as backward. This was certainly been the case with the Luddite movement[100]. The Luddites opposed the introduction not so much of machinery, but a particular kind machinery, that which condemned large sections of the rural population to poverty and misery. They broke the machines that broke prices. Their resistance, we may

---

[100] According to Thompson (1965) the Luddites were inventing new legislation from old. On the one hand they were guildsmen, looking back to old customs and the legislation of the past. On the other hand, many of their demands – for example, for an honest price, and later against child-labour, for a minimum wage, and the right to have trade unions – pointed forward towards what we now consider to be achievements of the labour movement.

say, was oriented to those machines that did not serve human beings, but rather made serfs from human beings, and in their struggle we see the clash of modernisations actually turn into a battlefield. A close look at this history may also lead us to reconsider the standard image of Luddites as 'anti-technology', and to appreciate them instead as (militant) re-constructionists.

3.1. Luddites

The Luddites were a well-organised, rural movement of machine-breakers that spread across southern England between 1811 and 1813. Although Luddites and the practice of machine-breaking have become synonymous, actually machine breaking had been a widespread practice in England long before the Luddite movement. Machine breaking started around 1640-1660, a period in which the attitude of the state towards machinery changed, and the former 'hostility towards devices which take the bread out of the mouths of honest men', gave way to encouragement of profit-making enterprise, at whatever social costs' (Hobsbawm 1964).

The Luddites smashed those machines they considered to be immoral, in particular the machines deskilling labour and operated by unskilled workers and child-labour. As a well-organised network of secret committees, they resisted new forms of domination symbolised by the factory system in textile production. New cotton mills had recently been established in South England, where the employment of child labour was used to lower wages and prices. (Thompson 1965: 548-549). This emerging factory production system undermined the then prevalent standards of craftsmanship and resulted in a beating down of the wages of workers and prices of self-employed artisans. The Luddites did not want to maximise returns and minimise expenditures regardless of social obligations and social costs. This destroyed the 'customs of trade' (often described as a moral economy), replacing them with a philosophy of profit-above-all and unrestricted competition. The new technology also transformed localities, where production had been community-controlled, into production-sites, where people became the labour-force.

The new factories were described as 'centers of exploitation, monstrous prisons', where hired workers were subject to harsh order and discipline. For the craftsmen, the difference between the working conditions they had been used to and the new subjugation was wide enough to cause them to rebel. As was written in a letter signed by Ned Ludd:

> 'We will never lay down Arms [until] The House of Commons passes an Act to put down all Machinery hurtful to [the common people], and repeal that [law] to hang Frame Breakers.'

Local communities sided with the Luddites. In public opinion, it was the factory owner, not the Luddites, who was considered as engaging in immoral and illegal practices. The Luddites resisted the politics of domination embedded in the new technology, and the community gave them its support. Between 1811 and 1813, thousands of machines and even complete factories were destroyed, sometimes by groups of as many as a couple of hundred masked and armed men (no women were allowed). However, their rage against the machine was by no means blind fury. The machines that were destroyed were, in Luddite terms, violating 'commonality'. In the words of the historian E.P. Thompson, 'every attack revealed planning and method' (Thompson, 1965: 554). A report on a machine-wrecking event reads:

> 'They broke only the frames of such as have reduced the price of the men's wages; those who have not lowered the price, have their frames untouched; in one house, last night,

they broke four frames out of six; the other two belonged to masters who had not lowered their wages, they did not meddle with.'

The Luddite movement came to an end in the second half of the 19th century, after a combination of severe repression and the transformation of secret committees dedicated to machine-breaking into legal organisations involved in the politics of social reform and labour unions.

Machine-breaking, and in particular the Luddites, has been largely side-lined in histiography. According to Eric Hobsbawm this is due to the fact that the movement and all it stands for did not fit the new emerging discourses of liberalism and socialism, the first praising the triumph of mechanisation and the pioneering-industrialists, the second rejecting the strong-arm methods and self-organisation in labour action. It became a dominant and widely shared view that machine-breaking was pointless, and that those who engaged in it were fighting with their backs to the future, facing only an inevitable defeat. The machine-breakers became a footnote in history. Although only a few historians have engaged in serious research on the phenomenon of machine-breaking, those that have done so were some of the most outstanding of their time, including Hobsbawm, Rudé and Thompson (Hobsbawm, 1964; Hobsbawm and Rude, 1969; Thompson, 1965). These studies reveal not only the popular character of the Luddite movement and its extensive repression – with mass-trials and death-penalties in 1813 along with the deployment of 12,000 troops, a number which exceeded that of the army Wellington had taken to the Iberian Peninsula in 1808 during the Napoleonic Wars – they reveal also the importance of the phenomenon of machine-breaking for the study of science and technology. The Luddites did not only engage in politics by wrecking, but as re-designers, a term used to indicate social practices directed against the politics of domination. The Luddites demanded a technology necessary for the realisation of a life worth living (Jongerden, 2006). Although one may argue that the Luddites were a peculiarly English phenomenon, machine-breaking as a form of collective bargaining or rejection, and other forms of violent resistance, are not (see for example Hobsbawm, 1959; Brown, 1977; Scott, 1985). At the heart of the machine question is a clash of modernisations, between a modernity regulated according to social principles, and a modernity in which human needs are subordinated to profits.

4. Instrumental or substantial rationality?

The Luddites dramatically raised the question of technology's rationality, the principles underlying the organisation of the man-machine relationship. Should it be considered in terms of efficiency or in terms of social justice? The Spinning Jenny replaced craftsmen with child-labour, and increased productivity as a result of which prices dropped. Should we consider the Jenny in terms of an instrumental rationality of efficiency in production, or a substantial rationality of social justice? Instrumental rationality is about the congruence between means and ends, and involves the best ways (efficiency, effectiveness) to realise an objective. Substantial rationality is about ends in themselves and involves the discussion and comparison of different kinds of ends, and different kind of means-ends systems.

Weber (1968) argued that modern society is characterised by an instrumental rationality. The prevalence of an instrumental rationality in the field of technology is inextricably related to the history of technology as constitutive of modernisation and the realisation of the modernist propject. The term 'technology' emerged at a time when industrial revolutions were overthrowing the production systems of guilds and craftsmen, and positivist science prevailed. Still, by the turn of the 19th century, 'technology' was not a current term. It did not begin to enter into popular usage until Jacob Bigelow at

Harvard College published his 'Elements of Technology' in 1829 (Meier, 1957: 618). The transferred sense of technology as the 'science of the mechanical and industrial arts' was first recorded in English in 1859 (Barnhart, 1995).

In the same period, the meaning of science started to change. Up to the 19th century, sciences were considered to be concerned with the ends and meaning of life. 'Men of science at such periods,' Alexis de Tocqueville observes with dislike, 'are consequently carried away towards theory; and it even happens that they frequently conceive an inconsiderate contempt for practice.' Their prototype was Archimedes who 'spent his talents and his studious hours in writing only of those things whose beauty and subtlety had in them no admixture of necessity' and considered the 'science of inventing and putting together engines, and all arts generally speaking which tended to any useful end in practice, to be vile, low, and mercenary' (De Tocqueville, 1997 [1840]).

De Tocqueville thought it characteristic for democratic societies that scientists would not devote themselves to abstract notions, but rather would be engaged in discovering practical sciences, the methods of application and the means of execution of knowledge. The new technological sciences should not only provide solutions for industry, by, for example disclosing the features of materials, but also provide rules and laws for the further development of industry[101]. An important forward thrust in technological science came with the creation of a separate community of practitioners, the engineers, with the establishment of the Rensselaer Polytechnic Institute in Troy, New York in 1824 an important milestone. In the course of the 19th century, science started to deal with the efficient organisation of life.

The development of applied science, a science which borrows its rationality not from conceptualisation but from industrial utility, deprives the (technological) sciences of a substantial rationality and leaves it with just an instrumental rationality, or worse, a normative and prescriptive one[102]. The transformation of science's rationality is the main subject of 'Eclipse of Reason', a remarkable book written in 1947 by Max Horkheimer, a leading figure of the so-called Frankfurt School. 'Eclipse of Reason' is a critical, not to say discerning examination of the transformation of thought in the epoch of modernity. An important argument is that the development of modernity transformed the world from one of ends into one of means. Horkheimer argued that modern thought is hardly concerned at all with ends in themselves; it is concerned, rather, with the coordination of means with ends as given. The emphasis is on its instrumental character, on the adequacy of procedures for ends that are more or less taken for granted. Little importance is attached to the question whether the ends as such are reasonable. These ends are simply considered a given end-state, they are not open for discussion. This stands in sharp contrast to pre-modern thought, which 'was intended to achieve more than the mere regulation of the relations between means and ends: it was regarded as the instrument for understanding the ends, for determining them' (Horkheimer, 2004: 7)[103]. The classical reference here is to Aristotle, whose philosophy sought out the ends that were good in themselves, rather than instrumental to some further end (and came up with one, that of happiness). The issue of reconstruction of technology is

---

[101] In the United Kingdom, Eaton Hodgkinson and William Fairbairn were pioneers in systematic experimentation in applied science. In 1854 William Fairbairn published his 'On the Application of Cast and Wrought Iron to Building Purposes', which immediately became a handbook for builders. In the United States, Alexander Bache did research on the strength of materials and the causes of steam boiler explosions.

[102] The distinction between instrumental and substantive rationality is to some extend a reformulation of Weber's rationalisation-theory (Löwy, 1996), but the terms as such were first used by Karl Mannheim in 'Man and Society' (Mannheim, 1940).

[103] Horkheimer concludes that functional rationality, the calculation of probabilities and convenient means for given ends, ceases to be a critical force because such a calculating functional reason destroys qualitative values.

contingent not on instrumental rationality, but on substantial rationality. It is dependent not just on the coordination of means for given ends, but on a discussion of different ends and means systems. It is concerned with the ends themselves.

4.1. Rationality and the study of technology

At the risk of simplifying to the level of caricature, one may distinguish between two dominant perspectives on the study of technology: technological determinism and social constructivism. I deliberately say perspectives, and not theories, for if we understand theory as an organised system of knowledge, then technological determinism must be disqualified as a theory. Technological determinism is not so much a theory, but a multi-headed notion of how technology induces social change. In general, the term is attached to the work of *others*, as some kind of accusation. I do not know of authors who would refer to themselves as technological determinists – with the possible exception of Marshall McLuhan, and one may question if he was. Social constructivism, on the other hand, may be considered a theory, which can be associated with groups of scholars: see, for example, the major reference books edited and written by MacKenzie and Wajcman (1985) and Bijker, Hughes and Pinch (1987) and Bijker and Law (1992).

A basic feature of technological determinism is that it seeks to explain social phenomena in terms of principal or determining factors, but it does so by differentiating technology from the social domain (Feenberg, 1999: 209). After separating technology from the social domain, the relationship of technology to the social is understood in terms of the interaction between domains external to each other. This results in two basic characteristics of the 'technological'. First, technology appears to be a world of its own. Second, technology appears as a reification, an object or arte-fact attributed human or living abilities. It is as if technology has a will of its own, outside human control, shaping society and changing under its own momentum (Chandler, 2001). Technological determinism is not so much a 'theory' for the study of technology, but a technology-oriented 'theory' of social change revealing the hidden logic of modernity. History is happening because of technology; technology is considered 'the prime mover' of history, releasing the forces of modernist progress

After a wave of technological determinism in the period 1950-1970 (also the most important years of the modernisation school in American and European scholarship) we witness a (renewed) interest in social constructivism[104]. Basically, constructivist approaches in studying technology offer a sociological approach to technology, holding that technologies are congealed moments in history of people acting together. Social constructivist approaches tend to bring in an actor-oriented perspective, distinguishing between a variety of what they call relevant social groups, competing to control a design. At the stage of design, constructivists argue, a technology is far from preordained. Each relevant social group has its own idea of the problem that the new artifact is supposed to solve and, in consequence, favors a distinctive technological design, including components and operational principles that may not be favored by other, competing groups. Eventually, one social group prevails over the others, its design prevails and the others are forgotten (Pinch and Bijker, 1984), or else two or more groups negotiate a compromise (Bijker, 1995). The most prominent case to which the approach has been applied has

---

[104] The idea of technology as a social construction may be traced back to the work of Karl Marx. In: *Grundrisse*, Marx writes: 'Nature builds no machines, no locomotives, railways, electric telegraphs, self-acting mules, etc. These are products of human industry; natural material transformed into organs of the human will over nature, or of participation in nature. They are organs of the human brain, created by the human hand; the power of knowledge, objectified' (Marx, 1973: 706).

been the social shaping of the bicycle (Bijker *et al.*, 1987). The conventional shape of the bicycle only acquired some permanency after decades of competition among diverse designs. While discussing the high front wheel bicycle and the safety bicycle – two low wheels, a diamond frame, a rear chain drive and pneumatic tires – Bijker introduces the concept of the relevant social group. Such groups might encompass consumers, manufactures and engineers with various ideas about the shape and features of the arte-fact, the bicycle. In the period of the development of the bicycle, Bijker distinguishes the relevant social group of the relatively well-off, young men, who preferred the bicycle to be a exclusive machine, which they would use to impress others. Others criticised that machine as unsafe and brought about the safety-bike, which permitted greater certainty of control. The differences created by agency of relevant social groups (the different interpretations of a thought artifact) Bijker refers to as the 'interpretative flexibility' of a technology[105]. The approach is subject to various critiques (see, for example, the Clayton vs. Bijker/Pinch debate), but the most substantial of which is that constructivists have narrowly confined themselves to the study of design and the winning of acceptance for particular designs of devices and systems, and lack any sense of a political context (Clayton, 2002; Bijker and Pinch, 2002; Feenberg, 1999: 11).

The two perspectives – technological determinism and social constructivism – are not 'interchangeable approaches', mainly because their objects of study are different: technological determinists study the social, or better, social change, while the social constructivists study technology. If we want to understand the variety of possible technologies and paths of modernity among which we can choose, critical theory has more to offer than determinism or constructivism, simply because critical theory does not only develop conceptual frameworks to analyse the new forms of oppression (associated with modernity and technology), but also argues that these can be challenged and alternatives are possible[106]. A wide ranging discussion of the relevance of critical theory to the study of technology is not possible here. Instead, I will focus on one particular aspect, and that is the rationality of technology.

5. Appropriating modernity and technology

The writer Bertolt Brecht argued that when people are struggling and change reality, they cannot do that by simply taking over existing works, or creating new works under the existing rules: 'People can only take over their cultural heritage by an act of appropriation' (Brecht, 2007: 81). Appropriation is an act of arrogation, confiscation or seizure in order to create something new[107]. An act of appropriation, Brecht argues, is an act of transformation. Just as work literally cannot be taken over physically, but needs an act of appropriation, so does technology need to be appropriated. According to Tracy B. Strong, 'I have appropriated something when I have made it mine, in a manner that I feel comfortable with, that is in a manner to which the challenges of others will carry little or no significance' (Strong, 1996).

---

[105] Bijker introduces two other concepts: closure and stabilisation. Closure refers to the process when the technology acquires a certain permanence, and stabilisation refers to consensus within a relevant social group.

[106] The term critical theory is linked with the Frankfurt School. Max Horkheimer defined critical theory in his 1937 essay 'Traditional and Critical Theory' (1972). The term critical refers to the use by both Kant and Marx of the term 'critique', i.e. meaning examining and establishing the limits of a phenomenon, especially through accounting for the limitations imposed by the system of which the phenomenon is a part. The term also refers to the need for an integrated approach to the study of society, echoed today in interdisciplinary approaches.

[107] To appropriate is to make something into one's own, and must be distinguished from expropriation, which is an act of deprivation. The distinction tends to be purely analytical, however, since an act of appropriation and of expropriation often occur simultaneously.

5.1. Appropriating modernity

'The great advantage of a tunnel vision,' James Scott muses, 'is that it brings into sharp focus certain limited aspects of an otherwise far more complex and unwieldy reality. This very simplification, in turn, makes the phenomenon at the center of the field of vision more legible and hence more susceptible to careful measurement and calculation [...] making possible a high degree of schematic knowledge' (Scott, 1998: 11). The disadvantage of a tunnel-vision is that it excludes to the extreme. Everything that does not fit the vision is ignored and marginalised to the level of invisibility. Concerned with the coordination of means to ends as given, the modernity project and its instrumental rationality has equipped us with a tunnel-vision focused on the assumed ends of history and provided us with schematic knowledge about the transformations needed to reach an end-state. Van der Ploeg (2003) refers to this modernity-project as a one-dimensional mega-project, directing and sanctioning actors to organise their activities towards the realisation of the end-state of a transformation of peasants into farmer-entrepreneurs, integrating their production-system in global production chains supplying commodities for international markets. Alternatives to this end-state are disqualified as inferior or non-desired and irrational because they are considered an intrusion on the mega-project, reducing the efficiency of the transformation process to the end-state. Teleological assumptions about the future function as a self-fulfilling prophesy.

Since modernity has never been unchallenged, it may be worth considering attempts to conceptualise alternatives, as well as critiques. It is striking that in spite of an abundance of critique, little effort is made to think about alternatives. Based on a brief review, I would say we can distinguish between two attempts to conceptualise alternatives, one parochial and exclusive, concerned mainly with the cultural dimension of modernity, the other libertarian and inclusive, concerned with a 'total' rethinking of the modernity project. Here, I will introduce the Turkish rural-sociologist Nusret Kemal Köymen and the Japanese novelist Junichiro Tanizaki as examples of a parochial and exclusive current, and Peter Kropotkin, Muray Bookchin and John Friedmann as examples of a libertarian current. The discussion will focus on the organisation of space, principally because of the mode of production that drained the countryside, turning the rural population into city-dwellers, also contributed to the growth of metropoles.

5.2. The parochial approach

By the 1930s, the Turkish sociologist Nusret Kemal Köymen had already developed a systematic critique of the concept of modernisation and settlement (Köymen, 1937). Köymen argued that the coincidence of urbanisation and industrialisation was not a law of history, but the biggest mistake in European history. It produced a scatter of desolate villages and pathologic cities. Köymen was fond of biological metaphors to describe the evil of the *megapolis*. He compared it with orthogenesis, the unstoppable growth of an organ damaging other organs and the body (the nation) in which it exists. Köymen argued that the first cities in history were relatively small and not the seats of production, like the modern factory-city, but rather administrative, cultural and economic service centers. In Anatolia, agriculture and industry had once been integrated, contended Köymen; villages at that time were the seats of a variety of industries, and many towns were little more than large industrial villages. The de-industrialisation of villages in Anatolia was an effect of the industrial revolution. Köymen declined the tendency to establish, under protection of the state, heavy and cumbrous machines, in itself unsuited for decentralised production, near urban centers, turning them into factory-cities.

Köymen considered the development of these factory-cities to be of the gravest danger for the nation, since the factory-city produced socio-cultural differentiation and class-society. Köymen's proposal was to decentralise industry and to establish a network of new settlements, created from an integration of rural and urban modules (founded economically upon an integration of agriculture and industry), which he named '*rurban*'. Köymen explicitly related the development of new settlements to a nationalist cultural program. The new environments created were supposed to assimilate populations within a national identity and contribute to the development of the nation. In this context, alternative modernity becomes a policy for nationalism and identity politics. Strikingly, in this idea of alternative modernity, technology was simply considered a tool, to be used in a good or bad way. The domain of science and technology was considered a universal, accessible to all nations – and Turkey should try to acquire and master the technologies developed in 'the West'. Culture, however, was seen as inextricably connected with the concept of the nation, marking the boundaries of nations, which Köymen thought to be the highest and most exalted form of political organisation.

In Critical Theory of Technology, Andrew Feenberg refers to a Japanese attempt to decline the modernity project and open up a world of technology entirely its own (Feenberg, 1991). Up to the beginning of the 20th century, Japan had committed itself to importing and manufacturing technology from the West, which met with severe criticism in the course of the 1920s, mainly from the cultural intelligentsia. In Japanese literary criticism this shift is referred to as *Nihon kaiki*, 'return to Japan'. The shift constituted a break with images of modernity (no longer considered universal, but depicted as Western), a rejection of Western technology and a conversion to a discourse of authenticity based on notions of ethnic purity and nationalism. One of the main figures of this *Nihon kaiki* was the novelist Junichiro Tanizaki. In the much celebrated essay *In'ei raisan* ('In Praise of Shadows', 1933), Tanizaki turns against the shape modern technologies had taken. He asserts the inappropriateness of western building materials and toilets and the unsuitability of contemporary versions of film, radio and phonographic recording to the Japanese arts. But the example I would like to cite here is related to pen and paper (Tanizaki, 1991: 18).

> 'To take a trivial example near at hand: I wrote a magazine article recently comparing the writing brush with the fountain pen, and in the course of it remarked that if the device had been invented by the ancient Chinese or Japanese it would have had a tufted end like our writing brush. The ink would not have been this bluish color but rather black, something like India ink, and it would have been made to seep down from the handle into the brush. And since we would have then found it inconvenient to write on Western paper, something near Japanese paper – even under mass production – would have been most in demand. Foreign ink and pen would not be as popular as they are; the talk of discarding our system of writing for Roman letters would be less noisy; people would still feel an affection for the old system. But more than that: our thought and our literature might not be imitating the West as they are, but might have pushed forward into new regions quite on their own. An insignificant little piece of writing equipment, when one thinks of it, has had a vast, almost boundless, influence on our culture. But I know as well as anyone that these are the empty dreams of a novelist, and that having come this far we cannot turn back.'

Despite this kind of observation, no alternative technologies were developed, and the *Nihon kaiki* essentially reflects and constitutes the development of a nationalist discourse, claiming ethnic, cultural, and technological difference.

5.3. The libertarian approach

Interestingly, the libertarian approach too tends to focus on the spatial and economic dimension of modernity: the rural-urban divide, and the farm-city/agriculture-industry divide, and the factory form. Different thinkers (Howard, 1902; Kropotkin, 1912; Mumford, 1961; Bookchin, 1974, Friedmann, 1996) have criticised the idea that modernity equals the coincidence of industrialisation and urbanisation, regarding this not as an unfolding law of history, but a false assumption and mistaken direction. I will mention some of the key thoughts of three of these.

Peter Kropotkin, in his classic study 'Fields, Factories and Workshops', argued that benefits could be derived from 'a combination of agriculture with industry, if the latter could come to the village, not in its present shape of a capitalist factory, but in the shape of a socially organised industrial production, with the full aid of machinery and technical knowledge' (Kropotkin, 1912). Kropotkin envisaged the development of a free association of people in 'industrial villages'. This suggestion follows his historical analysis of societal space and the relations of production:

> 'There was a time, and that time is not so far back, when both were thoroughly combined; the villages were then the seats of a variety of industries, and the artisans in the cities did not abandon agriculture; many towns were nothing else but industrial villages. If the medieval city was the cradle of those industries which bordered upon art and were intended to supply the wants of the richer classes, still it was the rural manufacture which supplied the wants of the million' (*ibid.*).

Concerning the relation between the city and the countryside, Murray Bookchin (1986) makes a similar observation in his book 'The Limits of the City'. Bookchin's main concern, however, is urbanism and the development of the city, arguing that the cities produced by early urbanism used to make people free (a theme he would continue in 'The Rise of Urbanisation and the Decline of Citizenship'), whereas those of contemporary urbanism aggregate the individual into a herd. Bookchin argues that there is a point beyond which a village becomes a city (which is difficult to determine precisely), and a point beyond which the city negates itself, becoming, in the words of Lewis Mumford, an 'anti-city' (Bookchin, 1986: 113).

Over a long process of development, Bookchin argues, the city created an universal terrain. It formed the arena for the emergence of a common humanity, rather than parochial tribal or kinship communities. 'The origin of the word 'civilisation' from *civitas* is not accidental: it authentically reflects the emergence of a distinctly human culture – universal in its scope – from city life as such' (*ibid.*: 7). In the modern city, however, Bookchin no longer recognises the values of *civitas*. Like Köymen, he considers the growth of the city pathological (*ibid.*: 112), explaining this pathology as the breakdown of 'self-constitutive restraints'. Moral economy had provided self-constitutive restraints, but the modern market economy, production for the sake of production, translated in urban terms, means the growth of the city for its own sake (*ibid.*: 88).

The spatial planner John Friedmann argues that where urbanisation and industrialisation coincide, villages and cities are produced as irreconcilable spatial formations. Friedmann's model of spatial integration seeks to transcend the rural-urban and agriculture-industry divides He looked to a new economic and spatial development of the countryside, based on a rethinking of processes of industrialisation and urbanisation. 'A central issue I wish to address is the rural-urban divide that pits cities against the countryside as two irreconcilable social, moral and physical formations. Since the early decades of the nineteenth century, overcoming the rural urban divide has been a persisting, but largely unrealised dream' (Friedmann, 1996: 129). To describe an alternative development model he introduces the terms 'agropolitan development' (Friedmann, 1988) and 'modular urbanisation' (Friedmann, 1996), a philosophy and a model for urbanisation without cities.

5.4. Appropriating technology

Much has been written about the transfer, adoption and distribution of technologies Comparatively little has been written about the appropriation of technology. A full and comparative study of attempts by social movements, groups and individuals to appropriate technologies is badly needed, not only as a part of writing history, but also to fully understand the idea of reconstruction and to be able to evaluate its value. How are technologies appropriated, in which webs of social relations, and what are the alternatives produced, technically and socially? I will briefly consider two possible cases of such a comparative study.

One possibly interesting case of appropriating technology is discussed by Kate Milberry in 'The Wiki Way' (Milberry, this volume). A wiki is defined by Wikipedia – the online encyclopedia iself a wiki – as a collaborative website which can be directly edited by anyone with access to it[108]. 'Most people, when they first learn about the wiki concept, assume that a website that can be edited by anybody would soon be rendered useless by destructive input. It sounds like offering free spray cans next to a grey concrete wall. The only likely outcome would be ugly graffiti and simple tagging, and many artistic efforts would not be long lived. Still, it seems to work very well'[109]. Milberry looked at Indymedia, a global network of activists from the global justice movement, able to publish their work as directly as possible, and by doing that reclaiming, even becoming the media.

In 'The Wiki Way', Milberry argues that the egalitarian structure of the wiki (and Indymedia) is based on the decentralisation of authority and a horizontal self-organisation. Since a wiki is administered by a group of people with equal rights who control each other and whose work and decisions are subject to all users, the gate-keeping power of editors and news producers to control the flow of information is obliterated.

> 'This egalitarian structure is characteristic of the GJM (Global Justice Movement), which eschews formal leadership and is configured rhizomatically in loose networks of autonomous nodes. A decentralisation of power is critical for the undermining of the social hierarchies common to modern capitalist societies, where the few rule over the many. In modern Western capitalism, this elite minority typically dominates the production of information (as well as technology), with the majority of citizens relegated to the passive, disempowered role of perpetual consumer. In a wiki, there are no access barriers: as with Indymedia, the producers of content are its consumers, and vice versa. [...] '[I]n technical terms, the wiki represents an advancement in digital communication; but in social terms, it both models and facilitates new modes of social organisation. [...] In their work, tech activists strive to reconnect technology with its logos – the

---

[108] See: http://en.wikipedia.org/wiki/Wiki

[109] Systems specialist Lars Aronsson cited from Wikipedia: http://en.wikipedia.org/wiki/Wiki

rationale for the good served. In doing so, they remind us that technology matters, that it is political, and that it is a scene of constant struggle. Does this indicate, or contribute to, a radical reform of the technical sphere? It remains to be seen. But it certainly offers hope that another world is possible' (Milberry, this volume).

Indeed, how the 'new modes of social organisation' fare over time in the context of existing power structures would form a major part of the evaluation of such efforts to reconstruct technologies, and the way forward for the reconstruction movement as a whole, if it could be termed such. To what extent, that is, can the new modes genuinely appropriate and challenge the dominant ideologies ('radical reform'), or at least provide real and accessible alternatives? Or do they become integrated into, consumed by and subjugated to them? Or does a symbiotic relationship or some other dialectic emerge? In the field of electronic media and information technology, one thinks of the poacher-turned-gamekeeper phenomenon of hackers now employed by large organisations to develop protections against future would-be hackers. Hackers, of course, are not generally technological reconstructivists – more techno-vandals perhaps – but the way in which their threat to the integrity of the technological system is turned to the advantage of that system is striking. Thus, policy decisions on the part of the managerial groups within the groups appropriating technology are inevitably examined hypocritically for any signs indicating a cultural shift away from an ends-oriented, substantive rationality to a means-to-ends, instrumental rationality. Wikipedia, for example, has to deal with tensions between open access (to contributing/editing) and product quality (of its articles), and faces issues related to matters such as editor restrictions and the rigidification of hierarchical structures. It is in this context that it has developed an open ('transparent') editorial policy peopled by contributors/editors acting as dispute arbitrators and as administrators, and, also, quotes its own co-founder Jimmy Wales voicing disquiet about the perceived status of administrators[110].

A group that has been exposed to the sharp end of the new modes/dominant ideologies relationship, the Critical Art Ensemble, provides another interesting example of appropriating technology and reconstructing the modern, this one in the area of biotechnology. The Critical Art Ensemble (CAE) is a collective 'dedicated to exploring the intersections between art, technology, critical theory, and political activism'. Information and communication technology (ICT) had previously been one of the focal points of the CAE, but gradually the group's work shifted towards biotechnology. CAE projects concerned with biotechnology are brought together in a theoretical framework named 'contestational biology'. Contestational biology aims to develop 'increasingly complex ways and means of slowing, diverting, subverting, and disturbing the molecular invasion through radical appropriation of knowledge systems and appropriation of the products and processes developed by imperial powers' (CAE, 2001: 12).

CAE criticises contemporary forms of biotechnology, among other reasons for the fact that public access to the processes of biotechnology is extremely limited. 'It is only the resultant product that appears as a commodity, resulting in misleading speculation, fear, disinformation and communicative disorder.' It is argued that contestational biology contrasts with the strategies of Luddites and bioluddites. While Luddites mainly threatened to demolish machines, or actually did destroy them, contestational biology aims to disrupt biotechnology by using biotechnology itself. It encourages people to gain knowledge about biology and to set up laboratories to carry out biotechnology experiments that can increase control over biotechnical developments. Biotechnology becomes a tool to contest biotechnology

---

[110] 'I don't like that there's the apparent feeling here that being granted sysop [systems operator = administrator] status is a really special thing' http://en.wikipedia.org/wiki/Wikipedia: Administrators

(Wietmarschen, 2007). CAE did not only develop a critique, however, but also became involved in projects doing biotechnology. CAE was involved in participatory performances that critiqued representations, products and policies related to biotechnology (CAE, 2001: 3). It developed a portable public facility to test foods for genetic modifications, placing labs in the public domain aimed to 'hurry the process of demystification', to do away with all the fantasy, myths, misleading speculations and disinformation around biotechnology. Two other CAE projects in this vain are concerned with genetic testing and the construction of GMOs (http://www.critical-art.net/).

In May 2004, when CAE co-founder, art professor Steve Kurtz, called the police following the sudden death of his wife, they became suspicious seeing the art works and scientific paraphernalia related to this work – including apparatus for the installation/performance/film projects 'GenTerra', 'Range Grains' and 'Marching Plague', which were related to the public GMO test, transgenic organism education and germ warfare demystification (this critiquing the Bush administration's earmarking of billions of dollars to erect high-tech, high-security laboratories to protect, supposedly, against the threat of biological warfare). The police called in the FBI to Kurtz's home, and the professor was detained and investigated for 'bioterrorism', with the Justice Department seeking charges under the US Biological Weapons Anti-Terrorism Act (even though the biological agents involved in the projects were only benign bacteria, such as a harmless form of gut *E. coli*). Although the bioterrorist charges seem to have failed, related court proceedings are still ongoing with a court trial expected next summer, and Kurtz still facing a maximum twenty-year jail sentence for charges now amended to 'mail fraud' and 'wire fraud' (http://caedefensefund.org/faq.html#law).

Toward the end of the second millennium, 'law enforcement' had meant the Luddites being confronted by 12,000 troops; as the third opens, a respected artist, academic and biotechnological activist seeking to reconstruct the dominant ideology has had his art and writings seized and suffers legal persecution. The context may have changed, but attempts at reconstruction can still come at a price. The experience of Steve Kurtz reminds us that the appropriation of (bio)technology does not only involve dominant ideologies of authoritarian modernism in the abstract, but very much takes places in the context of the realities of capitalist state politics.

6. Concluding remarks

In this article three interrelated issues – modernity and the peasantry, the rationality of technology, and the appropriation of technology – have been considered. First, the peasantry and modernity was discussed, with an emphasis on the destruction of peasant production systems in the 19th and 20th centuries. This was followed by a brief introduction to what is named a clash of modernisations, illustrated with the example of the Luddites. The Luddites functioned also as a hinge to a discussion on the rationality of technology. In the final paragraph, appropriation has been considered in relation to the general concept of modernity (rethinking modernity) and a more tangible appropriation of technology.

It has been argued that the destruction of peasant production systems was target and outcome of the modernity project. Successive generations of social scientists (from neo-liberals to orthodox Marxists) have portrayed the peasantries as emblematic for the world of the past, with those surviving in the present world seen as a residual of history, deemed to extinction, which did not need any further research. This position was seriously criticised from the 1970s onwards. Even though actual developments were complex and contradictory, the general idea had been that the peasantry, as a class, would decompose:

the historical persistence of the peasantry, however, was now used as an argument for its permanency (Kloppenburg, 2004: 28). Both positions are problematic. The first – the idea of an inevitable and irreversible decomposition of the peasantry – is hopelessly teleological. The second – the peasantry as a permanent category – is inadmissibly static.

A more fruitful position is one that considers the peasants as a social force, acting upon and resisting against the forces that try to eliminate them – and in so doing, reinventing themselves. Henry Bernstein rightly observes that 'the peasantries... that inhabited "the world of the past"... are indeed destroyed by capitalism and imperialism' (Bernstein, 2001: 45) Heather Johnson is probably right when she says that 'this process has resulted not in the disappearance of the peasantry, but in its redefinition. Today's peasantry is a population struggling for survival, clinging to control over the means of production [...] and in search of a sustainable livelihood' (Johnson, 2004: 54). The peasantries that inhabit the world of today are not mere survivors, whose fate has been decided upon. They occupy and expand social and economic spaces, and challenge the control of global capitalism over land and labour (Bernstein, 2001: 46). These struggles are contesting the modernity project as is. A connection between reconstruction and the peasantry may be considered an irony of history, yet only to the extent that one seriously believed in the claim that modernity demands the destruction of the peasantry.

It must be argued that a reconstructionist approach is in principle a substantial approach. It is not based upon a pre-conceived picture of development and the articulation of means with ends as given, not with effectivity and efficiency alone, but with the wider question of which ends should be reached by what means. This makes a reconstuctivist approach simultaneously a theory of social and societal change (development) and of the social shaping of technology, going beyond the notion of technological determinism, which is a technological theory of social change, and social constructivism, which is a social theory of technology. It has been argued too here that a reconstructionist approach implies a fundamental rethinking of technology and modernity.

Finally, when I started to think and work on this paper, I had something very different in mind than what finally emerged and I have presented here, at this conference. This was clear from the working title I had been using, and which had the words 'reconstruction' and 'biotechnology' in their title. The paper I had in mind was going to reflect on experiences that we, as a research group, are involved in, experiences with the reconstruction of biotechnology in countries like India, Ghana, Ecuador and Cuba. Yet, gradually, step by step, I ended up by writing a paper on modernity and technology. I think that, in spite of critical studies, a fundamental demerit in technology studies has been the absence of a critical evaluation of our society's constitution, or, in other words, the modernity project. I think, like others (see, for example, James C. Scott's 'Seeing Like State', on how certain schemes to improve the human condition have failed, or Jan Douwe van der Ploeg's 'The Virtual Farmer', that a rethinking of technology desperately needs a rethinking of the modernity project. This links to Adorno's assertion that a critical theory 'in the end negates the whole sphere it moves in' (Adorno, 2005: 197).

References

Adorno, T. (2005). *Negative Dialectics*. New York & London, Continuum.
Araghi, F. (1995). Global Depeasantization, 1945-1990. *The Sociological Quarterly* 36: 337-368.
Barnhart, R.J. (1995). *Barnhart Concise Dictionary of Etymology*. Harpercollins.
Benjamin, W. (2003). *Selected Writings, Volume 4, 1938-1940*. Cambridge & London, Harvard University Press.

Bernstein, H. (2001). The peasantry in global capitalism: who, where and why? In: *Working Classes, Global Realities: Socialist Register 2001*, (Eds. L. Panitch, C. Leys, A. Greg and D. Coates), New York, Monthly Review Press.

Bijker, W.E. (1995). *Of Bicycles, Bakelites and Bulbs: Toward a Theory of Sociotechnical Change*. Cambridge, MA, MIT Press.

Bijker, W.E., Hughes, T.P and Pinch, T.J. (Eds.) (1987). *The Social Construction of Technological Systems: New Directions in the Sociology and History of Technology*. Cambridge, MA, MIT Press.

Bijker, W. and Law, J. (1992). *Shaping Technology/Building Society: Studies in Sociotechnical Change*. Cambridge, MIT Press.

Bijker, W. and Pinch, T. (2002). SCOT answers, Other Questions, a reply to Nick Clayton. *Technology and Culture* 43 (2): 361-369.

Bookchin, M. (1986) [1974]. *The Limits of the City*. Montreal & Buffalo, Black Rose Books.

Bookchin, M. (1987). *The Rise of Urbanization and the Decline of Citizenship*. San Francisco, Sierra Club Books.

Bookchin, M. (1990). *Remaking Society: Pathways to a Green Future*. Cambridge, South End Press.

Bookchin, M. (1992). *Urbanization Without Cities: The Rise and Decline of Citizenship*. Montreal & Buffalo: Black Rose Books

Brecht, B. (2007). Against Lukacs. In: *Aesthetics and Politics*, (Ed. F. Jameson), London, Verso, pp. 68-85,

Brown, G. (1977). *Sabotage: A Study of Industrial Conflict*. Nottingham, Spokesman Books.

Chandler, D. (2001). *Technological or Media Determinism*. Aviabable at: http://www.aber.ac.uk/media/-Documents/tecdet/tecdet.html (accessed 23-07-2007).

Clayton, N. (2002). SCOT: Does it Answer. *Technology and Culture* 43 (2): 351-360.

Critical Art Ensemble. (2001). *The Molecular Invasion*. New York, Autonomedia.

De Tocqueville, A. (1997) [1840]. *Democracy in America* (volume 2). University of Virginia.

Feenberg, A. (1991). *Critical Theory of Technology*. Oxford, Oxford University Press.

Feenberg, A. (1999). *Questioning Technology*. London & New York, Routledge.

Frank, A.G. (1967). Sociology of Development and Underdevelopment of Sociology. *Catalyst* 3: 20-73.

Friedmann, J. (1988). *Life Space and Economic Space*. Transaction Books, Brunswick.

Friedmann, J. (1996). Modular cities: Beyond the rural-urban divide. *Environment and Urbanization* 8: 129-131.

Gellner, E. (1983). *Nations and Nationalism*. Oxford, Basil Blackwell.

Hardt, M. and Negri, T. (2004). *Multitude, war and democracy in the age of empire*. New York, Penguin Press.

Harvey, D. (1989). *The Condition of Postmodernity. An enquiry into the origins if cultural thought*. Oxford, Blackwell.

Hobbelink, H. (1987). *Biotechnology: New Hope or False Promise? Biotechnology and Third World Agriculture*. Intl. Coalition for Development Action, Brussels.

Hobsbawm, E. (1959). *Primitive Rebels, banditry, mafia, millenarians, anarchists, Sicilian fasci, the city mob, labour sects, ritual, sermons & oaths*. Manchester, Manchester University Press

Hobsbawm, E. (1964). *Labouring Men, studies in the history of labour*. London, Weidenfeld and Nicolson.

Hobsbawm, E. (1995). *The Age of Extremes: The Short Twentieth Century, 1914-1991. New York, Penquin*.

Hobsbawm, E. and Rude, G. (1969). *Captain Swing*. London, Lawrence and Wishart.

Horkheimer, M. (2004) [1947]. *Eclipse of Reason*. London& New York, Continuum.

Horkheimer, M. (1972). Traditional and Critical Theory. In: *Critical Theory; selected Essays*. Herder and Herder.

Howard, E. (1902). *Garden Cities of To-Morrow* (London). Reprinted (1946), edited with a Preface by F.J. Osborn and an Introductory Essay by L. Mumford. London, Faber and Faber.

Johnson, H. (2004). Subsistence and Control. The persistence of the peasantry in the developing world. *Undercurrent* 1 (1): 55-65.

Jongerden, J. (1996). *Patronen van Verscheidenheid* (Patterns of Diversity). Wageningen, Wetenschapswinkel.

Jongerden, J. (2006). Luddites, or the Politics in Technology – An Introduction. *Tailoring Biotechnologies* 2: 63-67.

Jongerden, J. (2007). *The Settlement Issue in Turkey and the Kurds, An Analysis of Spatial Policies, Modernity and War*. Leiden & Boston, Brill Academic Publishers.

Kemal, Y. (1976). *The Legend of the Thousand Bulls*. London, Collins and Harvill Press.

Kloppenburg, J.R. (1988). *First the seed. The political economy of plant biotechnology 1492-2000*. Cambridge, Cambridge University Press.

Kloppenburg, J.R. (2004). *First the seed: the political economy of plant biotechnology, 1492-2000*. Wisconsin: University of Wisconsin Press.

Köymen, N.K. (1937). Village, the Unit of Societal Organization. Master's thesis. Wisconsin, Wisconsin University.

Kropotkin, P. (1912). *Fields, Factories and Workshops or Industry Combined with Agriculture and Brainwork with Manual Work*. London, Edinburgh, Dublin and New York, Thomas Nelson & Sons.

Löwy, M. (1996). Figures of Weberian Marxism. *Theory and Society* 25: 431-446.

MacKenzie, D. and Wajcman, J. (1985). *The Social Shaping of Technology*. Philedelphia, Open University Press.

Mannheim, K. (1940). *Man and Society in an Age of Reconstruction, Studies in Modern Social Structure*. Harvest Book/Harcourt.

Marx, K. (1973). *Grundrisse, Foundations of the Critique of Political Economy*. New York, Vintage Books.

Marx, K. and Engels, F. (1848). The Communist Manifesto. Available at: http://www.marxists.org/archive/marx/works/1848/communist-manifesto/index.htm

Meier, H.A. (1957). Technology and Democracy, 1800-1860. *The Mississippi valley Historical Review* 43: 618-640.

Migdal, J. (1974). Why Change? Towards a theory of change among individuals in the process of modernization. *World Politics* 26: 189-206.

Mumford, L. (1961). *The City in History*. New York, Harcourt, Brace and World.

Paine, T. (1791-1792). *The Righs of Man*. Groningen, Department of Alpa-informatica, University of Groningen

Pinch, T. and Bijker, W.E. (1984). The Social Construction of Facts and Artefacts: or How the Sociology of Science and the Sociology of Technology Might Benefit Each Other. *Social Studies of Science* 14: 399-441.

Puente, D. (2007). Redesigning the Production of the *Bacillus thuringiensis* Bio-Pesticide within the Context of Subsistence Agriculture in Andhra Pradesh, India. *Asian Biotechnology and Development Review* 9 (3): 55-81.

Ruivenkamp, G. (1989). *De invoering van biotechnologie in de agro-industriële productieketen: de overgang naar een nieuwe arbeidsorganisatie* [The Introduction of Biotechnology in the Agro-Industrial Production Chain: The Transition to a New Organization of Labor]. PhD thesis, Faculteit politieke en sociaal-culturele wetenschappen, Amsterdam, Universiteit van Amsterdam, 354 pp.

Schumpeter, J. (1975) [1942]. *Capitalism, Socialism, and Democracy*. New York, Harper.

Scott, J.C. (1985). *Weapons of the Weak: Everyday forms of Peasant Resistance*. Yale University Press.

Scott, J.C. (1998). *Seeing Like a State, how certain schemes to improve the human condition have failed*. New Haven and London, Yale University Press.

Steinbeck, J. (2000) [1939]. *The grapes of wrath*. Penguin Classics.

Strong, T.B. (1996). Nietzsche's Political Misappropriation. In: *The Cambridge Companion to Nietzsche*, (Eds. B. Magnus and K.M. Higgins), Cambridge, Cambridge University Press.

Tanizaki, J. (1991) [1933]. *In Praise of Shadows*. London, Jonathan Cape.

Thompson, E.P. 1965. *The Making of the English Working Class*. London: Victor Gollanzc.

Van der Ploeg, J.D. (2003). *The Virtual Farmer. Past, present and future of the Dutch peasantry*. Assen, Van Gorcum.

Weber, M. (1968). *Economy and Society: An Outline of Interpretive Sociology*. (Eds. G. Roth and C. Wittick. New York, Bedminster Press.

Wietmarschen, H. (2007). Turning Tools for Democratizing Biotechnology in a Bioterrorist Threat. *Tailoring Biotechnologies* 3 (2): 25-29.

Wiskerke, H. (1997). *Zeeuwse Akkerbouw tussen Verandering en Conituiteit, een sociologische studie naar diversiteit in landbouwbeoefening, technologieontwikkeling en plattelandsvernieuwing*. PhD Thesis Wageningen University.

Wolf, E.R. (1969). *Peasant Wars of the Twentieth Century*. New York, Evanston and London, Harper & Row.

Reconsidering agricultural modernisation: three dimensions of questioning and redesigning biotechnologies for international agricultural development

Wietse Vroom

1. Introduction: biotechnologies for international agricultural development

We are living in an age of Millennium Development Goals. International policy makers, philanthropists and scientists are bundling their efforts in setting up programs to alleviate global hunger and poverty, and to find a cure for some of the most devastating diseases plaguing humanity, especially in the poorer regions of this world. The Millennium Development Goals (MDGs) themselves – a series of eight concrete objectives for global development[111] – are strong symbols of a contemporary international modernisation project, focused on bringing global food production and health care from the shadows of underdevelopment into the light of modernity.

Agriculture is widely acknowledged to play a key role in the economic development of less developed countries (Thirtle *et al.*, 2001; Dorward *et al.*, 2004; Diao *et al.*, 2007). Therefore, agricultural modernisation is considered to be a vital element in reaching the Millennium Development Goals. Such visions of an organised global agricultural modernisation are far from new, and go back in time at least a couple of decades to the first ideas of the Green Revolution. This project, which revolved around the introduction of high yielding varieties of several cereals, resulted in strong productivity increases and growth of incomes of large groups of farmers in less developed countries (Pingali and Heisy, 1999; Evenson and Gollin, 2003). Indeed, the Green Revolution is slowly regaining credibility among the most prominent development policy makers[112], after a period of rather harsh criticism in the 1980s and early 90s[113]. However, the precise effects (let alone benefits) of such agricultural modernisation processes are still ground for debate, especially in the dawning light of a potential new Gene Revolution in international agricultural development (Ruttan, 2004).

Various types of questions have been raised in response to pro-poor agricultural modernisation, and more recently in response to the role of molecular biology and transgenics in agricultural development. The focus in addressing such concerns is very often on the policy level, in terms of national agricultural policies and international trade relations between north and south. However, much of the contemporary controversy in international agricultural development revolves around specific (bio)technologies like the high yielding varieties of the green revolution, and the transgenic crops of today.

---

[111] http://www.un.org/millenniumgoals/

[112] E.g. Prabhu Pingali, Director of the Agricultural and Development Economics Division of FAO explicitly highlighted the advances made during the Green Revolution at the BioVision 2007 Conference in Lyon, France, March 2007. Similarly, Ismael Serageldin, chairman of CGIAR 1993-2000 has been an explicit defender of the rationale and results of the Green Revolution. See also the recent publication of Alston *et al.* (2006) for IFPRI.

[113] Some of the key criticisms of the Green Revolution have been that it largely left Sub Saharan Africa behind, that it has been environmentally unsustainable, and that it exacerbated income differentiation in less developed countries. See Vandana Shiva for one of the sharpest criticisms of the Green Revolution (Shiva, 1993), or Freebairn for a wider overview and discussion of the different perspectives on the effects of the Green Revolution (Freebairn, 1995).

On the level of technologies for international agricultural development, different types of concerns and categories of arguments can be distinguished. One set of arguments – in a rather straightforward way – focuses on the appropriateness of agricultural technologies in a different climatic, geographic, socio-economic or cultural context. This refers to a rather straightforward alignment of technological development with local priorities and conditions within which a technology will have to function.

A second category of concerns is somewhat more complicated and focuses on the mode of modernisation and industrialisation of agriculture. The core of this argument goes back to critical studies of biotechnology in the 1980s and 90s that treated (bio)technology development not just as a tool for development, but as an essentially social and political process that involved the restructuring of social relations of production (Ruivenkamp, 1989). The issue here is not so much to what extent improved crop varieties will grow in less developed countries, but the extent to which agricultural biotechnologies are instrumental in exporting (or imposing) a particular system of production and innovation which modernises farming, but takes away the control of a farmer over his own means of production. This general trend of a loss of farmer' autonomy as part of agricultural modernisation has not been uncontested. Today, proponents of an organic farming style, and especially supporters of the global 'Food Sovereignty' movement seem to focus heavily on self-sustainability and farmers' independence from multinational corporations (Rosset, 2006; Quaye, 2007)[114]. The rationale of such movements cannot be reduced to a single set of arguments, but they do represent a clear response to an increasingly undesired entanglement of farming in a global web of capitalist exchange, and an increasing convergence of farming styles as the result of international agricultural modernisation.

This paper cannot evaluate whether organic farming or Food Sovereignty will provide better models for agricultural production in less developed countries, than mainstream approaches. However, it does take the articulation of alternative visions on agricultural modernisation serious, and wonders what kind of technological development would be appropriate according to such visions of agricultural modernity. With that additional dimension of 'questioning' technological change in mind, a number of case studies have been carried out looking for instances where technological change was sensitive to its restructuring dimension in social relations of innovation and production. Specific examples were found where biotechnologies allow for a very different process of modernisation in which farmers were empowered, rather than deskilled. However, next to these examples, instances of rejection and redesign were encountered that could not be aligned with instrumental problems, or with problems associated with a loss of independence in production. A third type of argument appeared to play a major role in technological controversies, which requires a very fundamental reconstruction and resignification of technologies. This type of argument is related to the notion of indigenous identity formation and a fear of cultural erosion as a result of the adoption of fundamentally westernised technological systems.

This paper takes up the question what the role of technologies is in agricultural modernisation, and how technological issues are intricately intertwined with social and political issues. The discussion of three concrete cases leads to a typology of three levels (or dimensions) of understanding technological change, resistance and reconstruction. The rationales for adapting and reconstructing technology in response to these three dimensions will be elaborated and leads to a discussion of the wider international social

---

[114] Food Sovereignty is an approach that is largely based on the Right to Food and a right to produce. In contrast to the concept of Food Security (the right to have access to sufficient food of good quality), Food Sovereignty can be interpreted as 'the right to produce your own food'.

organisation of public sector agricultural research and its position *vis-à-vis* informal innovation and production systems.

2. Instrumental adaptations of technologies in a new context; the appreciation of laymen expertise

In debating agricultural technologies for development, the first and most dominant category of arguments focuses – in a rather straightforward way – on the appropriateness of modernisation tools in a different climatic, geographic, socio-economic or cultural context. For example, heavy machinery (tractors) on small scale, muddy fields, or in mountainous areas is simply not useful; especially if the capacity to maintain or repair such machines is not available. Crops that need a lot of fertiliser and irrigation to reach their productivity potential, or pesticides for protection against local insects, may in practice be environmentally unsustainable, and therefore undesirable in the long run.

The concerns are clear and generally straightforward, and have been a classic argument to involve local stakeholders and end-users in processes of technological innovation. Not only does this allow a better attuning of technologies with local conditions, it also allows a more sophisticated understanding of the specific needs and expectations of the future users (farmers) of the technology. Traditionally, plant breeders have been trained to focus on a specific set of traits which are considered to be important for agricultural production. These traits however are only valuable with respect to a specific production system. The value of certain varieties in an intercropping system, or the value of residual material as animal fodder are classical traits that have often been neglected by formal plant breeding programs, but are highly relevant in some local farming systems. The same goes for culinary value, cultural meaning or simply personal preferences of appearance which are not always quantifiable for an external observer[115].

The idea of participatory priority setting and evaluation of new crop varieties or other technologies did have one important condition: *the appreciation of laymen expertise*. While farmers have frequently been considered to be illiterate and therefore incapable of having sophisticated ideas about agricultural development, taking these people on board in an innovation process implies that their expertise and perspectives are taken seriously. In a recent paper, Dominic Glover discusses the transformation from a top-down extension system of 'Training and Visit', to more interactive models of participatory technology development (Glover, 2007). Chambers' *'Farmer First'* (Chambers *et al.*, 1989) and Brian Wynne's study of Cumbrian sheep farmers (Wynne, 1996) are frequently mentioned as seminal works in involving farmers in technological design and technical decision making. The key point is that scientific expertise may be a special category of expertise, but not necessarily more useful or appropriate in solving certain problems than other types of expertise[116].

As a result of such debates, notions of participation and bottom-up priority setting have become firmly embedded in all international agricultural development projects. The significance of this development for other types of technological reconstruction will be discussed later in this paper, but the point is clear that successful agricultural development cannot be achieved without including local stakeholders. The

---

[115] See (Scott, 1998: 273-301) for a discussion of the intricacies of polycultures in agriculture and the difficulties for external observers to appreciate the roles different crops with their traits play in such a system.

[116] However, also note the conceptual discussion on the nature of expertise in technical decision making that was initiated by Collins and Evans, demonstrating that the precise role of 'uncertified expertise', and other types of expertise remains far from clear from a theoretical standpoint. (Collins and Evans, 2002)

extent to which this participatory mode of innovation extents to the private sector is doubtful as some studies stress the increasing attention for 'open innovation' and 'user-producer interactions'[117], while others observe an interest of companies to target small scale farmers, but a failure in actually involving them in upstream technology development (Glover, 2007; Vroom, 2007). Nonetheless, it is widely acknowledged that a transfer of technology approach has to be complemented with a local component of development and adaptation of technologies[118].

The technical and pragmatic adaptation of technologies to specific circumstances in developing countries, belongs to the kind of problems that we are slowly learning to deal with. The raise of participatory methodologies has been a vital element of that process. In addition, increasing attention goes out to the building of a good innovation system, which ensures a linking of technology developers and farmers/end-users, and therefore different types of expertise that are all relevant and necessary to reach a useful new agricultural technology. However, the adaptation of technologies to a specific context in less developed countries does not end here.

3. Technologies restructuring social relations of innovation and production; the political dimension of agro-biotechnologies

A second type of concerns regarding agricultural development can be identified. Here the issue is not so much to what extent improved crop varieties will grow in less developed countries, but the extent to which agricultural biotechnologies are changing the agricultural innovation and production systems in a specific way. This perspective essentially refers to a political dimension in technological design, considering that it links technological development with changes in social structures.

Classical works in this tradition are the study of Goodman, Sorj and Wilkinson, who focused attention on processes of substitution and appropriation in the industrialisation of agriculture (Goodman *et al.*, 1987), and '*First the Seed*' of Jack Kloppenburg, who described the political economy of plant biotechnology, the commoditisation of seed and the new division of labour between the public and private sector in plant breeding (Kloppenburg, 1988). In a similar vein, elsewhere in this conference volume Joost Jongerden refers to the transformation of a peasant mode of production to a farmer mode of production ('First the peasant'). The central element in this transformation as discussed by Jongerden is the gradual loss of control over the means of production by the farmer.

A lot of the traditional critique of agricultural biotechnology has been illustrated by studies on the introduction of hybrids in maize breeding (Kloppenburg, 1988; Van de Belt, 1995). Like other cereals, maize was conventionally grown by farmers by replanting a portion of their harvest. This way, maize had a dual function (meaning) as both grain for consumption and marketing, and seed for the next growing season. The introduction of hybrids was primarily legitimised by their increased yield potential, based upon the 'heterosis' effect that occurs when different parents produce strongly heterozygous hybrid

---

[117] Note the more recent trend of 'Open Innovation' among companies that are explicitly looking for interaction with end-users to improve their innovative capacities. See (Chesbrough, 2003) and (Von Hippel, 2005) for a brief discussion of the relevance of the concept in the context of agro-biotechnology.

[118] See (Vroom *et al.*, 2007) for a discussion of the need for a transformation from a 'transfer of technology' model to an 'endogenous development' model.

offspring[119]. However, many critical scholars have observed that maize hybrids at the same time changed the production system, because the offspring of hybrid seeds is very diverse and therefore of diminished value for agricultural production. This implies that farmers would buy new seed every year, instead of saving a portion of their own harvest. In a sense, hybrids provided a biological patent that implied a very strong incentive to go back to the seed producer, and to refrain from 'copying' the seed through its biological potential for propagation. *Hybrids not only increased yields, they also fundamentally changed the agricultural production chain, and firmly embedded public or private external seed suppliers in the line of production.* Goodman *et al.* (1987) noticed how this was essentially an act of appropriation of a domain of production that used to be in farmers' control. Kloppenburg (1988) described the same mechanism in terms of commoditisation of previously common goods (seed).

A remarkable observation in studies of the emergence of hybrid seed was that while the 'biological patent' that hybrids introduced was clearly primarily beneficial for private sector seed breeders, its development and deployment was supported by public sector scientists. According to Kloppenburg, to them it only seemed fair that private sector companies would get some return on investment of their research into new varieties (Kloppenburg, 1988: 99). The ideological hegemony of a specific and perhaps arbitrary model of agricultural development was supported by both public and private sector scientists. This demonstrates that the source of funding itself, and the public mandate of public sector institutes does not guarantee that they are always independent in developing a conceptualisation of agricultural development[120].

Next to the introduction of hybrid seeds, other critical scholars identified other ways in which modern biotechnologies restructured production chains. Pesticides and chemical fertiliser stimulated a disconnection from agricultural system and the natural environment, often leading to unsustainable farming practices in the long run (Pretty, 2002). Various kinds of enzymes allowed a disconnection between agricultural product and food product (Ruivenkamp, 1989). This in turn allowed a interchangeability of various 'carbohydrate' or 'oil' resources, and potentially a substitution of agricultural products by chemical products. The classic example here is the exchangeability of sugar cane from the Philippines for corn fructose syrup from the United States (leading to large scale unemployment in the Philippines), and the potential substitution of sugar by artificial sweeteners (Ruivenkamp, 1986; Hobbelink, 1991). However, especially the example of hybrid maize still remains very powerful because of the very visible and direct relation between a new type of agricultural technology (seed) and a new structure of the production and innovation system.

This raises an important conceptual question. While the relation between a new technology and a restructuring of relations of production is very clear in the case of hybrid seed, the question is whether this political effect is embedded in the material design of the seed, or in the strategies of seed companies trying to sell more seed. What is the relation between technical design and the wider context in which it emerges?

---

[119] Kloppenburg challenges this notion that hybrids are responsible for yield increases, by arguing that the development of alternatives such as open-pollinated varieties was largely abandoned, and therefore does not provide a fair comparison. See (Kloppenburg, 2004: 92 -).

[120] Feenberg also discusses this hegemony of technical rationality (Feenberg, 1999: 86-87)

4. The politics of technologies

These questions seem to hit a core problem in recent Science and Technology Studies (STS) and the Philosophy of Technology. On the one hand, these traditions stress the way in which technical design reflects social norms, and that technological design and application is thoroughly contingent upon social, political and institutional context (e.g. Bijker, 1995; MacKenzie and Wajcman, 1999). But a number of scholars has also very explicitly drawn attention to the coercive power of technological artifacts, and the notion of 'the politics of technologies'. Langdon Winner addresses this issue in his article 'Do artifact have politics?' (Winner, 1985). He argues that technological design can 'solve' certain social problems, or can at least be highly complementary with certain types of social organisation (e.g. hierarchical or centralised management). A few years later, Madeleine Akrich and Bruno Latour have discussed the delegation of morality to artifacts in terms of a 'script of technologies' which allows apparently mundane artifacts (doorcloser, seat belt) to implicitly enforce certain social norms (Akrich, 1992; Latour, 1992). Also Andrew Feenberg builds upon this tradition when he introduces a notion of a 'technical code' which brings technological artifacts in accordance with the social meaning they have acquired. Technologies are argued to materialise social norms and ideologies, which become embedded in the material design of the technology. This implies that technologies can also be prescriptive in the kind of social relations they mediate. As far as Feenberg is concerned, this has an important implication for the democratisation of technological design. Not only should technological institutions be governed in a democratic way, also the material design of technological artifacts should be opened to reconstruction in the course of a democratic involvement of citizens in the socio-technical shaping of our world, ultimately leading to what Feenberg calls a 'Deep Democratisation' (Feenberg, 1999: 142-147) [121].

But there are two problems with this understanding of the politics of technological artifacts. One is conceptual, the other pragmatic. First, there is a potentially problematic interpretation of the concept of a 'technical code' of technologies, since it may seem to suggest that a technical object has a specific technical configuration, which leads to a specific social effect, in isolation of the wider social and economic context in which the technology functions. Winner already countered that interpretation by mentioning that *'A ship out at sea may well require a single captain and obedient crew. But a ship out of service, parked at the dock, needs only a caretaker'* (Winner, 1985: 39). In other words, context of application matters. Latour similarly stresses that the architecture of buildings from the Belle Epoque successfully separated the servants from the bourgeois in the house. But the same building today has the *'perverse tendency to force the students inhabiting its coveted 'chambre de bonnes' to climb six stories through a steep and narrow staircase, while the happy owners of the flats are allowed to glide through a comfortable lift inserted inside a wide staircase.'* (Latour, 2004). This of course is a totally unintended effect of the architecture from an entirely different epoch. Historical context matters.

In addition, Brian Pfaffenberger makes a powerful and useful argument on the importance of culture in the way technologies gain a specific coercive force. To him, technological artifacts have affordances which are inherently multiple, depending on the perception of users or affected stakeholders. Pfaffenberger argues that it is discourse ('ritual') that privileges and legitimises a specific interpretation of technologies, thereby constituting a political effect. 'The artifact embodies political intentions, but these intentions do not come to life in the absence of ritual' (Pfaffenberger, 1992: 294). He illustrates

---

[121] Finally, consider Jaap Jelsma, who goes even one step further and discusses the possibility to actively structure the socio-technical landscape around us by consciously inscribing technologies with, for example, more environmentally friendly scripts. (Jelsma, 2003)

his point with reference to the plain Victorian hallway bench on which servants had to await the master of the house, in the nineteenth century. A myth of hygiene mystified and legitimated the use of a plain and hard bench, which was supposedly not to humiliate the servant class, but because they would only soil nicer benches with filth from the streets. Today, many antique collectors place Victorian hallway benches in their homes, but with a very different intent. According to Pfaffenberger: 'What made the hallway bench into a political artifact in the nineteenth century was the ritualisation of the hallway space: Profound decorum standards called for members of the masters' class to be admitted straightaway into the interior of the house, while members of the servant's class were seated on the bench, signifying their inferiority' (Pfaffenberger, 1992: 294). So not only historical context matters, ritual matters too. The conceptual conclusion must be that technologies are profoundly political, but their political meaning depends upon the social and historical context they are part of, and the rituals and discourses they are surrounded by. This takes us away from an essentialist understanding of the political nature of technologies, in the sense that we don't see technologies as having only one particular political function or 'meaning'.

But as announced, a second –more pragmatic- problem with our understanding of the 'politics of technologies' emerges in the context of this article. The conceptual understanding of the coercive and prescriptive force of technology is directly related to a very concrete, material object with a specific technical design. In discussing agricultural modernisation and the role of modern biotechnologies, we don't always discuss a specific technical object, but rather a research trend, or breeding strategy. This first means that we explicitly have to consider what the material dimensions are to contemporary biotechnology or genomics development. But it also means that we have to look further than concrete technical objects, and need to consider, for example, the 'politics of breeding strategies'.

The challenge is to study the role of technological development as part of a specific modernisation project, including its discourses, projections and expectations. In relation to the issue of agricultural modernisation, the question becomes what models of agricultural modernisation are implicit in contemporary attempts to contribute to agricultural development in developing countries. Secondly, in what ways do technical objects and research methodologies that are being used reflect this model of modernisation? And, can we find instances where agricultural development is sensitive to the potentially problematic aspects of modernisation, and tries to mitigate this? Finally: how is that reflected in the technical design or the research methodologies that are developed and adopted?

These issues are taken up in a study of the work of The International Potato Centre (CIP) in Peru for resource poor potato farmers. This provides an empirical basis to discuss the relationship between technological development and different views on agricultural modernisation. CIP has been specifically chosen as case study because of the very specific requirements the Andean context poses for any breeding program, as will be elaborated below.

5. CIP's challenge: balancing modernisation, biodiversity and traditional production systems

The International Potato Centre (Centro International de la Papa, CIP) in Lima (Peru) is one of the international public sector research centers of the CGIAR[122] that works on a specific set of mandate

---

[122] CGIAR = Consultative Group on International Agricultural Research; http://www.cgiar.org/

crops (potato, sweet potato and a collection of Andean roots and tubers) and attempts to develop 'global public goods' primarily through releasing improved breeding material. The location of the centre in Peru, the centre of potato domestication and home to some 3,500 potato varieties (Brush *et al.*, 1995), is by no means coincidental. Nonetheless, the primary target areas and people for CIP are not necessarily in Peru or even South America. Potato and sweet potato are cultivated in large parts of Africa and Southeast Asia too, where the potential impact of improved varieties may be much bigger than in the highly diversified Andean potato systems. Nonetheless, CIP greatly benefits from the natural potato biodiversity of the Andean region and appears to feel committed to returning a favor. The institute is involved in a number of programs to alleviate poverty and hunger among potato producers in the High Andes, as well as in projects aimed at *in situ* conservation of agricultural potato diversity.

This work for Andean potato producing communities provides some fundamental challenges. In his book 'The botany of desire', Michael Pollan provides a strong image of the very basic challenge that contemporary plant breeders are facing when wishing to improve upon the existing potato production system in the Andes (Pollan, 2001). Peru, being the centre of domestication of potatoes, and featuring an immense agricultural biodiversity, has a long history in potato farming, which goes back to 5,800 BC (Pickersgill and Heiser, 1978; Weatherford, 1988). The harsh conditions in the high Andes, in terms of weather conditions as well as the virtually vertical pastures, have traditionally led to a potato production system that is more than anything geared towards resilience (Ortega *et al.*, 2005). In these traditional production systems, an immense degree of agricultural biodiversity makes sure that at least a significant portion of the harvest survives climatic peculiarities of every year, while at the same time providing the dearly wished variation in the menu which in some regions is very strongly dominated by the potato. Pollan is quick to notice how fundamentally this system differs from contemporary western production models where agricultural conditions are essentially controlled in order to fit the most desired crop variety. Many modern breeding schemes are essentially geared to such a western style production system in which the most desired variety is chosen in terms of productivity and quality characteristics, and the growing conditions can be largely controlled. An additional very important characteristic of this type of farming today is the social insurance and subsidy systems which allow farmers to take a bigger risk in cultivation, leading to higher productivity, and providing a backup in case of harvest failures. This is a luxury most farmers in less developed countries do not have; changing the priority from the highest possible productivity to a resilient system in the first place. As a result, the traditional Andean potato system primarily relied on a high degree of biodiversity to cope with an environment that is very difficult to control[123].

It should be noted that the traditional Andean potato production system as described here has largely disappeared in many areas in the Andes. Many Andean potato farmers have lost much of their potato biodiversity as market preferences have led to the dominance of only a few varieties. These include both 'cosmopolitan' landraces, as well as improved varieties from potato breeding programs. In large parts of Ecuador, Bolivia and Peru, today farmers mainly cultivate only 1 or 2 potato varieties. A notable exception includes the Peruvian region Huancavelica where still a large on farm biodiversity can be found[124].

---

[123] Which by the way does not imply that the ancient Peruvians did not attempt to control or modify their environment. The extensive terracing of the Andean slopes bears witness to the degree to which farmers were in fact able to adapt their natural environment, making it more fit for agriculture.

[124] Personal communication with Graham Thiele.

This loss of crop genetic diversity in large parts of the Andes presents a challenge to any institute that aims to improve agricultural production by releasing new improved varieties, since they tend to displace landraces. This would be undesirable since crop genetic diversity is considered to be an important common resource for future breeding, as well as a private resource for Andean potato farmers who value the variation in diet these landraces provide. Moreover, potato landraces are used as exchange material and gifts, and are as such associated with indigenous Quechua identity. Hence, crop genetic diversity is important; but more in general the question is how external interventions (by means of releasing improved potato varieties) stand in relation to the traditional extensive informal seed systems that farming communities have developed. For example, does increasing the quality of seed potatoes imply that seed potato production itself is externalised from the domain of farmers themselves, to specialised institutes?

Such tensions between modernising agriculture, and conserving a valuable resource like crop genetic diversity provides an interesting locus to think about (and witness) technological reconstruction. The question then is whether and how these tensions are being reflected in the breeding programs that form the core of research at international agricultural research centers such as CIP, and the varieties it releases. This level of thinking about technological design does not strictly relate to an instrumental level of adapting crop improvement and biotechnology development to a local context (as previously discussed in paragraph 2). CIP has many years of experience with all kinds of production systems in South America, Africa and Asia, allowing its researchers to have a good insight in the priority traits for different regions, and the response of potato varieties to a range of different agro-climatic conditions. However, the tension between releasing improved potato varieties and potentially reducing agricultural biodiversity demonstrates that also the basic model of Andean potato production itself is at stake here. This makes the way in which improved potato varieties and other agricultural technologies developed by CIP fit into local innovation and production systems, and potentially restructure them, an important object for analysis.

As will be demonstrated by the examples presented here, CIP is not entirely capable to resolve all tensions and paradoxes it encounters. However, a number of elements in its breeding strategy do demonstrate the sensitivity of CIP to these tensions, and to the importance of linking up with the valuable parts of traditional potato production systems, rather than replacing them.

5.1. Case vignette 1: breeding for biodiversity?

The contemporary re-evaluation of agricultural biodiversity as key source for international food production and plant breeding (Hoisington *et al.*, 1999), has changed the perspective on agricultural development. From a focus on the introduction of a limited number of improved varieties and a homogenised production system (in terms of genetic diversity and cultivation practices), concern is going out to the replacement of landraces by improved varieties, and the in situ conservation of landraces is gaining in importance (Brush, 2000). This means that instead of a strategy of replacement and industrialisation, the current strategy which is increasingly in vogue today is one of adding to the existing gene pool, and making use of the informal seed systems that already exist (Thiele, 1999).

This focus on agricultural development is influencing the breeding work at CIP, though in a subtle way. A number of observations can be made. First of all, CIP inhabits a position rather upstream in the innovation system, which means that the institute focuses on the improvement of breeding populations by introgressing relevant resistance, processing or other agronomic traits into existing potatoes. These improved populations are further developed into concrete varieties by national research systems (NARS) in several countries, including INIA in Peru. At that more downstream level, breeding may in fact focus on a very limited number of varieties as outputs, and interview data suggest that this is often the case. The traditional preference for producing a widely popular variety has its roots in an incentive structure in which a breeder would be credited for the popularity and wide adoption of 'his/her' variety.

However, both CIP and INIA are involved in participatory variety selection trials, which allow for somewhat more diversity in the varieties that are released. In these trials, farmers get to participate in the evaluation of a number of new varieties with a (number of) specific trait(s). In such cases, farmers receive e.g. 10 varieties that they can try, and choose perhaps 2 or 3 for official variety release. Interestingly, the other varieties that are not officially released, commonly stay in use by the communities involved in evaluation, and after a couple of growing seasons, they may prove to have other useful traits which make them more popular. Moreover, it is quite common for such trials to lead to different 'best varieties' selected in different communities. In other words, the system explicitly allows for regional and cultural differentiation in preferences of new potato varieties.

One of the important pragmatic challenges to participatory variety selection is the lack of credibility that non-standardised farmer evaluations have in scientific circles and among policy makers. To address that issue, a methodology of 'Mother-Baby' trials has been developed in which a centralised trial under reproducible conditions is complemented with a series of satellite trials by farmers. The variety of different evaluations from the farmer trials are backed up by statistically sound data from the Mother trial, which allows the formal release of newly selected varieties. Although the methodology can be applied in different ways with different outcomes, this kind of breeding strategy potentially allows for a much wider range of new varieties that will be released, in a much less prescriptive manner, than when a single best-performing variety is released[125].

Secondly, in catering for the needs of farming communities in the high Andes, breeders have started to consider a 'multiline' approach as a way to improve quality without reducing biodiversity. The general idea is that clones are identified and released which have complementary characteristics. This work was initially informed by nutritional studies in which the complementarity of micro-nutrients from different crops was at stake. However, a similar concept could apply to other traits of potato varieties that serve different needs to a farmer. This idea is currently still very much in a conceptual stage, but it demonstrates the possibility to reshape breeding strategies to address the concerns of biodiversity loss.

Thirdly, within CIP a complementary project has been set up to revalue indigenous landraces for market based production. One of the questions that CIP is in fact struggling with is the extent to which farmers do still maintain a large variation in landraces in certain areas (e.g. Huancavelica), and what the main drivers for this conservation of diversity are[126]. Some preliminary finding indeed suggest that farmers like to cultivate many different landraces because of the variation in the diet, and because landraces are

---

[125] There's one important catch about participatory varietal selection that should not be left unnoticed. While farmers may be very good in evaluating what varieties are useful for them, sometimes a trait like disease resistance may look very appealing during a trial, but may not be very durable in the field. Breeders can make this distinction between vertical (strong but in general not very durable) resistance and horizontal (much more durable) resistance. For farmers it's in practice impossible to make such a distinction. This illustrates how farmer selection is absolutely important, but also has its limits. A combination of breeders expertise with selection by farmers may be expected to work best in really improving agricultural production.

[126] This is ongoing research by Stef de Haan.

associated with quality and indigenous Quechua identity[127]. However, the Papa Andina project[128] that CIP is coordinating is an attempt to also attach market value to some of the traditional landraces as a way of making these varieties commercially attractive[129]. As part of the project, traditional landraces are marketed in supermarkets, or processed to potato crisps. This attempt to attach some market value to the traditional crop genetic diversity of potato in the Andes is only limited in scope. But it is interesting as a complementary activity to CIP's breeding work, in trying to offer something useful to Andean farmers, while not totally loosing the existing potato diversity.

All in all, it is a difficult question whether these approaches or initiatives are really successful in stopping the loss of potato biodiversity in the Andean region. Apart from the fact that biodiversity can be defined and measured in different ways, the patterns of change are different across the Andes[130], and are determined by a range of other factors next to CIP's breeding strategies. Especially the degree of market integration can be expected to have strong correlations with the specialisation on 1 or 2 improved varieties that fit the needs of urban consumers or processing industry. And even then, it is reported that the same communities that cultivate improved varieties for the market, also cultivate a range of traditional landraces for their own local consumption. While commercial farming in large areas of the Andes reduces the area in which landraces are grown, this combination of farming for the market and for local consumption does limit the overall reduction of crop genetic diversity (Brush *et al.*, 1992). In fact, this would be an example of a hybridisation of modern and traditional farming styles, as previously discussed in relation to the effects of the Green Revolution by Stephen Brush (Brush, 1992).

All in all, this paper can't make any definite comments on the success of CIP's breeding strategy. Instead, the observation that is most relevant for this research is that the technology development and breeding itself can respond to the tension between providing useful improved varieties, and running the risk of destroying crop genetic diversity. While a range of factors determines the successful hybridisation of traditional and modern farming styles, with their different crop genetic resources, breeding can potentially be flexible enough to allow such a hybridisation. This is in contrast with a vision of agricultural modernisation that per se involves the replacement of landraces by modern varieties and the destruction of traditional farming styles.

5.2. Case vignette 2: enabling farmers' seed potato production?

The virus-free multiplication of potato seed is a vital element in ensuring high productivity levels, for any potato variety (Haverkort, 1986). Since many viruses are transmitted to new generations via tubers, the specialised virus free multiplication is generally considered to be an essential part of seed potato systems. The most notable exception is the Andean region where at high altitudes the degeneration of potatoes is slower[131]. In other regions like Africa and Eurasia, the focus of development has generally been on the introduction of improved formal seed systems (Gisselquist *et al.*, 1998); an endeavor which

---

[127] Personal communication with Stef de Haan.

[128] See http://papandina.cip.cgiar.org/index.php?id=3&L=1

[129] Similar approaches exist elsewhere, like in Bolivia where Whipala potatoes are marketed as a mixture of indigenous landraces (see Puente, 2007).

[130] Personal communication with Graham Thiele.

[131] At high altitudes, temperatures are lower and therefore few aphids are found, which are the main vectors for the most damaging viruses (Thiele, 1999).

has not always been successful[132]. The externalisation of seed production can be problematic for a whole range of factors (see e.g. Cromwell, 1991), but one important element in the case of potatoes is that the crop is vegetatively reproduced, which always allows farmers to use a part of their own harvest for seed, making a monetary investment in formally produced seed potatoes less attractive, and risky.

Recently, CIP has started to explore the possibility to entirely reverse conventional wisdom in this area, by suggesting that empowering informal seed multiplication by farmers may be a better idea than setting up external systems of seed potato production. The key elements of such a strategy would consist of the production of virus resistant potato varieties, cheap and easy diagnostic kits for viruses, and 'positive selection' methodologies for seed potato selection. Virus resistant[133] potatoes would significantly reduce the degeneration of potatoes that are saved from the harvest, and would prevent the build up of virus infections over the years (leading to the need for virus free seed production). The use of diagnostic kits would allow farmers to measure the actual level of virus infection in their field, allowing a much more educated decision as to invest in commercially produced (virus free) seed, or not. Moreover, it would allow some sort of quality control of potatoes which may replace the current centrally organised quality system of formal seed producers. Thirdly, the 'positive selection' practice implies that healthy looking potato plants are selected before harvest. The tubers from these plants can then be used as seed potatoes for the next growing season. This contrasts with the common practice of selecting the smallest tubers after the harvest, without consideration of which plants were in fact visibly infected by a virus. In practice, since virus infection tends to reduce tuber size, selection for the smallest tubers in fact implies a selection for virus infection as well. This can largely be prevented by a positive selection practice as described.

The interesting aspect of this case is that the production of virus resistant potatoes constitutes an example where biotechnology in breeding has an entirely new structuring role in the social relations of production. Where biotechnology has been argued to be instrumental in the industrial appropriation of breeding and multiplication of seed management, in this case it can be instrumental in bringing it back into the domain of farmers' management[134].

A comparison with apparently similar examples of host-plant resistance illustrates the importance of the precise material shape (and therefore technical code) of resistant varieties. In principle, host plant resistances can make farmers less dependent upon agricultural inputs such as pesticides or herbicides. However, in the case of potato, the virus resistance goes hand in hand with the vegetative reproduction of the crop by farmers themselves. The opposite is generally happening in (e.g. transgenic) host plant resistances that are marketed in hybrid seeds (e.g. of vegetables or cereals). In that case, the introduction of a new variety (with a useful resistance trait) often goes along with a transformation from farm-saved seed to commercially propagated seed, and therefore an externalisation of seed production[135]. The

---

[132] Interview data with CIP scientists

[133] Resistant, not tolerant. While a tolerant potato may still produce in spite of infection, the marketing and transport of infected potatoes might in practice spread virus diseases instead of containing them.

[134] Which is not to say that the occasional injection of quality seed from an external source wouldn't still be a very good idea. Rather than arguing that farmer seed production is per se better for farmers than external seed multiplication, the issue here is that farmers can be involved again in their seed production with the help of modern plant breeding. It's the reversing of a trend of externalisation that is the key observation here.

[135] See (Vroom, 2007) for a discussion of the introduction of pest-resistant cabbages, which simultaneously reinforces the externalisation of seed production and pest control in that production system.

nature of potato as clonally propagated crop has important consequences for how apparently similar traits have an effect on the social relations of seed production.

It is to be expected that – apart from the technical challenges of actually producing potatoes with sufficient and appropriate virus resistance – this strategy of empowering informal potato seed systems will run up against some serious institutional and regulatory challenges. One of the components of agricultural policy in many countries is the certified production of high-quality seed. Certification of formal varieties provides a mechanism of trust within the seed system, because the quality of the seed is guaranteed. While intended as a measure to protect farmers from bad-quality seed, it also strongly prescribes the mode of agricultural production, since it basically prevents farmers from exchange of their own seeds[136]. Instead, it introduces a strong bias for the external, industrialised production of a limited number of improved varieties, the quality if which can be easily tested, monitored and guaranteed. While this model in potato production has been problematic, the transformation from a certified seed system, to a system that grants farmers the possibility to produce their own, but uncertified seed, will be difficult. In informal seed systems in the Andes, it is a social system that creates trust and so-called 'neighbourhood certificates' for seed sellers. A similar mechanism to address the 'trust component' in seed systems needs to be addressed in some way. The development of intermediate certificates of e.g. 'informal seed, but of tested quality' is suggested by CIP scientists as a way forward.

6. The third dimension: culture and identity

This article explores how different perspectives on agricultural modernisation materialise and are reflected in concrete material technical design, and breeding strategies. As elaborated, both a instrumental-pragmatic dimension of questioning technology can be identified, as well as a dimension that relates to the structuring of social relations of innovation and production. But there appears to be more. Not all controversy over biotechnologies can be explained with reference to a simple 'non-functioning' of these technologies, or by a diminished autonomy or sovereignty of farmer production systems.

Illustrative in that respect is that proponents of the Food Sovereignty movement, or of organic farming, are generally hostile to genetic engineering technologies (Rosset, 2006), while in terms of pure autonomy of production such technologies may in fact increase farmer sovereignty. Similarly, indigenous farmer communities and a whole range of NGOs in Latin America are highly critical regarding the development and deployment of transgenic crops in e.g. Peru. Next to the perspectives already elaborated, understanding this controversy requires the introduction of at least one other perspective on technological modernisation; one that is related to issues of identity formation and a fear for cultural erosion. This perspective appears to be especially vital in understanding the controversies around genetically modified crops.

6.1. Case vignette 3: re-signification of molecular biology: from 'improvement' to 'mapping and protecting'

The use of modern breeding technologies and especially transgenics in the context of native potatoes with strong cultural and traditional symbolic values is problematic. With a long history of colonialism, exploitation, and deterioration of natural living environments, indigenous communities have a reason to

---

[136] See for example recent discussions on the Indian Seed Law (Madhavan and Sanyal, 2006). See also the special issue of *Seedling* on seed laws in 2005: (GRAIN, 2005)

be suspicious of external influences that may 'pollute' the very genetic basis under a natural resource that is considered to be a vital heritage from their Inca ancestors. This is where the rejection or reconstruction of technology can no longer be understood in merely technical terms, nor in terms of the (re)structuring of social relations of production and innovation. Instead, the technological controversy primarily takes place at a discursive level where different social groups (in support, or opposition of the technology) attempt to embed the technology in different symbolic discourses, or myths.

Pfaffenberger has discussed this dynamic as part of a 'Technological Drama', in which a design constituency creates technology in accordance with its own political aims, but at the same time constructs a discursive, cultural context in which the technological artifact gains meaning and is legitimised. He calls this legitimisation of technological development in line with a specific vision of progress: 'regularisation' (Pfaffenberger, 1992). According to Pfaffenberger, impacted or affected groups can challenge this socio-technical shaping of the world in a number of ways, one of which is 'countersignification':

> 'Countersignification is an act of mythos substitution that decomposes and rehistoricises the meanings embodied in artifacts. In so doing, it creates a conspiracy theory of regularisation...' (Pfaffenberger, 1992: 300)

NGOs engaged in the protection and conservation of the cultural and agricultural heritage of indigenous groups in the Andes are involved in the countersignification of transgenic technologies. As part of that process, a discourse is created in which transgenic crops are connected to the invasion of foreign multinational companies and the erosion of an indigenous identity based upon a cultural and technical heritage. This resonates with different meanings and values attributed to the existing crop genetic diversity of traditional potato landraces. While biodiversity is primarily treated as a valuable common resource for future breeding by scientists and breeders, for indigenous communities it has strong connotations with their Quechua identity. The tension between maintaining traditional crop genetic diversity and introducing improved varieties is thus countersignified (or somewhat more constructively: re-signified) into a tension between traditional Quechua identity and external capitalist intervention; most powerfully represented by genetically modified crops. This type of 'resignification' of technological development cannot simply be addressed or reverted by making it technically more appropriate to a certain geographic region, by participatory methodologies, or by renegotiating social relations of production. The controversy simply does not take place at that level. Rather than the technical functioning, the 'meaning' of the technology is at stake.

However, that doesn't mean that some elements of the technology cannot be useful in an entirely different way. Countersignification can lead to acts of redesign, making the technology appropriate to this new cultural discourse it is embedded in. This is what has been happening in the efforts to publish a catalogue of native potato varieties in Peru (CIP and FEDECH, 2006). Concretely, molecular biology as starting point for improving crops (or 'polluting traditional genetic diversity', depending on your perspective) can be reframed into a method for categorising and cataloguing existing biodiversity. This in turn potentially allows an improved use of the biological richness that is available. Next to phenotypical and taxonomical classification, native potatoes have also been 'fingerprinted' using a number of SSR markers[137]. This makes the potato clones recognisable and traceable and provides another opportunity to map diversity, and to prevent duplications in the catalogue (or seed bank).

---

[137] SSR = single sequence repeats; also known as microsatellites. These are DNA markers that allow the tracing and identification of specific individuals by their genetic composition.

The use of molecular biology as tool for mapping and categorising is not reported to be controversial among indigenous communities. In order to further bridge the gap between scientific understanding of potato diversity, and the indigenous communities, a system has been developed to represent molecular fingerprints in Kipu-like diagrams (Figure 1). The Kipu[138] is an ancient, precolombian system of ropes with knots to represent or register information regarding harvests, exchanges, taxes, etc. The representation of SSR markers in the shape of a Kipu diagram is a very interesting and apparently successful strategy to present the fingerprinting technology as something that can be easily incorporated into the vision of native potatoes as indigenous material, free from external interference. In terms of Pfaffenberger, the fingerprinting technique is a counterdelegation (adjustment strategy) in addition to the countersignification which problematised the dominant understanding and discourse of molecular biology as key element in agricultural modernisation.

What's more is that the categorisation and cataloguing of native potato varieties is an important tool in keeping this germplasm in the public domain. In the light of international debates on biopiracy, and unclear access and benefit sharing mechanisms, there are important reasons to make sure that indigenous material cannot simply be appropriated by private companies. Once potato varieties are characterised by molecular markers, and published as such, their presence as traditional Peruvian landraces is proven, which prevents any future private appropriation. Molecular biology is in that sense reconstructed from a tool for modification of germplasm, to a tool for mapping and protecting native biodiversity.

---

[138] Or Quipu, see (Jacobsen, 1964). See also http://en.wikipedia.org/wiki/Quipu

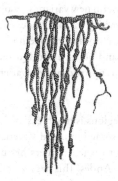 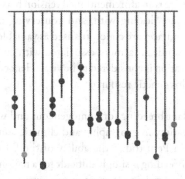

Figure 1. A precolombian Kipu (left) and a Kipu-like molecular fingerprint diagram (right). Reprinted with permission from the 'Catálogo de variedades de papa nativa de Huancavelica – Perú' (CIP and FEDECH, 2006: 49).

7. Reconsidering agricultural development by CIP: three key tensions to deal with

What has become clear from the sections above is how molecular biology and breeding can have very different roles with respect to how they support agricultural biodiversity, farmers' autonomy in seed production, and the categorisation and protection of native potato varieties. If anything, this makes clear how flexible and malleable biotechnologies are in practice, depending on the project they are part of and context they are applied in. Although Goodman *et al.* (1987) and Ruivenkamp (1989) could formulate theories about a number of general trends that were apparent in global agricultural industrialisation and modernisation, the examples described here demonstrate that agricultural modernisation is surely no unilinear path all over the world, and that variations can be made. This itself is consistent with the views of Sorj and Wilkinson who themselves mentioned the 'polyvalency' of biotechnology (Sorj and Wilkinson, 1990). Also Ruivenkamp is explicitly looking for potential to redesign technologies, which implies both a flexibility in design and social meaning, as well as a certain degree of human agency in technology development (Ruivenkamp, 2005).

So the question is not so much whether biotechnologies can have different meanings, or structure life in different ways, but what contextual, historical factors will allow biotechnology development to engage with the needs and perspectives of peasant communities like those in the Andes. Why has CIP been looking for innovative approaches in potato breeding and development which move beyond the mere instrumental adaptation of technologies in a new geographic and institutional context? And to what extent can we expect similar things to happen in other institutes or projects? These are difficult questions to answer, but one important observation is that CIP has to deal with a number of tensions in its work, which call for innovative solutions.

First of all, there is an aforementioned tension between providing useful new varieties to solve problems that farmers are dealing with, and conserving crop genetic diversity in the Andes. Whether the traditional landraces are conceptualised as valuable common resource for future breeding, or as private resource for Andean farmers, associated with culinary variation and quality, and Quechua identity, their conservation is considered to be vital. Hence, CIP's concern over providing new varieties, without destroying this genetic resource.

Secondly, plant breeding for homogeneous and well-controlled regions is much easier than breeding for a great variety in geographic and climatic conditions. While some improved varieties are widely distributed in Peru today, the ability of the formal seed system to reach farmers has been limited sometimes. Breeding is simply difficult for a region as diverse as the Andes. This has naturally increased the interest in involving farmers in participatory varietal selection, and linking formal with informal seed systems. As it appears, informal farmer selection, storage and distribution strategies are quite good to serve their specific and highly varied needs. Moreover, as Graham Thiele put it in 1999: '*improved seed was not as good (and farmers' seed not as bad) as had been thought. In Ecuador and the central highlands of Peru, studies in the 1970s found that yields from formal seed were not significantly higher than yields from farmer seed*' (Thiele, 1999: 86). It is in this light that relatively recent CGIAR initiatives can be understood that have been dedicated to an improved understanding of how research and development can in practice lead to impacts in farmer fields, and how different institutions and organisations together contribute to actual innovations (to be distinguished from inventions). Andy Hall has written extensively on this work, especially for the CGIAR, and e.g. published 'Innovations in Innovation' on the way in which notably ICRISAT[139] in India was reconsidering its role in the Indian innovation system (Hall *et al.*, 2004; Hall, 2005). In a similar vein, researchers at CIP have been engaged in research into Agricultural Knowledge and Information Systems (AKIS) (Ortiz, 2006). The contemporary conventional wisdom is that farmers are engaged in an intricate and diverse 'agricultural knowledge and information system', which is constituted by a whole range of institutions, NGOs and persons (Berdegué and Escobar, 2001). And in the same vein, Graham Thiele wrote his article for CIP on 'Informal potato seed systems in the Andes', which clearly highlights the strong points of existing potato innovation, multiplication, exchange and production systems, which may be helped by CIP research, rather than replaced (Thiele, 1999).

The third and final tension that can be pointed out here, is related to changes in the funding structure of international agricultural research centers (IARCs) like CIP. Historically, CGIAR centers have been dedicated to producing 'global public goods', which serve as a basic resource for agricultural development all over the world, without being restricted by intellectual property rights (non-exclusive, non-rival). Next to the public nature of these products, the limited funding and research capacity of the IARCs, combined with the ambition to produce goods of global relevance, has led to a policy of prioritising research that focuses on strategic key improvements in a relatively upstream domain, rather than focusing on solving very localised problems. However, the problem with investing in upstream research is that it is very difficult to document or measure impacts on actual development; which is an increasing demand by many international donor agencies and governments. Decades of international development aid with often disappointing results has created an atmosphere of accountability of development projects and a quest for efficiency and impact maximisation. It is concrete 'Research for Development' that is becoming more popular as a result, rather than strategic upstream research.

CGIAR centers are forced to follow such donor preferences or demands, but it does imply that the global public good mandate, and the research-for-development projects are pulling the research programs in two different directions. On the one hand, CGIAR centers need to invest in strategic upstream research, and on the other in development oriented, applied projects with a clear local impact. Bringing those two dimensions together in a coherent research program is obviously challenging, but at the same time the result is interesting as it is. The combination of downstream research for development and upstream strategic research has the potential (at least) to firmly ground the strategic research in actual priorities that are recognised in the partnership programs. Moreover, the availability of improved germplasm with new traits can serve as a potential valuable resource for some downstream programs that bring together a number of elements, technologies and partners, but among which the influence of improved varieties can be very significant.

8. Closing remarks

The elaboration of a number of specific research initiatives of CIP has demonstrated that agricultural modernisation can take different shapes, as reflected in breeding and fingerprinting strategies by CIP. This adaptation of technological development trajectories to three key tensions can take place in different dimensions. The article has distinguished between an instrumental-pragmatic dimension, a dimension related to the social relations of innovation and production, and a third dimension related to the cultural connotations of potato biodiversity. The significance of mentioning and elaborating these analytically

---

[139] ICRISAT = International Crops Research Institute for the Semi-Arid Tropics; http://www.icrisat.org/

different dimensions of questioning and adapting technologies, is that it may help in deconstructing some of the public debate and controversy around agricultural modernisation. International debates have had a strong focus on the instrumental dimension of making biotechnologies 'appropriate' to small scale farmers in developing countries. Critical scholars in the late 1980s and early 1990s have stressed the second dimension of technological design: the potential to restructure social relations of production. If this dimension is ignored, technological development runs the risk of becoming controversial because it may implicitly prescribe changes in production systems, consolidation unequal power relations. This has been forcefully pointed out by the aforementioned scholars such as Kloppenburg, Feenberg and Ruivenkamp. However, also the third dimension, focusing on the 'meaning' of technology, rather than its technical 'functioning' is significant. This implies that a categorical rejection of biotechnologies has little to do with the actual political functioning of many contemporary biotechnologies. Cultural arguments are sometimes much more important, and it is wise to treat them as such; not to mask them by reframing them in technical or socio-economic arguments. That does not necessarily make the solution to the controversy much easier, but at least saves one the trouble of proposing alternatives that do not address the concerns that are voiced by critical groups today.

Finally, the three tensions that were indicated for an institute like CIP provide an insight into why and how an institute like CIP is challenged to reconsider agricultural modernisation and the role of the technologies and breeding strategies it develops as part of that project. All three tensions (how to conserve biodiversity, how to link formal breeding with farmers' needs, how to reconcile different donor requirements) are existing in a wider context than just for CIP. It is to be expected that all CGIAR and other public sector institutes to some extent have to deal with these issues in their strategies and breeding activities. It is perhaps important to note that the public or private nature of funding for this kind of research, is not expected to be of primary importance. Both public and private sector institutes have in the past pursued a strong top-down strategy of agricultural development, in which (for example) little initial concerns over the loss of landraces was present. Today, the main differences in strategies may be expected to be related to the kind of farmers and region that is targeted. A seed breeding company or public institute producing for relatively large scale potato producers in coastal areas of Peru, or for large scale rice farmers in the Punjab in India, will not encounter the tensions mentioned here. In contrast, companies or institutes trying to develop seed for farmers in more marginalised and diversified areas do run into these issues and may be expected to adapt their strategies accordingly. In conclusion the difference has less to do with the nature of funding, than with the intended end-user of new varieties and biotechnologies.

Acknowledgements

I would like to thank Guido Ruivenkamp, Joost Jongerden, Conny Almekinders, Graham Thiele, and Merideth Bonierbale for comments to earlier drafts of this paper. In addition, I'd like to thank all interview respondents for their invaluable contribution to this research.

References

Akrich, M. (1992). The de-scription of technical objects. In: *Shaping technology/Building society*. (Eds. W.E. Bijker and J. Law), Cambridge, Massachusetts, The MIT Press, 205-224.

Alston, J.M., Dehmer, S. and Pardey, P.G. (2006). International initiatives in agricultural R&D: The changing fortunes of the CGIAR. In: *Agricultural R&D in the developing world: Too little too late?* (Eds. P.G. Pardey, J.M. Alston and R.R. Piggot), Washington, DC, IFPRI.

Berdegué, J.A. and Escobar, G. (2001). Agricultural knowledge and information systems and poverty reduction. AKIS discussion paper. World Bank.

Bijker, W.E. (1995). *Of bicycles, bakelites, and bulbs: toward a theory of sociotechnical change*. Cambridge, MIT Press.

Brush, S., Kesseli, R., Ortega, R., Cisneros, P., Zimmerer, K. and Quiros, C. (1995). Potato Diversity in the Andean Center of Crop Domestication. *Conservation Biology* 9(5): 1189-1198.

Brush, S.B. (1992). Reconsidering the Green-Revolution - Diversity and Stability in Cradle Areas of Crop Domestication. *Human Ecology* 20(2): 145-167.

Brush, S.B. (2000). The issues of in situ conservation of crop genetic resources. In: *Genes in the field: On-farm conservation of crop diversity*. (Ed. S.B. Brush), Rome, Italy, International Plant Genetic Resources Institute.

Brush, S.B., Taylor, J.E. and Bellon, M.R. (1992). Technology Adoption and Biological Diversity in Andean Potato Agriculture. *Journal of Development Economics* 39(2): 365-387.

Chambers, R., Pacey, A. and Thrupp, L.A. (Eds.) (1989). *Farmer First: Farmer innovation and agricultural reserach*. London, Intermediate Technology Publications.

Chesbrough, H.W. (2003). *Open Innovation: the new imperative for creating and profiting from technology*. Boston, Massachusetts, Harvard Business School Press.

CIP and FEDECH (2006). *Catálogo de variedades de papa nativa de Huancavelica - Perú*. Lima (Peru), Centro Internacional de la Papa (CIP), Federación Departamental de Comunidades Campesinas (FEDECH).

Collins, H.M. and Evans, R. (2002). The third wave of science studies: Studies of expertise and experience. *Social Studies of Science* 32(2): 235-296.

Cromwell, E. (1991). Sustaining the sustainable. *Seedling*: October 1991.

Diao, X., Hazell, P., Resnick, D. and Thurlow, J. (2007). The role of agriculture in develompent. Implications for Sub-Saharan Africa. Research Report 153. Washington D.C., IFPRI.

Dorward, A., Kydd, J., Morrison, J. and Urey, I. (2004). A policy agenda for pro-poor agricultural growth. *World Development* 32(1): 73-89.

Evenson, R.E. and Gollin, D. (2003). Assessing the impact of the Green Revolution, 1960 to 2000. *Science* 300 (5620): 758-762.

Feenberg, A. (1999). *Questioning technology*. New York, Routledge.

Freebairn, D.K. (1995). Did the Green-Revolution Concentrate Incomes - a Quantitative Study of Research Reports. *World Development* 23(2): 265-279.

Gisselquist, D., Kampen, J., Sykes T. and Alex, G. (1998). Initiatives For Sustainable Seed Systems In Africa. Seed policy and programmes for sub-saharan Africa, Abidjan, Côte d'Ivoire, FAO. 23-27 November 1998. Available at: http://www4.fao.org/cgi-bin/faobib.exe?database=faobib&rec_id=392717&search_type=ef_copy&de_ worksheet=ORDER&de_copy_init=ECORD&de_mail_pft=mailo&lang=eng

Glover, D. (2007). Farmer participation in private sector agricultural extension. *IDS Bulletin* 38(5): 61-73.

Goodman, D., Sorj, B. and Wilkinson, J. (1987). *From farming to biotechnology. A theory of agro-industrial development*. Oxford, Basil Blackwell Ltd.

GRAIN (2005). Seed laws: imposing agricultural apartheid - Editorial. *Seedling*: 18 July 2005.

Hall, A. (2005). Capacity development for agricultural biotechnology in developing countries: an innovation systems view of what it is and how to develop it. *Journal of international development* 17: 611-630.

Hall, A., Yoganand, B., Sulaiman, R.V., Raina, R.S., Prasad, C.S., Naik, G.C. and Clark, N.G. (Eds.) (2004). *Innovations in Innovation: reflections on partnership, institutions and learning*. Patancheru, New Delhi, CPHP, ICRISAT, NCAP.

Haverkort, A.J. (1986). Forecasting National Production Improvement with the Aid of a Simulation-Model after the Introduction of a Seed Potato Production System in Central-Africa. *Potato Research* 29(1): 119-130.

Hobbelink, H. (1991). *Biotechnology and the future of world agriculture*. London, New Jersey, Zed Books Ltd.

Hoisington, D., Khairallah, M., Reeves, T., Ribaut, J., Skovmand, B., Taba, S. and Warburton, M. (1999). Plant genetic resources: What can they contribute toward increased crop productivity? *Proceedings of the National Academy of Sciences of the United States of America* 96(11): 5937-5943.

Jacobsen, L.E. (1964). The Ancient Inca Empire of Peru and the Double Entry Accounting Concept. *Journal of Accounting Research* 2(2): 221-228.

Jelsma, J. (2003). Innovating for sustainability: involving users, politic and technology. *Innovation* 16(2): 103-116.

Kloppenburg, J.R. (1988). *First the seed. The political economy of plant biotechnology 1492-2000*. Cambridge, Cambridge University Press.

Kloppenburg, J.R. (2004). *First the seed. The political economy of plant biotechnology. Second edition*. Madison, Wisconsin, The University of Wisconsin Press.

Latour, B. (1992). Where are the missing masses? The sociology of a few mundane artifacts. In: *Shaping technology/ Building society*. (Eds. W.E. Bijker and J. Law), Cambridge, Massachusetts, The MIT Press, pp. 225-258.

Latour, B. (2004). Which politics for which artifacts? Available at: http://www.ensmp.fr/~latour/presse/presse_art/GB-06%20DOMUS%2006-04.html.

MacKenzie, D. and Wajcman, J. (Eds.) (1999). *The social shaping of technology*. Buckingham, Open University Press.

Madhavan, M.R. and Sanyal, K. (2006). PRS Legislative Brief: Seed Bill 2004. Available at: http://www.indiatogether.org/2006/jun/law-seeds.htm.

Ortega, R., Halloy, S.R.P., Yager, K. and Seimon, A. (2005). Traditional Andean cultivation systems and implications for sustainable land use. *ISHS Acta Horticulturae 670: I International Symposium on Root and Tuber Crops: Food Down Under* CAB Abstracts: 31-55.

Ortiz, O. (2006). Evolution of agricultural extension and information dissemination in Peru: An historical perspective focusing on potato-related pest control. *Agriculture and Human Values* 23(4): 477-489.

Pfaffenberger, B. (1992). Technological Dramas. *Science Technology & Human Values* 17(3): 282-312.

Pickersgill, B. and Heiser, B. (1978). Origins and distributions of plants domesticated in the New World tropics. In: *Advances in Andean archaeology* (Ed. D.L. Browman), The Hague, Mouton, pp. 133-165.

Pingali, P.L. and Heisy, P.W. (1999). Cereal crop productivity in developing countries: past trends and future prospects. CIMMYT Economics paper 99-03. Mexico D.F., CIMMYT.

Pollan, M. (2001). *The botany of desire: a plant's-eye view of the world*. New York, Random House Inc.

Pretty, J.N. (2002). *Agri-culture. Reconnecting people, land and nature*. London, Earthscan.

Puente, D. (2007). Searching democratic trajectories for the deployment of genomics within the potato crop systems in the Bolivian Andes. DSA annual conference 2007: Connecting Science, Society and Development, Sussex. 18-20 September 2007.

Quaye, W. (2007). Food sovereignty and combating povety and hunger in Ghana. *Tailoring Biotechnologies* 3(2): 69-78.

Rosset, P. (2006). Agrarian reforms and food sovereignty: Alternative model for the rural world. *55th Annual Conference of the Center for Latin American Studies: Alternative Visions of Development: The Rural Social Movements in Latin America*. Gainesville, Florida, USA.

Ruivenkamp, G. (1986). The impact of biotechnology on international development: competition between sugars and sweeteners. *Vierteljahresberichte des forschungsinstituts der Friedrich Ebert Stiftung* 103 (March): Special issue on New technologies and Third World development.

Ruivenkamp, G. (1989). *De invoering van biotechnologie in de agro-industriële productieketen: de overgang naar een nieuwe arbeidsorganisatie*. PhD thesis, Faculteit politieke en sociaal-culturele wetenschappen, Amsterdam, Universiteit van Amsterdam, 354 pp.

Ruivenkamp, G. (2005). Tailor-made biotechnologies: between bio-power and sub-politics. *Tailoring Biotechnologies* 1(1): 11-33.

Ruttan, V.W. (2004). Controversy about agricultural technology: lessons from the green revolution. *International Journal of Biotechnology* 6(1): 43-54.

Scott, J.C. (1998). *Seeing like a state. How certain schemes to improve the human condition have failed*. New Haven & London, Yale University Press.

Shiva, V. (1993). *Monocultures of the mind: perspectives on biodiversity and biotechnology*. Atlantic Highlands, NJ, Zed.

Sorj, B. and Wilkinson, J. (1990). From Peasant to Citizen - Technological-Change and Social Transformation in Developing-Countries. *International Social Science Journal* 42(2): 125-133.

Thiele, G. (1999). Informal potato seed systems in the Andes: Why are they important and what should we do with them? *World Development* 27(1): 83-99.

Thirtle, C., Irz, X., Lin, L., McKenzie-Hill, V. and Wiggins, S. (2001). Relationship between changes in agricultural productivity and the incidence of poverty in developing countries. DFID Report No. 7946 27/02/2001.

Van de Belt, H. (1995). How to critically follow the agricultural technoscientists: Kloppenburg versus Latour. Agrarian Questions: The politics of farming anno 1995, Wageningen (The Netherlands). May 22-24, 1995.

Von Hippel, E. (2005). *Democratizing Innovation*. Cambridge, Massachusetts & London, England, The MIT Press.

Vroom, W. (2007). From rejection and regulation to redesign: making pro-poor biotechnology responsive to social needs. Connecting Science, Society and Development. Development Studies Association Annual Conference 2007, University of Sussex. 18-20 September 2007.

Vroom, W., Ruivenkamp, G. and Jongerden, J. (2007). Articulating alternatives: biotechnology and genomics development within a critical constructivist framework. *Graduate Journal of Social Sciences* 4(1): 11-33.

Weatherford, J. (1988). *Indian givers: How the Indians of the Americas transformed the world*. New York, Crown Publishers.

Winner, L. (1985). Do Artifacts Have Politics? In: *The social shaping of technology* (Eds. D. MacKenzie and J. Wajcman), Buckingham, Open University Press, pp. 26-38.

Wynne, B. (1996). Misunderstood misunderstandings: social identities and public uptake of science. In: *Misunderstanding Science? The public reconstruction of science and technology* (Eds. A. Irwin and B. Wynne), Cambridge, Cambridge University Press.

Ethicisation of biotechnology research, politicisation of biotechnology ethics

Shuji Hisano

1. Introduction

Public distrust of science and technology in general and agricultural biotechnology in particular has rapidly increased. The latter is largely because of hasty introductions on the market of new technologies like GMOs and nanotechnology in spite of pervasive public concerns. It is also prompted by a spate of food scandals and government mismanagements of food safety issues like BSE.

Such distrust has grown to the extent that science communities, and business and government sectors that have a stake in science and technology development, cannot ignore and have to deal with it. As a way to regain the public confidence, a growing attention tends to be drawn to ethical considerations on science and technology[140]. This *might* be an important step forward to increase the 'social robustness' of science and technology and re-embed it into society However, it is open to question whether and to what extent such a growing attention to ethics really addresses socio-ethical questions enough to make science and technology socially robust. If we can say that science and technology must be seen as inextricably tied into broader social, cultural and political contexts (Bijker, 2001), and raising important socio-ethical questions in such contexts (MacKenzie and Wajcman, 1999), what we should expect from ethics is not just an instrumental tool for judging something about values in abstract terms, but a discursive and communicative tool for reflecting on what values for whom in more concrete context. The purpose of this paper, therefore, is to examine the efficacy and limitations of ethics as it is, and to give some implications for critical reflections among scientists by expanding and politicising ethical discourses on science and technology.

In so doing, we will first look at the so-called *ethicisation* of science and technology, and then some criticism against its 'slippery' aspects. We will use the term of ethicisation to describe an accelerating tendency to introduce ethical considerations into an institutional setting or discourse concerning a decision making. When it comes to science and technology, the term of ethicisation is referred to a tendency to incorporate ethical considerations into all areas and stages of science and technology, from scientific research to technology development, to risk assessment, and to science and technology policy. However, the term is, especially in our context, meant to imply that it is just a tendency or an attempt, and not necessarily addressing ethical issues enough to make those engaged in science and technology self-critical and reflective on their practice and its outcomes. Indeed, ethical discourses are often used by proponents of a certain technology (e.g. GMOs) to justify the application and commercialisation of the technology, on the ground that, for example, GM technology enables us to produce more food with less external resources even in developing countries, without critical and empirical evaluations. This makes a critical analysis of ethical discourses such as 'biotechnology for the poor' quite relevant (Hisano, 2005).

---

[140] Ethics is defined as the study of values and customs, whether individual or collective. It covers the analysis and employment of concepts such as right and wrong, good and evil, and responsibility, especially when we should deal with moral problems and dilemmas.

Second, we will review some attempts to bridge the gap between traditional bioethics and social sciences so as to critically reconstruct bioethics. The term of *politicisation* will be used to describe such a process of reconstruction of bioethics. The term of politicisation is borrowed from Bijker (2001), who defines it, in the context of socially constructed technology, as showing hidden political dimensions, putting issues on the political agenda, and opening issues up for political debate.

Third, while ethicists and social scientists have been struggling to commit to decision making processes of research, development and evaluation of science and technology, several agricultural scientists are getting actively involved in *ethical reflections on their own practice*. By referring to some of these, we will examine the possibility of bringing scientists into a process of critical reflections on the social meaning of agricultural science and technology by use of ethical terms. This will be a clue for critical reconstruction of agricultural science and technology for alternative development, given the important role of scientists in the process of social transformation of technological hegemony.

In light of the understanding that agricultural science is a *social learning process*, we will finally argue for the need of *interdisciplinary approaches* in the context of scientists' critical reflection and the expectations for *ethical education* in the context of interdisciplinary approaches.

We fully agree with Ruivenkamp (2003), when he argues that it is necessary to reflect on which role (genomics) researchers can play in the reconstruction process taken in consideration that they must operate within the hegemonic political, economic, legal, and ideological setting of biotechnology development. At the same time, he stresses that it is possible to develop an 'alternative coalition of social and scientific elements' in (genomics) research, by uncoupling it from its one-dimensional association with industrialising agriculture and simultaneously inter-relating it with initiatives working towards endogenous innovations. What we need in order for it to take place is to set up new research networks in which researchers gain a view of their possibilities for social choices in their research as an 'endogenous catalyst and facilitator for endogenous development' (Ruivenkamp, 2003).

Then, why should we shed light on the role of ethics, not directly on critical social sciences that have revealed the 'social constructedness' of agricultural science and technology as well as bioethics? Although important role of critical social sciences in the process of reconstructing agricultural science and technology is needless to say (Kloppenburg, 1991), critical social sciences are, in our view, too subversive to be a common platform or discursive tool for those engaged in agricultural science and technology to interact with ethicists, social scientists and a range of stakeholders. On the other hand, ethics has been, more or less, mainstreamed and institutionalised in research and technology development activities and policies. Few scientists and technical professionals now disagree on the need of ethical consideration, regardless of the extent to which it is thoroughly carried out. We should start with this fact – ethics as a relatively easy 'entry point' – to involve scientists into the process of reflection and reformation of themselves and to develop a discursive and institutional setting for reflective researchers to gain moral and intellectual control over the social support networks and material resources for the alternative goal, i.e. reconstruction of agro-biotechnologies. This can be a part of the 'counter-hegemonic tactics from within to democratise technology' (Hisano, 2005).

2. Ethicisation and institutionalisation of ethics

It should be noted that a number of natural scientists and engineers have begun to recognise that the technologies resulted from their scientific research can have an array of impacts, good and bad, on our society and the environment, and therefore recognise the scientists' social responsibilities. One of well-referred statements to this effect is of Dr. Shirley Ann Jackson, delivered as a presidential address at the American Association for the Advancement of Science (AAAS) 2005 Annual Meeting. She urges scientists and engineers to renew their commitment to public engagement and serving humanity: 'We must address the ethics of the application of science in key areas' (Jackson, 2005). While this is a rather classical approach to ethics, which is just applied to the application of science and technology that is assumed to be neutral as discussed in the following sections, Dr. John Ziman, a prominent theoretical physicist, is going step further. He contributed an essay to *Science* even earlier, arguing that: 'Even the "purest", "most basic" research is thus endowed with potential human consequences, so that researchers are bound to ask themselves where all the goals of the activity in which they are engaged are consistent with their other personal values' (Ziman, 1998).

Apart from these noticeable remarks, it is also important to know that, since as early as 1980s, there have been dialogues among agricultural scientists, ethicists (or philosophers), and social scientists on the ethical aspects of practice and policy of food, agriculture and natural resource management. Some of these attempts and outcomes have been shared within the relevant journals such as *Agriculture and Human Values*. Our concern is to what extent these early messages have been spreading and shared among mainstream scientists to produce results, and if not, why, and what can we do. Winner (1990) actually noticed the gap between the ideal and the real, the attempts and the achievements:

> 'Recent attempts by American colleges and universities to teach ethics for scientists and engineers deserve strong praise. They represent a shift away from the idea that questions about ethics and morality are best left to humanists or to elder statesmen of science, a recognition that such matters ought to be an important part of education in the technical professions. One can hope that through these efforts a new generation of men and women will obtain a firm grounding in the ethical aspects of their vocations early enough to make a difference. Despite these admirable aims, however, the approach often used to teach ethics to scientists and engineers leaves much to be desired' (Winner, 1990).

Meanwhile, the so-called *institutionalisation of ethics* has been taking place (Paula and Van den Belt, 2006). There are different ways in which ethics can be institutionalised or incorporated in scientific research and technology development: (1) through ethics committees and other institutional bodies functioning as an instrument to frame and address public concerns as ethical issues by providing the analysis and vocabulary to do so; (2) through organised public debates and other participatory technology assessment initiatives functioning as a forum supposed to articulate and address public concerns as ethical issues by use of stakeholder dialogue or broader citizen consultation; (3) through regulation or legal frameworks to govern the introduction and management of new technologies on the ethical grounds, such as biosafety, labeling and co-existence measures; and (4) through ELSI/ELSA[141] programs with the view that every major research project should be accompanied by research on its ethical, legal and social implications/aspects, and for that purpose a substantial amount of research budget has been spent.

---

[141] ELSI stands for the 'Ethical, Legal and Social Implications' of research and technology, mainly used in the U.S., while ELSA, the 'Ethical, Legal and Social Aspects' of research and technology, is used in Europe.

Especially, the ELSI/ELSA approach marks a milestone in the institutionalisation of ethics. The first of such ELSI/ELSA programs was established in 1990 by the U.S. Department of Energy and the National Institutes of Health, who devoted 3 to 5 percent of their annual Human Genome Project budgets to the ELSI program. In Europe, bioethics forms an important part of the European Commission's Research and Technological Development (RTD) activities since under the 4th RTD Framework Program (1994-1998), when bioethics was extended to include fisheries and agriculture research as well as biomedicine and health research. Now, under the concept of ELSA, experts in ethics, law and social sciences are, whenever relevant, encouraged to participate in the research projects (NGI, 2004).

The U.N. organisations are no exception. The UNESCO, for example, created an International Bioethics Committee in 1993, and the World Commission on the Ethics of Scientific Knowledge and Technology in 1998 to give an ethical reflection on science and technology and its applications; and WHO has hosted several external committees that have addressed issues related to medicine and medical genetics. Because of its mission and actual involvement into global food security, management and utilisation of natural resources, sustainable rural development, and food safety issues, the ethical role of the FAO has long been interpreted broadly by member countries. But, the FAO's explicit commitment to an 'ethics initiative' was started in 1998, when the Director-General has launched a process of reflection within the Organisation, by establishing an internal inter-departmental Committee on the Ethics of Food and Agriculture. Since 2000, the Panel of Eminent Experts on Ethics in Food and Agriculture has been commissioned to advise the Organisation and raise public awareness of ethical considerations.

All of these settings are expected to encourage scientists to critically, not only individually but also collectively, reflect on their practice and its outcomes enough to make science and technology socially robust. So, it might be appropriate to say that ethics can be a sort of *nexus* of science and society, a *discursive tool* to stimulate dialogues among scientists as well as between scientists and the public at large, or a *platform* for scientists to reflect on the nature of their practice.

But, the institutionalisation of ethics itself does *not* necessarily mean that critical self-reflections among scientists on the nature of science and technology they are engaged in are automatically facilitated. There is sufficient evidence to raise a question as to whether a particular body such as an 'ethics committee' is really doing ethics (Levidow and Carr, 1997); Wynne, 2001); whether public concerns are properly framed as socio-ethical issues through participatory approaches (Stirling, 2005; Wynne, 2001); and whether ELSI/ELSA programs can provide scientists with increasing and substantial opportunities to interact and mutually understand with those in social sciences and humanities (Tait *et al.*, 2002; Bruce *et al.*, 2004). As briefly reviewed below, there are a plenty of sociologists' criticism against the role of ethics played in some institutionalised ethics bodies and fora.

3. Criticism of ethicisation

Social scientists, especially those in science and technology studies, cast critical doubt on whether an ethics committee is really doing ethics.

3.1. End-of-pipe commitment

Levidow and Carr (1997) heavily criticise the role of official bioethics (committee), saying that it has been devised to judge where to 'draw the line' between science and ethics in applying biotechnological knowledge, possible risks of which are *assumed* to be dealt with only by objective science. Such an objective risk assessment is *assumed* to be beyond ethical scrutiny. This separation between science and ethics is largely due to 'scientific values', the values commonly shared among scientists (Burkhardt, 1999). One of those values, implicitly or explicitly, shared among scientists is the belief that the science itself is value-free, whereby science is regarded as an independent, detached and objective observation of the world, whether natural, human, or social. Therefore, the agenda setting of scientific research, or the process of technology design, is *assumed* to be beyond ethical scrutiny. In this way, bioethics is not really incorporated into the process of biotechnological knowledge production, but rather complemented at the end-of-pipe as a token consideration.

3.2. Narrow utilitarian framing

Levidow and Carr (1997) also point out that state-sponsored, expert-led bioethics may judge how to 'balance' risks and benefits, *as if* their definition were not an issue. Even when risks are potential enough to cause public concerns or to be perceived inevitable, the role of ethics is narrowly reduced to the marginal role of risk-benefit analysis, at most on the application of biotechnological knowledge, *as if* its production were neutral. At issue here is a very basic utilitarianism. A typical example can be seen in Straughan (1992, cited in Levidow and Carr, 1997) and Reiss and Straughan (1996), who, according to Levidow and Carr (1997), reduce bioethics to a simplified treatment within a narrow utilitarian frame. Although utilitarianism seems to be preferably applied by many regulators of science and technology, as in established approaches like economic cost-benefit analysis and scientific risk assessment, it has got a lot of criticism (Paula and Van den Belt, 2006). Not only critical social scientists and philosophers, but even agricultural scientists raise a fundamental question about utilitarian ethics.

In deciding whether certain developments of technologies are 'good', we look at outcomes as benefits. But, Chrispeels and Mandoli (2003) question that: 'How should these benefits be calculated? Benefits for whom and for how long?' They thus conclude that the utilitarian ethics has its limits. A weed scientist, Radosevich and his colleagues (1992) also cast doubt on risk-benefit analysis, most commonly attempted by weed scientists to assess the relative merits of its tools such as herbicide, tillage, and herbicide tolerant crops. Based on their expert experience, they say that this approach, with its roots in the utilitarian ethics, employs a 'calculus' or scaling of the potential goods and harms of a tool or activity, and would simply express all values in economic terms (e.g. yield, productivity, profitability, efficiency), in spite that many of the values of a tool or activity are economically intangible (i.e. externality). Another weed scientist Zimdahl (1998b) raises a question about the primacy of instrumental reason when weed scientists think about their role and about what research should be done. Instrumental reason is evident in the frequency with which technological solutions are chosen in the utilitarian terms and with the almost automatic assumption of their appropriateness in the weed science community. As a result, there is a tendency to rely on technology to overcome problems it has developed, and that reliance may lead to ignoring the fragile and intangible values. In short, when a narrow utilitarian frame is applied, it is probable that a risk-benefit analysis is carried out without thoroughly examining socio-economic and environmental benefits and costs, and therefore, there is a strong tendency to overestimate benefits of new technology while underestimate its (external) costs.

As a consequence, state-sponsored and expert-led bioethics is quite likely to be used as a way to legitimise the contentious values implicitly embedded in biotechnological knowledge production, and then to pacify opposition and pubic concerns in the name of ethical judgement (e.g. policy statements issued by scientific communities and professional societies; ethical commitments of private companies such as CSR policy statements, among others). On the other hand, it is not improbable that a narrow utilitarian frame is also applied 'against' a certain technology.

It is not only those bioethics narrowly applied by a state-sponsored expert committee that should be critically examined. More importantly, theoretical frameworks of bioethics in general, and bioethics education in special, are not without problems. The main complaints about bioethics are (1) its almost exclusive focus on personal moral responsibilities; and (2) its empirical weakness.

3.3. Personal moral responsibilities with universal principle

First, bioethics tends to position individuals as the sole judge in ethical decision-making, in which an ideal of universal ethical principles is presupposed to govern individual moral practice, with the view that there is a single, correct solution for each ethical problem, largely independent of person, place, or time. Ethical discourses, therefore, tend to ignore the role of social and cultural factors (Hedgecoe, 2004). The orientation of ethics toward personal conduct is also a concern of Winner (1990). He's made out that, in the way the topic is usually presented, personal responsibilities are situated in extremely limited, asocial contexts. He describes there is a logical juncture 'where ethics finds its limits and politics begins'.

In response, engineering ethics, for example, has come up with several categories to address the deficit of its individual orientation and to broaden its scope by analysing three frames of reference from which engineering ethics can be viewed – individual, professional and social – then grouping ethics into 'microethics' and 'macroethics' (Herkert, 2005). Other relevant categories include 'technical ethics', 'professional ethics' and 'social ethics' (McLean, 1993) and 'ethics in engineering' and 'ethics of engineering' (DeGeorge, cited in Roddis, 1993). But, it is still an open question to what extent the situation has changed, since it seems that most research and teaching in engineering ethics still has a 'micro' focus concerned with ethical decision making by individual engineers and the engineering profession's internal relationships, rather than 'macro' focus referring to their collective social responsibility and to societal decisions about technology:

> 'The role of engineers and technical professionals is crucial. They are intimately involved in maintaining key social patterns and in inventing new ones as well. In that work they become, in effect, unelected delegates and representatives of the rest of us, charged with the work of building basic structures of our social and political future ... The lack of an ability among engineers and other technical professionals to engage in political reflection is by now much more than a small failure in education; it threatens to become a great tragedy .... What is needed is less the study of pre-packaged ethical dilemmas than the cultivation of two crucial skills: political savvy and the capacity for political imagination ... Political imagination [means] the ability to envision the contributions of one's own work to society as a whole, to the quality of public life' (Winner 1990).

If engineering ethics is insufficient to provide engineers and technical professionals with such skills (political savvy and the capacity for political imagination), Winner argues, it is the disciplines in humanities and social sciences that should be brought into educational programs for engineers and technical professionals since these disciplines have 'important resources to offer, resources that at present remain largely untapped' (Winner, 1990).

3.4. Empirical weakness

Furthermore, the bioethics literature sometimes contains testable, potentially disputable statements, presented as 'fact'. Although social scientists are not immune from overstating their positions and going beyond their empirical evidence, they give greater importance to a commitment to root one's ideas in social reality (Hedgecoe, 2004). We can analytically (empirically) evaluate the quality of knowledge and information used as a source of moral judgement by ethicists, and then we can challenge and re-frame the ethical framing assumed with no doubt by ethicists.

It should not be presumed, however, that we need a sort of the division of labour between ethics and social sciences, in which the latter provide the empirical data upon which the former make judgements. We need to go beyond the presumption of such a 'linear relationship' (Nelson, 2000, cited in Haimes, 2002) or 'ethical reductionism'[142]. On the one hand, descriptive facts social sciences can provide are inherently tied to theories and methodology of knowing the world[143]. So, the contribution of social sciences to ethical discourses is not limited to providing descriptive facts and information. On the other hand, there is a growing interest within bioethics in conducting empirical investigations (Haimes, 2002). There is also some interest for bioethicists in seeing themselves in a social context (they are actually getting subject to an increased scrutiny and criticism): therefore, there is a potential for a fruitful collaboration between ethicists and social scientists (Haimes, 2002).

Due to the differences in styles of reasoning, however, there is very few collaboration or interaction between ethicists and qualitative social scientists, and there seems even a growing antagonism and disrespect between them to the extent that can be described as 'ethics wars' (Hoeyer, 2006). According to Hoeyer, who seems to stand largely for bioethics contrary to Hedgecoe and Haimes, it is largely because social science studies employ either (1) a deficit model (social science perspectives accommodate the sense of context that bioethics lacks); (2) a replacement model (social scientists have found a better way of doing bioethics); or (3) a dismissal model (bioethics should be abandoned all together as a misconstrued veil of power). It is less important here to ask whether his claims are appropriate or not, than listening to his point for productively overcoming 'ethics wars'. His observations show that the antagonism between different approaches to studying moral issues might reflect the fact that each tradition works with different criteria, research questions, and modes of reasoning. In this regard, he suggests that we should start with this fact in order to engage fruitfully with other forms of reasoning, rather than simply assuming either form will be able to embrace all questions or generate all answers. What is needed is to 'assume a more modest mode of inquiry embracing reflexivity concerning its own limitations' (Hoeyer, 2006).

Thus, instead of denying the role of bioethics on the ground of its empirical weakness, and instead of denying the role of social sciences on the ground of their normative weakness, it is possible to incorporate social science research into bioethics, at the same time firmly rooted in philosophical bioethics (Hedgecoe, 2004). But how?

---

[142] Following Thompson (1995), Alrøe and Kristensen (2002) clearly describe ethical reductionism is the idea that: 'there is a sharp and inviolable distinction between facts and values, and that this distinction entails that science is neutral and has no implications for ethics (and *vice versa*)'.

[143] This is also true of natural sciences, whereby we clearly hold 'neutral science' to be faulty.

3.5. Politicisation of ethics

From a point of view of qualitative social sciences, sociology of ethics (or politicisation of ethics) is needed to critically analyse and reveal the socio-political construction of ethical discourses and to reconstruct ethics to make it more socially robust. In social science understandings, ethics is shaping and being shaped by major social forces and therefore being historically and culturally located and embedded within the social world (Haimes, 2002), exactly the same as natural science and technology. Social sciences have a broad range of theoretical and epistemological resources that can be used to assess: (1) the locations and institutions in which ethical dilemmas arise; (2) the actors involved in the activities and discussions around these dilemmas; (3) the diverse range of definitions and meanings that underpin both the dilemmas and the answers proffered; and (4) the broader socio-cultural frameworks.

According to Herkert (2005), who endorses the division of ethics into 'microethics' and 'macroethics' as mentioned above, mechanisms for incorporating macroethical perspectives include: (1) integrating engineering ethics and STS, in which the latter is expected to broaden discussions of ethics to include ethical implications of institutional and political aspects of engineering; (2) closer integration of engineering ethics and computer ethics, in which the latter has practical experience and knowledge necessary for in-depth understanding of ethical problems; and (3) consideration of the influence of professional societies and corporate social responsibility (CSR) programs on ethical engineering practice. Although his argument for the positive role of professional societies and CSR-kind programs seems too naïve to draw engineers' attention to an 'inconvenient truth' about their practice, his implications for requiring appreciation of ethics in historical, societal, political and institutional contexts is important.

Hedgecoe (2004) borrows the term of 'critical bioethics' from Parker (1995) to imply the need of critical self-reflection on the nature of bioethics and the decisions it supports. He characterises critical bioethics as being (1) empirically rooted; (2) theory challenging; (3) reflexive; and (4) politely skeptical. First, critical bioethics is rooted in empirical research. The problems, dilemmas and controversies analysed come from looking at a particular setting, not from the standard bioethics debates. Second, the results of such empirical research would feed back to challenge, and even undermine, an analyst's theoretical frameworks. Thus, critical bioethics tests its theories in the light of empirical experience, and changes them as a result. Third, reflexivity is about acknowledging one's personal context. It is not just about an individual's self-awareness, but a critical reflection on the nature of knowledge production itself. It is also about acknowledging the social and cultural influences that make us who we are. Finally, being polite means one does not necessarily dispute matters with scientists. Being skeptical means accepting that the truth is often more complex than scientists claim. Such an approach allows bioethicists to unpack ethically controversial topics, and analyse the interests that underpin some of the claims made about the objectivity and reliability of scientific results.

Through challenging (politicising), bioethics will be reconstructed and integrated into a broader framework that includes the social context of science and technology it concerns.

4. Ethical reflection among agricultural scientists

Although the term of 'philosophy' is used instead of 'ethics', Soda (2006) [2000] describes the philosophy of agricultural science as the field of academic study that questions the place of agricultural science in the modern world. This definition derives from his predecessor Sukekata Kashiwa, who defines the philosophy of agricultural science as a field of academic research that attempts to answer the question about what agricultural science is; the area of study that questions the basis of agricultural science. It is therefore a process of the self-criticism and self-reflection on agricultural science. We can trace the basis of their discussions back to early Japanese philosophers such as Kitaro Nishida, Hajime Tanabe, and Kiyoshi Miki, who elaborated on the philosophy of science. According to them, the philosophy of science is an enquiry into the knowledge that lies at the heart of science; it examines the basis for and significance of scientific cognition, and asks about its raison d'être; and therefore it is 'the reflection of scientists about their scientific pursuits' and 'self-critique of science'. With these definitions in mind, we will take a look at several agricultural scientists, who have been concerned with ethical considerations, or self-criticism and self-reflection, about what they do in a broader social context.

Among backgrounds of the increasing ethicisation of agricultural science and technology, we should be reminded of a history of social criticism against the modernisation and industrialisation of agriculture and its adverse effects on human health and the environment. We can go back to Rachel Carson's *Silent Spring* (1962), among others. Since then, our rich experiences of social and political discussions on agrarian and ecological issues have enabled us to expand the literature of agricultural ethics. Additionally, food safety issues derived from agricultural and food technology are now included in the list of the literature. It is interesting that, while ethics in other disciplines such as engineering ethics and medical ethics is largely limited to micro-oriented or professional ethics, the discourse of agricultural ethics includes a much broader range of subjects from the beginning. One of those who have contributed to the development of agricultural ethics is Paul B. Thompson, who puts an emphasis on 'social and political ethics' as one of four domains of ethics[144]. Similarly, there are several definitions of agricultural ethics, commonly given from socioeconomic, political and cultural perspectives:

- Kristin Shrader-Frechette in *Encyclopedia of Ethics* (1992, 2001)
- Richard P. Haynes in *Encyclopedia of Bioethics* (1995)
- Ben Mepham in *Encyclopedia of Applied Ethics* (1998)
- Paul B. Thompson in various articles and books (1995, 1998, 2001)

If these discussions and messages have been passed on to agricultural scientists and others engaged in agriculture, we would have socially robust agricultural science and technology, notably agro-biotechnology. It seems, however, that critical reflections and ethical discourses have not yet been shared within mainstream agricultural science communities, as if ethical considerations of several agricultural scientists reviewed below are just a heroic effort. But still, it is worthwhile looking into these outstanding examples.

A weed scientist Zimdahl, the author of *Agriculture's Ethical Horizon* (2006), has been raising reflective questions as to the ethical foundation of agricultural science, based on his expert experiences in weed science. Because many aspects of agricultural practice have ethical dimensions in terms that they affect all of society, they demand ethical consideration. Because agriculture is an essential human activity and is the largest and most widespread human environmental interaction, those in agriculture have a special responsibility to address the ethical dimension (Zimdahl, 2000).

While it is true that maintaining or increasing food production has been the primary goal, and it is still a good one, it must be remembered that public has a right to hold agricultural scientists responsible

---

[144] Others include religious and metaphysical ethics, environmental ethics, and professional ethics (Thompson, 2001).

for positive and negative results of research (Zimdahl, 1998b). What makes it challenging to bring agricultural scientists into critical reflection on their activities, however, is the fact that those engaged in agriculture, whether practitioners, research scientists, extension agents, technology developers or technology suppliers, share a definite but unexamined moral confidence or certainty about the goodness and correctness of what they do (Zimdahl, 2002). Weed scientists, for example, explicitly or implicitly, presume that their activities are to meet basic human needs through improved food production, promoting the common good through abundant food, improving people's lives through efficient production of safe food, and increasing farmers' profit. If the primacy of these values attached on production and profitability were not questioned, weed scientists would be justifiably proud of their research on use of herbicides and the ensuring reduction in labour required weeding farm crops. In this regard, we should consider broadening our concept of productivity and efficiency, he contends, to include basic environmental, resource, ecological, health, social, and political processes and the costs and benefits of weed control technology (Zimdahl, 1998a).

Between May 2003 and March 2004, a special series of essays on agricultural ethics was organised in the journal of *Plant Physiology*. In an 'informal introduction' to the series, two biologists Chrispeels and Mandoli (2003) describe that ethics is about choices, and agricultural ethics is about choices for people engaged in agriculture. 'Although all of us make choices, few of us actively engage in an ethical analysis of our actions or can provide reasons for the choices we make'. Based on this critical self-reflection, the series was designed to encourage professionals in plant biology to think about the wider implications that their work has on society and examine their own beliefs, values, and morals and then reconsider their choices from this new perspective.

Just like Zimdahl (2002), Chrispeels and Mandoli also acknowledge the prevailing moral confidence, saying that: 'people engaged in agriculture, whether as producers, scientists, administrators, legislators, or protestors, all tend to believe that they are on high moral ground. Because they are part of that most noble of human endeavors, to feed the people of the Earth, they have a 'moral confidence' in their profession and often fail to see the need to examine their choices' (Chrispeels and Mandoli, 2003). They put an emphasis on engaging in the dialogue with people who have different visions from the prevailing paradigm, 'production agriculture'. Such an experience is expected to make research scientists aware that the concept of 'production agriculture' is not the only possible vision for agriculture. For instance, there is an ethical debate in the context of sustainability versus production agriculture: 'Is the health of rural communities in developed countries and the desire to please certain consumers more important than the food security of millions in developing countries?' However, we can put it differently. While it is absolutely true that agricultural sustainability in the developed world may depend also on increasing food production in the developing world, we should be reminded that, 'we still invest only modestly in research that will push agriculture toward sustainability, and we do not use our subsidies to bring about this transformation.... In addition, we could invest in developing agricultural systems for tropical and semi-tropical regions that would raise productivity there in a sustainable way'. Likewise, when it comes to GMOs, the ethical issues that face plant scientists are much broader than 'GM or Not GM', but pertain to the entire complex of agricultural practices and to the transition from production to sustainability. They therefore conclude that there is plenty of room for ethical choices in this area (Chrispeels and Mandoli, 2003).

Lockwood (1997) is another example. He, as an entomologist who has been working on biological control and has all the time argued for taking agricultural ethics into agricultural science research and education, served the guest editor for a special issue of *Agriculture and Human Values* on the theme of the ethics of biological control. In an introductory essay, he claims that biological control has the demonstrable potential, with no doubt, to dramatically and meaningfully decrease harm to human and environmental health relative to many pesticides, but 'technologies are not often, if ever, good in-and-of-themselves' (Lockwood, 1997). If we simply replace chemical control technologies with biological ones, then it seems that little moral progress has been made. Because the question of how we apply this powerful set of ecological technologies is a complex matter, he insists, rigorous ethical analysis is strongly required.

Lastly, we will listen to what Radosevich and his colleagues have argued for, also in the field of weed science. Already in a paper of *Weed Technology* published in 1992, they said that weeds, and concomitant decisions about weed control, involve a complex web of biological, economic, social, and aesthetic factors. 'It is, therefore, almost impossible to discuss objectively the introduction of a new tool for weed control without considering each factor separately and collectively' (Radosevich *et al.*, 1992). Because of a lot of uncertainties in such broad terms, and because of a lack of a framework that links scientific disciplines together within the arenas of, in the case of weed science, weed biology, weed technology, and ethics of weed control, our discussion of herbicide tolerant crops, for example, is incomplete. When they revisited the same questions as to potential benefits and concerns about herbicide tolerant crops after the ten year period, they had to admit that: 'It was suggested at that time that if such questions were answered, HTC development would proceed on a more solid foundation, whereas continued uncertainty and criticism would probably result if the questions were not answered ... Questions and concerns posed on decade ago are still pertinent, but current knowledge is still insufficient to address them' (Martinez-Ghersa *et al.*, 2003).

In sum, there are several common points of concerns and perspectives among those reviewed above. First, they acknowledge that agricultural scientists carry a particularly heavy burden of *moral responsibility* for what they do. This is not only because science and technology have significant impacts on the environment and society we (and next generations) inhabit, but also science and technology reflect scientists' values and beliefs, which must be made explicit so that we can understand and, if necessary, dispute them. Second, they agree that science and technology *need to be evaluated in the social, economic and political context*. This is what they mean by saying ethical considerations, which is never limited to utilitarian ethics such as risk-benefit analysis. They are rather critical of unexamined utilitarian calculus. Instead, they insist that ethical considerations need to include a complex of components of relevant disciplines. Third, in order to make such ethical considerations possible, they agree that scientists *need to be involved in dialogues* with those in social sciences and ethics, and with stakeholders in the society. And finally, but not the least, they recognise that most of agricultural scientists and practitioners share a sort of *moral confidence*, which can and should be critically examined. In turn, by overhauling moral confidence, and by probing explicit and implicit goals and values presupposed as being ethical by those scientists, we can overcome the limitation of ethics as is often used without careful consideration.

What remains to be answered is how and on what conditions these agricultural scientists could be able to reflect on the role of their own expertise. A survey conducted by Zimdahl in the US in 1998 and 1999 shows that agricultural scientists have not engaged in ethical reflection as have their colleagues in other disciplines. They have not placed ethics at the core of the curriculum, and dealt with ethics at the margin, or not at all. Agriculture has therefore suffered because of the lack of a carefully articulated ethical foundation (Zimdahl, 2000). As his new book (2006) makes clear, positivist or productionist values are still quite prevalent in agricultural sciences, whereas debate, discussion, teaching and thoughtful deliberation on the ethical dimensions of agricultural science and technology are still exceptional.

In his report on a symposium on agricultural ethics that he helped organise in 1988, Lockwood (1988) said that: 'If the field of agricultural ethics is to realise its potential and if the agricultural and philosophical communities are to address the impending changes in world food production, there is a need for education in public, government, and academic arenas. The development of a symposium on agricultural ethics is an effective method for 'raising awareness' of the imminent need for a consolidation of philosophical and agricultural expertise'. Actually, his goal as the editor for the 1997 special issue was to assist scientists and philosophers in developing a foundation for a continuing, productive, and respectful definition and resolution of ethical issues associated with biological control (Lockwood, 1997). Here is a clue for critical reflection and reformation of themselves, while, as Zimdahl laments, it is still arguably challenging to bring those engaged in agricultural science and technology into dialogue with others who have different views on agriculture. In the next section, we will look into the so-called interdisciplinary approach, not just in research but also in education, how it has been come up with and put forward as a way to make science and technology more reflective on itself.

5. Social learning process, interdisciplinary research, and ethics education

5.1. Agricultural science as a social learning process

Interdisciplinary research, or a fruitful interaction between the different kinds of scientific disciplines, is hampered by the idea that reductive science is more objective and more scientific than the less reductive sciences of complex subject areas. In this regard, Alrøe and Kristensen (2002) attempt to establish a systemic and inclusive research methodology and framework of science through rethinking the role of values in science. Their conceptual analysis can be summarised as follows.

Because agricultural practice involves both social and ecological systems, agricultural research into these socio-ecological systems faces dual challenge of understanding complex agro-ecosystem interactions, and understanding the practices of human actors in social systems. Therefore, agricultural research is inherently framed in a social context, and necessarily involves questions concerning different interests and values in society as well as different structures of rationality and meaning, as endorsed by aforementioned reflective agricultural scientists. Characterising agriculture and agricultural research as such, Alrøe and Kristensen stress a need for rethinking the approaches, processes, and institutional structures of agricultural research. According to them, agricultural research is a systemic science, a science that influences its own subject area, and therefore values play an important role in science. When science in turn plays such a role in the society, the traditional criterion of objectivity becomes problematic. Thus, we need to analyse agricultural science as an interactive, social learning process, in which contextual values and goals (visions) are integral to the very process of scientific knowledge and technological production. Science as a social learning process consists of two meanings.

First is a cognitive process of self-reflexive circle, in which the cognitive process starts from the inside viewpoint, or stance, of the 'actor' within the system; then it moves to the detached viewpoint where the 'observer' views the system from outside; and the observation made from the outside point of view can take effect upon returning to the first viewpoint. Because the outside stance always rests on a specific inside point of view (values), any research that considers itself only a detached 'observer' of the world, without also being an 'actor' to be influenced by and influence on the system, is blind to parts of its own function. In this respect, self-reflection allows research to make a description of the social and intentional, value-laden context of the observations, and when this self-reflection would be made 'objectively', i.e. inclusively and communicatively exposed, we can name it as 'reflexive objectivity'[145]. Second meaning of science as a social learning process is the relation between the different kinds of science. In order for science to function as a common learning process, there is a need for a common framework. By so doing, science can achieve good and valid communication and critique of the results. Given an ongoing, unfruitful disparity between different modes of knowledge production (reductionistic, less reductionistic, or holistic), a reflexive awareness of the social, intentional and observational context of science is crucial. In order to locate different kinds of science for functioning within a transdisciplinary work, Alrøe and Kristensen suggest a distinction between a normative interest and an empirical interest as commonly shared across natural, social and human sciences despite of their different modes of knowledge production. Importantly, they are called two 'aspects' of science, because the one builds on the other and *vice versa*: 'The different kinds of research have the same potential for doing good science, and this view of science can therefore serve as a basis for promoting transdisciplinary systemic research' (Alrøe and Kristensen, 2002).

In a similar manner, Soda (2006) [2000] emphasises that, if agricultural science is to solve the various problems of modern agriculture and in real life, its raison d'être rests in the kinds of values (social goals) it pursues and their validity in the actual social context. Because agricultural science is a value-seeking practical science as such, we should conduct research with the clear specification of its value objectives. In this regard, agricultural science must be transformed from a mere study of biological production to a holistic science of the life. There is a need to build a new mutual relationship between the three scientific and technological spheres of (1) human sciences that strive to understand humans and their societies, (2) the deepening natural sciences that strive to understand the natural world, and (3) the development of practical science based on the progress of the former two and used for the purpose of creatively constructing the human-nature relationship.

5.2. Growing interdisciplinary drive

Meanwhile, interdisciplinary approaches and collaboration among institutions and researchers with different disciplines are encouraged at international (e.g. EU Framework Programs) as well as national (e.g. ESRS Genomics Network in the UK) levels. It is often assumed that interdisciplinary approaches will contribute to more effective innovation and enhanced competitiveness, while they are also expected as a means of solving complex socio-scientific problems and answering questions that cannot be addressed sufficiently by a single discipline (Bruce *et al.*, 2004). Although in this sense, interdisciplinarity is not necessarily regarded as a way to make science and technology more socially robust through getting ethical reasoning incorporated into the scientific knowledge production, and although the role of ethics in the European research policy has served as a source of disagreement (Abels, 2003), practical experiences of interdisciplinary collaboration in European RTD programs might be useful to understand the nature of interdisciplinary integration within and between natural, social and human sciences.

---

[145] This concept also implies that research that operates only as an involved actor fails to be scientific, and therefore any practical idea of the good must build on empirical knowledge of the possible.

There are three comparable categories: multidisciplinary, interdisciplinary, and transdisciplinary approaches (Tait *et al.*, 2002; Bruce *et al.*, 2004; Gibbons *et al.*, 1994). Multidisciplinary research approaches an issue from the perspectives of a range of disciplines, but each discipline works in a more or less autonomous way with little cross-fertilisation among disciplines. Therefore, it involves low levels of collaboration and does not require any changes in the academic worldviews of the researchers themselves. Interdisciplinary research similarly approaches an issue from a range of disciplinary perspectives, but the contributions of the various disciplines are integrated to provide a holistic or systemic outcome, giving rise to intellectual and practical changes among the researchers and their institutions. Although transdisciplinary research, in which interdisciplinary work is accompanied by a mutual understanding of disciplinary methodology and theory, might be an ideal platform for self-reflection on their role and transformation of their disciplines involved (Alrøe and Kristensen 2002), it seeks to transcend the academic disciplinary structure and therefore makes the approach quite exceptional and difficult.

As a matter of fact, it is disappointingly concluded that there are few projects in the EU's 5th Framework Program (FP5) program that seemed by the above-mentioned criteria to be clearly interdisciplinary, particularly in crossing the boundary between natural and social sciences (Bruce *et al.*, 2004). It is stressed that interdisciplinary research does not occur automatically simply by bringing together several disciplines in a research project. There are a lot of difficulties identified, including, among others, the use of different 'languages' (methodologies and theories) that causes problems in communication and mutual understanding. A conscious and time-consuming effort to develop 'discursive collaboration' is not always successful especially due to a 'disciplinary asymmetry' (Marzano *et al.*, 2006), in which social scientists find their involvement just a token in a science-led project and are unhappy with a view among some natural scientists that social science methods are unscientific and inferior. Nevertheless, there is no established training, rewarding and evaluating systems for interdisciplinary research. In spite of the poor achievements of interdisciplinary research projects in the FP5, however, Bruce *et al.* (2004) expect that given continuity of interdisciplinary drive in the FP6 and subsequent programs and increasing interdisciplinary experiences in the research community, the understanding and attempts of interdisciplinarity will be getting spread and the quality of synergetic outputs will increase.

5.3. Expectations for ethics education

Ethical consideration and critical self-reflection might be stimulated through these interdisciplinary research experiences, even if many of the constraints on effective and fruitful communication between the researchers remain to be overcome. In order for interdisciplinary collaboration to work successfully, what need to be enforced further are not only the will and commitment of researchers and their communities (as well as policy makers), but also interdisciplinary education, including ethical reflection, for the next generation of researchers, i.e. students, to get integrated into interdisciplinary culture.

For example, Zimdahl has developed a keen interest in ethics education for students of agricultural sciences, while he himself as a critical agronomist has been struggling for raising awareness of ethical dimensions of weed science and technology. He argues that: today's students are tomorrow's farmers, business people, professors, and policy makers; therefore, they need to understand the ethical dimension of what they do now and may do in the future. If we fail to include true ethical study in their education, they will be still defensive when confronted with an ethical issue and unable to respond except with assertions or 'moral confidence' based on the production agriculture paradigm (Zimdahl, 2000).

A compatibility between the nature of education and ethics is also a strong rationale. Education is not solely about the transmission of information, knowledge or skills. It is as much about the transference of values, about individual teachers shaping, transforming, enriching, and disciplining young people's whole lives (Burkhardt, 1999). Based on experience with teaching a subject of 'engineering for sustainability' and working with adult educators, Yasukawa (2006) has become convinced that developing a capacity for critical reflection on the ethics *of* engineering and technology, not just the ethics *in* engineering and technology, is crucial; however, it is a lifelong and lifewide endeavor, linked very closely with one's identity formation. 'This is why the fundamental question of one's relationship to technology and humanity needs to be asked and reflected on in an ethics education for technologists and engineers so that the future engineers and technologists know not just how to complete an engineering project ethically, but can express who they want to be and how they want the world to be through the work they do'. This is arguably true for those engaged in agricultural science and technology. Her observation is also resonated with Winner's argument for education programs, in which the disciplines in humanities and social sciences are incorporated so as to bring the skill of political savvy and political imagination into engineers and technical professionals (Winner, 1990). Any effort to teach engineering ethics must produce a vital, practical and continuing involvement of scientists and technical professionals in public life, not just as an observer, but as an actor who is shaping and being shaped by the social world. This is really a process of self-formation.

Contrary to many scientists' perception, the teaching of values, and various kinds of moral education take place, not only in ethics courses but also in science classes, although the values instruction in science classes is usually not acknowledged by the teacher and, as a result, students are left with the understanding that the values they are taught are not values (Burkhardt, 1999), while being paradoxically given an unexamined moral confidence. As mentioned before, in the mainstream scientific community, and even in the official ethics community, there is a prevailing idea that ethical considerations are at best extra-scientific, or end-of-pipe, if not simply irrational. Indeed, to attempt to include ethical considerations in routines would require a significant paradigm shift in people's lives, practices, and institutions (Burkhardt *et al.*, 2005). Therefore, teaching ethics should not be just extrapolated to existing curricula, but fully incorporated within education programs. In so doing, we believe, interdisciplinary approaches both in research and education, involving both researchers and students (and public citizens in the case of adult education), will overcome the limitation of ethicisation.

6. Conclusion

Our primary purpose in the paper was to examine and reflect on the role of scientists and technical professionals, while focusing on the role of ethics for bringing them into the process of critical reflection on what they do. This does not mean 'First the Scientist' at all. For scientists to get reflective and critical on the social meaning of their practice, it must start with the social reality at the local, interacting and communicating with those to whom science and technology is to contribute. It is noteworthy in this regard that Soda (2006) [2000] emphasises, in a similar way as many researchers in rural sociology studies (Van der Ploeg and Long, 1994), development studies and science and technology studies (Leach *et al.*, 2005), that the key to resolving many of the problems of today depends on concrete and practical actions in which actors of each local community are aware of the problems and then, by unifying the knowledge derived from their daily experiences and that derived from science and technology, create new schemes, while joining together and cooperating to recreate the local community. In this sense, we have to admit that our focus on research and education programs, whether they are fully interdisciplinary and ethically reflective or not, is just a partial (but still crucial) attempt to reconstruct science and technology.

Finally, as Ruivenkamp (2003) clearly warns, scientists are deeply embedded in the system and would be faced with visible and invisible pressures to operate within the system, while any kinds of education take place within the system, too. In order to reconstruct research and education in the direction of alternatives, we must fight the current hegemonic institutional setting, or 'Biotechnology Project' in the words of Schurman and Munro (in this volume). Gramsci (1971) argues that hegemony is where a politically dominant class maintains its position not simply by force or the threat of force, but also by an ideology to win a sort of consent among various social groups to the social order maintained under the intellectual and moral leadership of the dominant. Hegemony as such is produced and reproduced through a network of institutions, social relations, and ideas outside the direct political sphere. This nature of hegemony makes the social meaning (or cultural horizon, according to Feenberg (1999)) behind a certain technology invisible once it is translated into technical terms. Formal education and academic science system is among others, which function as a hegemonic strategy. Our focus on critical reflection and education for scientists needs to be understood in the context of hegemony and counter-hegemony. Reformation of interdisciplinary research activities and ethics education is an attempt of *counter-hegemonic tactics at the epistemic level* (Hisano, 2005), which must also be complemented with *counter-hegemonic strategy at the institutional level*.

Acknowledgements

The author would like to thank the participants, especially Guido Ruivenkamp and Joost Jongerden for their comments.

References

Abels, G. (2003). The European Research Area and the Social Contextualization of Technological Innovations: The Case of Biotechnology. In: *Changing Governance of Research and Technology Policy* (Eds. Edler, J., Kuhlmann, S. and Behrens, M.), Cheltenham, UK, Edward Elgar, pp. 311-332.

Alrøe, H.F. and Kristensen, E.S. (2002). Towards a Systemic Research Methodology in Agriculture: Rethinking the Role of Values in Science. *Agriculture and Human Values* 19: 3-23.

Bijker, W.E. (2001). Social Construction of Technology. In: *International Encyclopedia of the Social and Behavioral Sciences*, Amsterdam: Elsevier.

Bruce, A., Lyall, C., Tait, J. and Williams, R. (2004). Interdisciplinary Integration in Europe: The Case of the Fifth Framework Program. *Futures* 36: 457-470.

Burkhardt, J. (1999). Scientific Values and Moral Education in the Teaching of Science. *Perspectives on Science* 7(1): 87-110.

Burkhardt, J., Comstock, C., Hartel, P.G. and Thompson, P.B. (2005). Agricultural Ethics. *CAST Issue Paper*, No. 29.

Carson, R. (1962). *Silent Spring*. Boston, Houghton Mifflin.

Chrispeels, M.J. and Mandoli, D.F. (2003). Agricultural Ethics. *Plant Physiology* 132: 4-9.

Feenberg, A. (1999). *Questioning Technology*. London, Routledge.

Gibbons, M., Limoges, C., Nowotny, H., Schwartzman, S., Scott, P. and Trow, M. (1994). *The New Production of Knowledge: The Dynamics of Science and Research in Contemporary Societies*. London, Sage Publications.

Gramsci, A. (1971). *Selections from the Prison Notebooks.* (Eds. Q. Hoare and G.N. Smith), NY, International Publishers.

Haimes, E. (2002). What can the Social Sciences Contribute to the Study of Ethics? Theoretical, Empirical and Substantive Considerations. *Bioethics* 16(2): 89-113.

Haynes, R.P. (1995). Agriculture. In: *Encyclopedia of Bioethics, revised ed.*, Vol.1, (Ed. W.T. Reich), NY, Simon & Schuster Macmillan, pp.101-108.

Hedgecoe, A.M. (2004). Critical Bioethics: Beyond the Social Science Critique of Applied Ethics. *Bioethics* 18(2): 120-143.

Herkert, J.R. (2005). Ways of Thinking about and Teaching Ethical Problem Solving: Microethics and Macroethics in Engineering. *Science and Engineering Ethics* 11: 373-385.

Hisano, S. (2005). A Critical Observation on the Mainstream Discourse of Biotechnology for the Poor. *Tailoring Biotechnologies* 1(2): 81-105.

Hoeyer, K. (2006). 'Ethics wars': Reflections on the Antagonism between Bioethics and Social Science Observers of Biomedicine. *Human Studies* 29: 203-227.

Jackson, S.A. (2005). The Nexus: Where Science Meets Society. *Science* 310: 1634-1639.

Kloppenburg, J. Jr. (1991). Social Theory and the De/Reconstruction of Agricultural Science: Local Knowledge for an Alternative Agriculture. *Rural Sociology* 56(4): 519-548.

Leach, M., Scoones, I. and Wynne, B. (Eds.) (2005). *Science and Citizens: Globalization and the Challenge of Engagement.* London, Zed Books.

Levidow, L. and Carr, S. (1997). How Biotechnology Regulation Sets a Risk/Ethics Boundary. *Agriculture and Human Values* 14: 29-43.

Lockwood, J.A. (1988). Taking Agricultural Ethics to the Forefront: A Practical Guide to the Organizational and Philosophical Issues. *Agriculture and Human Values* 5(4): 96-101.

Lockwood, J.A. (1997). Competing Values and Moral Imperatives: An Overview of Ethical Issues in Biological Control. *Agriculture and Human Values* 14: 205-210.

MacKenzie, D. and Wajcman, J. (Eds.) (1999). *The Social Shaping of Technology*, 2nd ed. Buckingham, UK, Open University Press.

Martinez-Ghersa, M.A., Worster, C.A. and Radosevich, S.R. (2003). Concerns a Weed Scientist Might Have About Herbicide-Tolerant Crops: A Revisitation. *Weed Technology* 17: 202-210.

Marzano, M., Carss, D.N. and Bell, S. (2006). Working to Make Interdisciplinarity Work: Investing in Communication and Interpersonal Relationships. *Journal of Agricultural Economics* 57(2): 185-197.

Mepham, B. (1998). Agricultural Ethics. In: *Encyclopedia of Applied Ethics*, Vol.1, (Ed. R. Chadwick), San Diego, CA, Academic Press, pp. 95-110.

McLean, G.F. (1993). Integrating Ethics and Design. *IEEE Technology and Society* 12(3): 19-30.

Nelson, J.L. (2000). Moral Teachings from Unexpected Quarters: Lessons for Bioethics from the Social Sciences and Managed Care. *Hastings Center Report* 30, no. 1 (January-February): 12-17.

NGI: Netherlands Genomics Initiative (2004). *Ethical, Legal and Societal Aspects of Genomics Research in the EU, Canada and the U.S.*, prepared by W.M. Beer and A.J. A. Jansen for the European Commission FP6-ERA-NET Specific Support Action.

Parker, L. (1995). Breast Cancer Genetic Screening and Critical Bioethics' Gaze. *Journal of Medicine and Philosophy* 20: 313-337.

Paula, L. and Van den Belt, H. (2006). *The Institutionalisation of Ethics in Science Policy: Practices and Impact.* A Report for Work Package 5, Ethics in Food Technologies.

Radosevich, S.R., Ghersa, C.M. and Comstock, G. (1992). Concerns a Weed Scientist Might Have About Herbicide-Tolerant Crops. *Weed Technology* 6: 635-639.

Reiss, M.J. and Straughan, R. (1996). *Improving Nature? The Science and Ethics of Genetic Engineering.* Cambridge University Press.

Roddis, W.M.K. (1993). Structural failures and engineering ethics. *Journal of Structural Engineering* 119: 1539-1555.

Ruivenkamp, G. (2003). Genomics and Food Production: The Social Choices. In: *Genes for Your Food, Food for Your Genes: Societal Issues and Dilemmas in Food Genomics* (Eds. R. van Est, L. Hanssen and O. Crapels), Rathenau Institute. pp. 19-41.

Shrader-Frechette, K. (1992). Agricultural Ethics. In: *Encyclopedia of Ethics*, Vol.1, (Eds. L.C. Becker and C.B. Becker), NY, Garland Publishing, pp. 30-33.

Shrader-Frechette, K. (2001). Agricultural Ethics. In: *Encyclopedia of Ethics*, Vol.1, 2nd ed. (Eds. L.C. Becker and C.B. Becker), NY, Routledge, pp. 45-48.

Soda, O. (2006) [2000]. *Philosophy of Agricultural Science: A Japanese Perspective.* Melbourne, Trans Pacific Press (translated from Japanese version, originally published in 2000).

Stirling, A. (2005). Opening up or closing down? Analysis, participation and power in the social appraisal of technology. In: *Science and Citizens: Globalization and the Challenge of Engagement* (Eds. M. Leach, I. Scoones and B. Wynne), London, Zed Books, pp. 218-231.

Tait, J., Williams, R., Bruce, A., Lyall, C., Grávalos, E., Rodriquez, P., Jolivet, E. and Jorgensen, U. (2002). *Interdisciplinary Integration in the Fifth Framework Program*, Final Report to European Commission, Contact No. SEAC-1999-00034.

Thompson, P.B. (1995). *The Spirit of the Soil: Agriculture and Environmental Ethics.* NY, Routledge.

Thompson, P.B. (1998). *Agricultural Ethics: Research, Teaching, and Public Policy.* Ames, IA, Iowa State University Press.

Thompson, P.B. (2001). The Ethics of Molecular Silviculture. In: *Proceedings of the First International Symposium on Ecological and Societal Aspects of Transgenic Plantations* (Eds. S.H. Strauss and H.D. Bradshaw), Oregon State University, pp. 85-91.

Van der Ploeg, J.D. and Long, A. (Eds.) (1993). *Born From Within: Practice and Perspectives of Endogenous Rural Development.* Assen, NL, Van Gorcum.

Winner, L. (1990). Engineering Ethics and Political Imagination. In: *Broad and Narrow Interpretations of Philosophy of Technology* (Ed. P.T. Durbin), Dordrecht, NL, Kluwer Academic Publishers, pp. 53-64.

Wynne, B. (2001). Creating Public Alienation: Expert Cultures of Risk and Ethics on GMOs. *Science as Culture* 10(4): 445-481.

Yasukawa, K. (2006). Engaging Students in the Ethics of Engineering and Technology. Working papers on learning 2006-2, Department of Education, Learning and Philosophy, Aalborg University.

Ziman, J. (1998). Why must scientists become more ethically sensitive than they used to be? *Science* 282: 1813-1814.

Zimdahl, R.L. (1998a). Ethics in Weed Science. *Weed Science* 46: 636-639.

Zimdahl, R.L. (1998b). Rethinking Agricultural Research Roles. *Agriculture and Human Values* 15: 77-84.

Zimdahl, R.L. (2000). Teaching Agricultural Ethics. *Journal of Agricultural and Environmental Ethics* 13: 229-247.

Zimdahl, R.L. (2002). Moral Confidence in Agriculture. *American Journal of Alternative Agriculture* 17(1): 44-53.

Zimdahl, R.L. (2006). *Agriculture's Ethical Horizon.* Amsterdam, Elsevier.

Part IV. Quality agriculture and networks

European quality agriculture as an alternative bio-economy

Les Levidow

1. Introduction

Agri-biotechnology has been a focus of intense conflict and broad public opposition in Europe. Agbiotech products have also faced a commercial boycott by food retailers, seeking to protect their reputations and sales. Widely known as 'GM food', agbiotech has been widely stigmatised as a symbol of multiple threats – economic globalisation, environmental risks, genetic pollution, biopiracy, etc.

These attacks had a basis in government efforts to promote agbiotech as an instrument of a specific political-economic agenda. European policies have promoted capital-intensive technological innovation as an essential means for more efficient production methods, more lucrative products, and thus a competitive advantage in the global economy. This agenda corresponds to a series of policy slogans – the 'Biosociety' in the 1980s, the 'knowledge-based economy' in the 1990s, and the 'Knowledge-Based Bio-Economy' (KBBE) in the current decade.

In this policy context, Europe has had little discussion of 'alternative agri-biotechnologies', much less as a means to assist the rural poor (the topic of this conference). Within the KBBE framework, 'alternative agri-biotechnologies' could denote non-food uses, e.g. the extraction of industrial or pharmaceutical substances from crops. European agri-food alternatives rarely use the language of technology, though there are exceptions (e.g. Almekinders and Jongerden, 2002: 19-22).

For food agriculture, current alternatives are promoted as quality products and cultivation methods. 'Quality' agri-food products relate to agri-ecological methods, specific territorial characteristics, and/or special aesthetic qualities such as distinctive taste or freshness. Conventional retail chains have increasingly incorporated 'quality' products as a means to appropriate the extra market value from food producers. Sometimes these have been called alternative *agri-food* networks, emphasising quality food products. Without truly alternative networks, quality products may lose their alternative character. Organic producers in particular have undergone such pressures (Smith, 2006; Rigby and Bown, 2007).

By contrast to those conventional markets, alternative agri-food *networks* have more directly linked consumers with producers, as a means to enhance and capture market value for food producers. Product 'quality' is defined by the networks which sustain those relationships, as much as by any tangible characteristics. Examples include the following:

- Quality foods, linked with specific localities, are promoted through retail systems such as *prodotti tipici* in Italy or *produits de terroir* in France.
- *Teikei*, meaning 'cooperation' or 'joint business', is a system of community-supported agriculture in Japan, where consumers purchase food directly from farmers. This model has been extended through distinctive local products (Ikejima and Hisano, 2007).
- Organic products promoted through special retail chains, as well as by the International Federation of Organic Agriculture Movements, www.ifoam.org.

As a short-hand, AAFNs will denote alternative agri-food *networks* in this paper. AAFNs have been promoted through Europe-wide organisations and networks, for example:

- Alimenterra, the European Network for Sustainable Food Systems, where 'sustainable' means diverse agri-food production linked to the seasonality and uniqueness of each locality.
- Coordination Paysanne Européenne (CPE), representing a collective identity as 'peasants' who develop relatively less intensive, more skilled production systems.
- Agri-food networks around the European Social Forum, raising the slogan, 'Another Agriculture is Possible'.

This paper discusses the following questions:

- In EU policy on the 'knowledge-based bio-economy' (KBBE), especially for agriculture, how are some knowledges favored over others?
- How do AAFNs develop resources for a different kind of knowledge-based bio-economy?
- How are different human and natural resources being commoditised in the two models? How do AAFNs provide an extra option – and perhaps an opposition?
- What underlies the conflict between these types of bio-economy?

Answers are brought together in the concluding paragraph and summarised in Table 1 there.

2. Contending bio-economies: analytical concepts

In answering the above questions, the paper will draw upon several analytical concepts: discursive frames, master narratives, naturalised futures and agricultural paradigms.

Controversies can be generally analysed as contending ways to frame issues and to define the problems that need solutions. As a link between ideas and practices, 'framing is a way of selecting, organising, interpreting, and making sense of a complex reality to provide guideposts for knowing, analysing, persuading and acting' (Rein and Schön, 1991: 263). Power can be exercised through discursive accounts of reality, by promoting one future vision over others. New story-lines can bring together actors into coalitions. 'Political change may therefore well take place through the emergence of new story-lines that re-order understandings' (Hajer, 1995: 55-56).

Likewise, master narratives frame societal problems in ways that promote particular solutions. Through such narratives, some possible futures become imaginable or even inevitable, while others are marginalised or excluded. Master narratives often conflate societal progress with technological advance. For example, it is asserted, Europe will fall behind globally in productivity gains unless we capitalise on new technological developments. Such narratives convey imagined and promised European futures, 'in order to justify interventions and pre-empt disruptive public responses', according to a critical report (EGSG, 2007: 75).

Some narratives represent science as the proper basis of government policy, thus equating science with official expertise. In practice, expertise is selected and guided according to dominant policy frameworks. For an alternative approach, a democratically-committed 'knowledge society' would work out how multiple social worlds and visions could creatively interact with a freer, more diverse science (*ibid.*: 78).

Illustrating a master narrative in European policy discussions, biotechnology has been elaborated as a solution to numerous problems since the 1980s. Societal problems were attributed to genetic deficiencies of human and crops, as a basis to propose remedies through a European 'Biosociety' (Gottweis, 1998: 228). European companies could not adequately compete in an increasingly global market, so they must be converted into competitive multinational companies. For this economic aim, alongside a more efficient use of resources, an essential tool would be the application of modern biotechnology to European agro-food industries (*ibid.*: 170).

Since the late 1990s biotech has symbolised a larger 'bioeconomy', likewise promoted as a response to the dual threats of biological vulnerability and economic competition. This bioeconomy depends on new types of capital flows and commodities, whose value derives from expected economic returns. Such biovalue originates from perceived failures of biological vitality; bioscience research identifies illness or morbidity which can be attributed to genetic deficiencies, thus warranting biotechnological treatments (Birch, 2007: 94).

Such narratives have been most prominent in biomedical science. Such research attempts to identify genetic bases for 'normal' variability in bio-vitality, e.g. longevity or sexuality, thus blurring the distinction between illness and health (Rose, 2001: 21). Analogous narratives also feature in agbiotech: genetic deficiencies explain crop vulnerability to pests, disease, adverse climates, etc., while bio-vitality can be found only in laboratory solutions.

Through promised remedies, capitalist accumulation mobilises biological resources, while testing the potential and limits of such appropriation.

> '...that capitalist promise is counterbalanced by wilful deprivation; its plenitude of possible futures [is] counter-actualised as an impoverished, devastated present, always poised on the verge of depletion' (Cooper, 2007: 28).

Natural resources are constructed in the image of laboratory solutions, through narratives of a deficient, vulnerable nature needing correction. That problem-diagnosis actively constructs scarcity (Xenos, 1989), thus stimulating new markets. Some biological knowledges are demarcated as economic assets that can be incorporated into current or new markets. At the same time, broader patent rights manufacture scarcity in these resources by favoring a research focus on patentable knowledge, while deterring further research which may be subject to patent disputes. Discourses of economic competitiveness naturalise research priorities which seek genetic solutions; this naturalisation provides a self-fulfilling prophecy through institutional changes which reinforce the priorities (Birch, 2006: 7-9).

Thus a particular type of 'bioeconomy' is turned into an objective imperative – necessitated by biological morbidity, genetic deficiency, and market forces. Given such problems and threats, solutions must come from laboratory-based knowledge. Otherwise Europe will suffer from productive inefficiency, poverty and backwardness. This combined threat and promise leaves no alternative: 'In this dominant political conception, the economism of globalisation discourse is combined with an authoritarian technological determinism' (Barben, 1998: 417).

Agbiotech research focuses on knowledge-intensive products, e.g. GM seeds, enzymes, fatty acids and amino acids produced by microbiological methods. This industrial appropriation facilitates control from a distance by multinational companies of agricultural activities, e.g. decisions over what crops are grown, how they are cultivated, etc. These knowledge-intensive inputs produce ignorance at farm level, thus creating a new market for information and inputs (Ruivenkamp, 2005: 13-14).

Master narratives promoting agbiotech have intersected with a prior conflict over agricultural futures, especially in Europe. This conflict has been theorised as 'food wars' between the 'Life Sciences Integrated' paradigm versus the 'Ecologically Integrated' paradigm (Lang and Heasman, 2004). In analogous ways, Marsden and Sonnino (2005) theories a competition among three paradigms. As the dominant one, complementing a neoliberal policy framework, the agri-industrial paradigm promotes globalised production of standardised food commodities for international markets. In the post-productivist paradigm, rural spaces become consumption spaces for urban and ex-urban populations. In the agrarian-based rural development paradigm, agricultural production is relocalised, by embedding food chains in highly contested notions of place, nature and quality.

Given these contending paradigms, 'rural space within Europe has become a 'battlefield' of knowledge, authority and regulation' (*ibid.*; also Sonnino and Marsden, 2006). At the same time, synergies can develop among rural activities such as quality agriculture, off-farm sales of products, agri-tourism, recreational activities, and agri-environmental schemes (Marsden and Smith, 2005: 441; Magnaghi, 2005: 96). Thus there can be a complementary relation between the post-productivist and agrarian-based rural development paradigms. Through quality agri-food chains, agrarian-based development has extended to rural-urban links through quality agri-food chains and even to urban agriculture (Jongerden, 2006).

Contending agricultural models can be seen as a specific case of EU narratives on the 'knowledge-based economy' (KBE). Through discourses of threat and opportunity, this concept helps to justify EU interventions into more policy areas, e.g. in the name of defending a European KBE from globalisation. The interventions empower some interests, while disorganising or demoting others (Burfitt *et al.*, 2006). Indeed, they favor neoliberal policy frameworks of economic competitiveness. At the same time, the KBE concept has scope for counter-hegemonic versions:

> '...once accepted as the master narrative with all its attendant nuances and scope for interpretation, it becomes easier for its neo-liberal variant to shape the overall development of the emerging global knowledge-based economy [...] This said, we should not neglect the scope for counter-hegemonic versions of the knowledge-based economy and for disputes about the most appropriate ways to promote it' (Jessop, 2005: 157).

By drawing on the above perspectives, this paper will analyse how the European agbiotech controversy extends prior conflicts over the agri-industrial paradigm. Master narratives promoting agbiotech encounter oppositional narratives of quality agriculture. This antagonism provides threats and opportunities for alternative agri-food networks.

3. EU policy for a neoliberal bio-economy

In European Commission policy, biotechnology has been long promoted as an essential tool for productive efficiency, economic competitiveness, wealth and the quality of life. Such benefits have been uniquely attributed to biotech, implying that only genetic modification techniques can provide them. This possible future has been naturalised as an objective necessity, while turning agbiotech into a symbol and instrument of a wider political-economic agenda.

The 1992 Maastricht Treaty promoted an integration of public and private sectors. By 1990 EC funds for biotech research became dependent upon industry partners committing resources to any project proposal. Research was given a clear economic function, especially in the private sector, with 'more careful attention to the long-term needs of industry ... The most vital resource for the competitiveness of the biotechnology industry is the capacity to uncover the mechanisms of biological processes and figure out the blueprint of living matter', according to managers of the DG-Research Biotechnology Division (Magnien and Nettancourt, 1993). Through such metaphors of mechanical engineering, GM techniques and molecular knowledge were naturalised as the basis of an entire industry.

Policy language blurred any distinction between conventional and GM techniques, as a discursive means to promote the latter. According to the European Commission's White Paper on *Growth, Competitiveness and Employment*, 'biotechnology' has a direct impact on sectors which comprise 9% of value-added in the EU. The global revenues of 'the biotechnology industry' was foreseen to reach 100bn ecu by the year 2000. Given these competitive pressures, 'perhaps only modern biotechnology has the potential to provide significant and viable thrusts....' The entire agro-food industry became discursively 'based on biotechnology' (CEC, 1993: 100-103). Within this narrative, all biological processes and products became economically dependent upon biotech for success in market competition. Therefore government had to remove any obstacles to the maximum exploitation of agbiotech.

In such ways, the 1993 White Paper counseled European adaptation to competitive pressures: 'The pressure of the market-place is spreading and growing, obliging businesses to exploit every opportunity available to increase productivity and efficiency' (CEC, 1993: 92-93). Moreover, Europe 'must exploit the competitive advantages associated with the gradual shift to a knowledge-based economy' (*ibid.*: 58).

This imperative was linked with technological innovations: 'The European Union must harness these new technologies at the core of the knowledge-based economy' (*ibid.*: 7). For the agricultural sector, priority was given to knowledge of genetic characteristics which could enhance productive efficiency or produce high-value substances. Institutional changes were promoted to turn that narrative into a self-fulfilling prophecy, towards making European agriculture dependent upon GM techniques and genetic knowledge.

Such knowledge was literally valorised by broader patent rights. Following a decade-long debate, the 1998 EC Patent Directive allowed broad patents on 'biotechnological inventions', even if they comprise new varieties of plants. Invention was broadly defined: 'Inventions which are new, involve an inventive step and are susceptible of industrial application are patentable even if they concern a product consisting of biological material' (EC, 1998). These broader patent rights gave financial incentives for public-sector research to seek lucrative genetic information and develop GM techniques for production of novel organisms.

A concept from the 1993 White Paper, the 'knowledge-based economy', was extended at the 2000 Lisbon summit of the EU Council. According to its 'Lisbon strategy', by 2010 Europe will be 'the globally most competitive knowledge-based region in the world, capable of sustainable economic growth with more and better jobs and greater social cohesion'. In this master narrative, research and scientific innovation would be the driving force behind wealth creation. Agbiotech has been promoted as a key element of that strategy. When the Commissioner for Enterprise and Industry presented the Commission's new biotech strategy in 2005, he stated that it was his most important goal: 'It is my objective to ensure that we create the conditions so that Europe becomes the natural home for biotechnological innovation'. Conversely, agbiotech has been naturalised as progress, potentially as a self-fulfilling prophecy, while pre-empting alternatives.

This naturalisation has been promoted through a master narrative of future Europe as a Knowledge-Based Bio-Economy, the topic of a major conference held in 2005. This narrative features terms such as Mother Nature's innovations, nature's toolset, biotech pistons, cell factories, food factory, nature's bounty, etc.; these metaphors link biology, mechanics and cornucopias. Industrialisation is attributed to inherent properties of Nature, while any economic activity involving biological material is classified as 'the bio-economy'. GM techniques are associated with specialty products: 'Biotechnology is opening up new possibilities in terms of tailor-made foods targeted at specific consumer needs'. The KBBE narrative has emphasised the extraction of 'renewable bio-resources', within a model similar to the agri-industrial food factory (CEC, 2005) – by contrast to the wider European debate over rural multifunctionality, imagining diverse socio-economic roles of agriculture.

The KBBE initiatives were applauded by the main industry organisation:
> 'Biotechnology can make a major contribution to Europe becoming more sustainable and economically dynamic. There is a whole new industry just emerging that can develop these clean and competitive materials' (EuropaBio, 2005).

Here 'clean' denotes an input-output efficiency of resource usage.

Around the same time as the KBBE 2005 conference, the OECD launched a global agenda for a bioeconomy. This meant economic activity 'which captures the latent value in biological processes and renewable bioresources to produce improved health and sustainable growth and development' (OECD, 2005: 5). According to a subsequent report from an expert group, the main driver is new biological knowledge: 'The bioeconomy is made possible by the recent surge in the scientific knowledge and technical competences that can be directed to harness biological processes for practical applications'. Potential benefits may be lost or delayed unless government decision-making procedures are adapted to those rapid advances; product safety remains the only grounds for any doubt to be overcome (OECD, 2006: 3).

Public need supposedly drives a bioeconomy in the agriculture sector. According to the OECD expert report:
> 'The main drivers for using biotechnologies in the agriculture sector are increasing population (in developing countries) and rising standards of living creating demand for higher input foods, products and services, along with urbanisation pressures on land availability and the negative side-effects of trying to increase production via traditional means. Biotechnologies allow crops to be grown in less favorable conditions, and help meet consumer demands on food quality, e.g. regarding pesticides and shelf life. Biotechnology can also be used to transform plants into 'factories' that can produce everything from modified foods to commodity chemicals' (OECD, 2006: 7-8).

From the problem-definition of sub-optimal productivity, traditional agricultural methods are portrayed as inefficient – by contrast to GM crops, which enhance productivity and add market value. 'Increasing value will be added at the level of primary production – though it is by no means certain whether the proportion of the value taken by the primary producer will also change...' (*ibid.*). This euphemistic phrase acknowledges doubts about the social distribution and benefits of extra market value, as if any societal inequities were incidental. OECD reports downplay commercial drivers – e.g. strategies for the agri-food chain to appropriate more value from farmers, even to control their production choices and methods.

The KBBE pervades the Commission's Framework Programme 7, featuring this definition:
> 'The term 'bio-economy' includes all industries and economic sectors that produce, manage and otherwise exploit biological resources and related services, supply or consumer industries, such as agriculture, food, fisheries, forestry, etc.'

The KBBE provides a general code for the thematic priority on 'Food, Agriculture, Fisheries and Biotechnology'. For example, GM crops for industrial uses are linked with trade liberalisation, thus modelling the global South as an agricultural factory for exporting bio-resources. New opportunities for GM crops are sought through non-food uses (CEC, 2006). Framework Programme 7 offers some funds for organic agriculture, seen as a minor exception to the dominant paradigm, though not for other alternatives.

As a policy narrative, then, the KBBE selectively favors some knowledges as if responding to societal needs or objective imperatives. This narrative gives priority to laboratory knowledge of genetic characteristics which can be marketed for higher-value inputs or outputs. Extending the earlier policy of the European Commission, the KBBE links agbiotech with neoliberal agendas – for extending industrial agriculture, commoditising natural resources through laboratory knowledge, marketising public-sector research and providing new regulations to facilitate such changes. The drive to maximise profits through intensification goes hand-in-hand with a flexible labour market in the agricultural sector, promoting super-exploitation of migrant labour in particular (e.g. Lawrence, 2004).

This agenda has appropriated the language of sustainable agriculture. GM crops will promote the 'sustainable intensification of agriculture', according to Novartis. This agri-industrial model emphasises input-output efficiency, as a basis for individual producers to compete more effectively in global markets. Regulatory standards have facilitated approval of GM crops which extend the normal hazards of intensive monoculture, especially by supplementing the 'pesticide treadmill' with a genetic treadmill (Levidow, 2005).

Such links between agbiotech and neoliberal agendas have provided a vulnerable target for European opponents (Levidow, 2008). Anti-biotech protest eventually limited the commercialisation of GM products, especially when European supermarket chains excluded GM grain from their own-brand products in the late 1990s. At the same time, opponents have demanded and promoted alternatives to intensive industrial agriculture.

4. European quality agriculture: drivers and constraints

Long before agbiotech became controversial in Europe, many farmers were developing 'quality' alternatives to the agri-industrial system of standard bulk commodities. These alternatives gained impetus from a general European crisis of conventional intensive, productivist agriculture. This crisis features various drivers for quality alternatives and constraints upon their development.

Farmers have been trapped in a technological-economic treadmill, through dependency upon a series of commodity inputs in order to maintain or increase productivity for competitive advantage. Farmers have submitted to this treadmill and/or had greater difficulty in maintaining their income. Peripheral, less favored rural regions have faced threats to their livelihoods, driving farmers into debt and poverty. A technological determinism drives farmers into dependence upon market pressures. By contrast, economic success has been gained by innovative peasant practices which maintain a distance from such pressures (Van der Ploeg, 2003).

As a positive basis for alternative markets, consumers have had many reasons to prefer 'quality' food. Globalised, agri-industrial economies have anonymised the relations of production. Consumer distrust has resulted from various food scandals – e.g. *Salmonella*, dioxin residues, pesticide residues, and especially the 'mad cow' crisis (bovine spongiform encephalopathy or BSE). Beyond consumer safety, there has been greater public sensitivity to environment, health and rural livelihoods – under threat from productivist, globalised food systems. Alternatives have gained greater interest and support, especially since the BSE crisis.

All these issues have intersected with a prior debate over how to reform the Common Agricultural Policy (CAP), which has generally spent about half the EU budget. Impetus for CAP reform came mainly from WTO rules, which prohibited agricultural subsidy linked with production. Until the early 1990s CAP funds were spent mainly on price support to guarantee farmers a minimum price for their products; the more they produced, the greater the subsidy they received. In 1995 the EU started to pay rural development aid, designed to diversify the rural economy and make farms more competitive. Since the late 1990s reforms have partly decoupled subsidies from production levels, instead linking payments to food safety, animal welfare, and environmental standards.

In a 1997 proposal for reform, *Agenda 2000*, the environment was broadly defined to encompass the overall socio-economic effects of agriculture, not simply agrochemical pollution. The proposal emphasised rural livelihoods, the quality of food production, and its 'environmental friendliness'. In passing, it suggested: 'The development of genetic engineering, if well controlled, could enhance production but may raise questions of acceptability to consumers' (CEC, 1997: 27, 29).

By the late 1990s *Agenda 2000* aimed to secure 'a multifunctional, sustainable and competitive agriculture', partly as a means to justify extension of subsidies. According to the Commission, agriculture can harm the natural environment, but 'abandonment of farming activities can also endanger the EU's environmental heritage through loss of semi-natural habitats and the biodiversity and landscape associated with them'. Consequently, 'The CAP's objectives include helping agriculture to fulfil its multifunctional role in society: producing safe and healthy food, contributing to sustainable development of rural areas, and protecting and enhancing the status of the farmed environment and its biodiversity' (CEC, 2003: 2; also in EU, 2000).

The Commission also made food quality the new key to farming policy. A rural development policy encouraged farmers to diversify their production and marketing: 'Europe is known for the diversity of its farming and its agricultural products, which derive from its natural environment and farming methods developed over centuries'. Commission policy has favored 'the freedom to farm to market demands', i.e. to produce according to consumer preferences. This means contradictory pressures; farmers are expected to be 'efficient and sustainable', while guaranteeing quality, food safety and environmental protection (CEC, 2004: 1).

Those policies indicate a shift from quantity to quality, potentially towards an agrarian-based rural development, yet the CAP still favors agri-industrial methods. Subsidies are still spent mainly on direct subsidies according to the land area and animal units in production. Farmers can also benefit from agri-environmental schemes, especially by removing the least productive or least profitable land from production, while continuing intensive cultivation methods on the rest. Subsidy goes mainly to the biggest farmers, i.e. large agribusinesses and hereditary landowners, who receive the largest proportion of the funds.

Consequently, the CAP reform maintains social and geographical inequities. Half of all direct payments go to the largest beneficiaries in the more productive and competitive areas, such as the Paris basin, Lower Saxony and East Anglia. These are mainly export-led farms, contracted to multinational firms (Sarasuà and Scholliers, 2003). Price support measures of the CAP give organic farms 20-25% less benefit than comparable conventional farms (Häring *et al.*, 2004). The demand exceeds supply for locally produced organic products, thus generating lower-cost imports.

Despite significant demand for quality agriculture, then, the agri-industrial paradigm still prevails in European policy. According to a proponent, farm subsidies should be reduced and transferred to 'funding to sustain and enhance the environment', while 'ensuring that environmental regulations do not stifle global competitiveness' (Haskins, 2002). Moreover:

> '... where European agriculture can be competitive, this competitiveness should, within environmental limits, be maximised. Where it cannot be competitive, farming per se should be downgraded behind good environmental husbandry as the linchpin of a subsidy/welfare system' (*ibid.*: 7-9).

In that scenario, subsidies would become conditional upon measures for environmental conservation beyond agriculture. As the EU lowers the support prices, farmers would need to produce competitively at world prices. In this way, trade liberalisation complements the further industrialisation of agriculture.

Small-scale producers have attacked that agenda, especially the plan for lower support prices. This policy would make farmers more dependent upon direct payments for their overall income. This 'will benefit agri-industry and the distribution sector rather than consumers', argued the Coordination Paysanne Européenne (CPE), representing farmers that rely on relatively less-intensive production systems. Under the reform, European taxpayers will continue to 'pay huge subsidies to huge farms, while driving the small ones out of business [and] support increasingly industrial farming methods, to the detriment of employment and the environment', according to the CPE (1999).

In the *paysan* view, the dominant reforms encourage large-scale farms to continue their intensive methods and (at best) to steward an 'environment' outside farming. This would 'accelerate the disappearance of multi-functional family farms'. *Paysans* advocate instead de-intensification measures, based on 'remunerative agricultural prices and sustainable family farming, with multiple benefits for society' (CPE, 2001). In other words, multi-functional skills should be sustained within farming communities, not simply outside agriculture. Thus the Coordination Paysanne Européenne counterposes a rural development paradigm to the agri-industrial paradigm.

Those proposals were echoed by the European coordination of consumer cooperatives:
> 'Social criteria for sustainable development means that the rural policies should take into account the issues of employment, cultural diversification, regional development, living rural areas, etc. From the Consumers' point of view, the social criteria also imply that the CAP has to serve as a way to improve food quality and safety' (Euro Coop, 2002).

Although the CAP reform claims to support 'sustainable agriculture', tensions continue between its incentives and quality products. In some countries, dependence on the CAP perpetuates industrial agriculture. In Italy, success in creating quality networks corresponds to the degree of distance from the CAP, its certification system and its bureaucratic criteria. Thus quality networks need to defend their spatial and social boundaries from the CAP (Marsden and Smith, 2005: 449).

The CAP reform extends a split between two agricultural models:
> 'It seems likely that we are headed for a two-track agriculture. One track is a largely "competitive" free-market industrialised agriculture that can only survive in the EU and in large parts of the U.S. with massive subsidies, which, to be acceptable to the WTO, have to be recast as direct supports, independent of production. The second track is a small, designer quality/organic sector that produces highly priced niche goods for those who can afford it and who wish, for health, environmental and taste reasons, to escape the dangers of mass-produced fare' (Kuper, 2007: 78).

Although these models may co-exist, the quality sector faces both threats and opportunities from the agri-industrial system, as analysed in the rest of this article.

5. Re-localising agri-production through quality

Since the 1990s European agri-food systems have developed an 'economy of quality', by analogy to the viniculture system of *appellation d'origine contrôlée*, whereby local characteristics provide a basis for long-distance markets. This economic model differentiates among qualities rather than increasing productive efficiency. Such innovations valorise diverse qualities whose market value has a symbolic, immaterial character as well as a basis in material resources; their market value remains dependent upon consumer satisfaction and thus vulnerable to reputational crises (Allaire, 2002: 172-176).

Diverse notions of 'quality' include the following: an identifiable place of origin, a regional reputation, aesthetic characteristics of a product, nutritional quality, social justice, etc. (Sonnino and Marsden, 2006). Through such qualities, small-scale producers attempt to differentiate their products as a means to obtain added value, rather than compete for ever-higher productivity of global commodities (Ilbery and Kneafsey, 2000).

Quality also relates to methods which re-peasantise European farming (Van der Ploeg *et al.*, 2000: cited in Goodman, 2004: 6-8). Retro-innovation appropriates traditional methods through a backward-forward technological adaptation. Sources include traditional types of animal slaughter, curing techniques, pest management, etc. (Marsden and Smith, 2005: 450). For example, agronomic changes can mean farming in more economically and ecologically ways, by using on-farm resources rather than commodity-inputs, thus also reducing costs. These methods link two aims that were previously competing – environmental protection and farmers' income. Not simply reducing pollution, this means recasting the 'environment' as cultural values, both within and beyond agriculture. As another strategy, farmers can also produce high-quality products which add value to the market price (Goodman, 2004).

Based on special cultivation methods and/or product characteristics, diverse 'qualities' are socially validated through alternative food supply chains (Murdoch *et al.* 2000: 122). These have linked producers more closely with consumers. In this way, products can gain market value and allocate this value to primary producers. Short supply chains depend upon direct local links (Renting *et al.*, 2003). As means to connect producers with distant consumers, alternative supply chains also depend upon territorial quality labels; they develop 'spatially extended networks, which are selling brands, labels and seriously commodifying their culinary repertoires' (Marsden, 2004: 138-39).

Short supply chains have been stimulated by local government policies. Many large cities (London, Rome, Munich, Copenhagen, New York, Vienna) or regions (Wales, Andalusia, Tuscany) have official strategies which link support for nearby farmers with general objectives for the whole population (e.g. water quality, biodiversity, health). Those policies share common principles: namely, that market signals from public or private sectors can effectively complement subsidies. Short supply chains are a vector for those signals, especially through regular box schemes, as well as public purchase for catering in schools, hospitals, elderly homes, etc. In a detailed case study, bread production has been analysed as a short-supply chain; this revealed the expansion, specialisation and concentration process within the Austrian food system, as well as its impacts on landscape (Penker, 2006).

'Local' food provides a general basis for short-supply chains. They have linked local with quality food, especially for three characteristics: freshness, taste and provenance. The latter relates to trusted local sources, which could be extended beyond a locality through trusted networks (Ilbery *et al.*, 2005b: 7).

Alternative agri-food networks (AAFNs) re-localise food networks by linking food more directly with local farming practices, rural nature, landscapes and resources. AAFNs are territorially based networks, dependent upon active involvement of diverse actors – farmers, food processors, local distributors, local NGOs, restaurants, consumers, local and regional authorities – who interact mainly through informal contact (Roep, 2002; Renting et al, 2003). In the UK such initiatives are located mainly in rural areas, though concentrated in specific areas near urban centers and access to particular trunk roads, as well as near a specific geography of farming types and alternative lifestyles (Ilbery *et al.*, 2005a). In Brittany AAFNs are located more in peri-urban areas, which indicate a new paradigm in town-countryside relations, as well as an experimental arena for sustainable development practices.

There are divergent, even competing definitions of quality, both between and within countries (Renting *et al.*, 2003). In France, Italy and Spain for example, AAFNs build on regional traditions and direct sales to consumers. In southern Europe more generally, quality is shaped by the production context, e.g. culture, tradition, terrain, climate, local knowledge system. In northern countries like the UK, Germany and the Netherlands, 'quality' emphasises environmental sustainability or animal welfare, as well as retailer-led forms of marketing (Sonnino and Marsden, 2006). In such northern European countries, AAFNs reconcile agricultural production and environmental protection along lines which can be theorised as ecological modernisation (Evans *et al.*, 2002). The latter can mean an emphasis on less-intensive methods rather than distinctive quality products.

Some alternatives illustrate a general strategy of territoriality, whereby a locality builds upon a distinctive history and identity, as an approach to sustainable development. This involves 'positive relationships between the three components whose reciprocal interactions produce territory: the natural environment, the urban environment, and the social and human environment'. In this strategy, agricultural producers expand their roles towards the production of common goods, e.g. hydro-geological conservation, land reclamation, rural tourism, craftsmanship, etc. (Magnaghi, 2005: 83, 87).

Some Italian localities illustrate convergence between agrarian-based and post-productivist development models. Organic farming has increased through agri-tourism, rural guesthouses, etc., as well as from urban interest in educational projects linking quality agriculture, local cultures and cultivars, agri-parks, etc. (*ibid.*: 96). A viniculture cooperative has returned to organic methods, which provide a basis for an agri-teaching project, agri-tourism and recreational activities. Italian livestock and milk producers have proposed a territorial strategy for alternative agricultures; this would develop short production-consumption chains, with quality labels indicating the place of origin, and animal breeding methods which respect animals. This means a real economy rather than a fictitious one (Dalla Costa and De Bortoli, 2005).

In all those ways, new socio-economic networks develop 'alternative geographies of food production' (Whatmore and Thorne, 1997; Maye *et al.*, 2007). Their success depends upon political alliances between farming and other interest groups in the agri-food chain, as well as with consumer and environmentalist interests around issues of food quality, safety and ecology. Such alliances have challenged the policy consensus around agri-industrial productivist systems (Whatmore, 1994: 59-60).

6. Oppositional roles of AAFNs

Through such alliances, alternative agricultures have been increasingly counterposed to industrial agriculture. In some cases, 'quality' schemes are promoted as a form of opposition.

> 'Here labels are employed to draw attention to the environmental, social and distributional processes associated with particular products, and to distance them from the perceived negative consequences of product standardisation, mass marketing, environmental degradation, and health and safety concerns. For example, organic, bio-dynamic or integrated farming and fair-trade labels, together with a number of "sustainable farming" schemes in Europe and North America, are driven largely by such concerns and are proselytised as alternatives to more classic forms' (Ilbery *et al.*, 2005a: 120).

In some EU member states, some farmers' organisations have played a major role in opposing agri-industrial systems. Europe-wide opposition to agbiotech came initially from the Coordination Paysanne Européenne, defending farmers' skills and quality production along lines much broader than organic agriculture (Seifert, this volume). *Paysan* activists have drawn analogies between *paysans* in Europe and the global South; they all seek independence from the global market for standard commodities, symbolised by GM crops (Bové and Dufour, 2001). That conflict was extended through a series of Europe-wide conferences linking regional authorities and stakeholder groups.

The first major conference on 'GMO-free zones' was held in Berlin in January 2005. Organised by the Foundation for Future Farming, the conference criticised the European Commission policy on the 'coexistence of GM, conventional and organic agriculture'. Several regional authorities there linked 'GMO-free zones' with food sovereignty, quality labels on food products and regional-cultural biodiversity. With the slogan, 'Our Land, our Future, our Europe', their charter identified GM crops as a threat to 'sustainable and organic farming and regional marketing priorities for their rural development' (FFA, 2005). In particular:

> 'Most European regions have made the promotion of sustainable and organic farming
> and regional marketing priorities for their rural development... Most Europeans don't
> want GM-food. To serve this demand is part of a region's food sovereignty and an
> important economic chance. Regional authorities must be able to protect quality labels,
> purity standards, organic production and designations of origin at competitive prices'
> (FFA, 2005).

At a subsequent conference in Florence, speakers more explicitly promoted a geopolitical alliance for a 'sustainable' future agriculture against agbiotech. As conference host, the Tuscany Regional President linked the precautionary principle, zero tolerance for the presence of GMOs, and uncertainty about their compatibility with environmental protection:

> 'We wish to avoid any standardisation of products which no longer have anything to do
> with their place of production. In Europe there must be room for a model of agriculture
> which is based on a genuine identity, cultural characteristics, high-quality GMO-free
> products' (AER/FoEE, 2005).

The Florence conference resulted in 'The charter of regions and local authorities of Europe on the coexistence of GMOs and conventional and organic crops', which in turn started the Network of GMO-free Regions. According to the charter, specific 'coexistence' plans would be based on in-depth feasibility studies examining the environmental, socio-economic and cultural impact of GMOs. Areas could be designated as 'GMO free' in order to protect any added value of certified quality products.

A larger conference broadened the network for alternative futures. Entitled 'Safeguarding Sustainable European Agriculture: Coexistence, GMO free zones and the promotion of quality food produce in Europe', the conference was sponsored by the Assembly of European Regions and Friends of the Earth Europe. It aimed 'to define the most appropriate EU legal framework for an efficient coexistence regime' (AER/FoEE, 2005). Moreover, local environments were framed as cultural-economic assets under threat from GMOs. In their declaration, the organisers sought:

> 'To allow regions to determine their own agricultural development strategy, including
> the preservation and development of regionally adapted genetic resources and the right
> to prohibit GMO cultivation' (*ibid.*).

At the conference numerous regional representatives elaborated that agri-development theme, which was counterposed to agri-industrial methods including GM crops. They described their 'natural' environment or special cultivation methods as a basis to market local products and services. According to a speaker from southwest England, for example, their local authority is committed to 'treating the environment as a highly valuable capital asset to be managed intelligently for long-term economic benefit' (FoEE, 2005). According to a report on the conference, the speakers had emphasised 'how their local specialised agriculture was a precious resource that plays a vital role in marketing their region' (*ibid.*: 15).

An even larger conference promoted 'GMO-free regions' to a broader public during International Green Week in January 2006 in Berlin. By this time, 'GMO-free' declarations had come from more than 160 regions, 3500 municipalities and local authorities, and tens of thousands of farmers in Europe. According to the Network, they were reclaiming their rights to local and regional self-determination – with regard to their landscapes, eco-systems, agricultural practices, food traditions and future economic development. Moreover, farmers and food processors cooperated to find sources of non-GM animal feed for their GM-free animal products, thus increasing the pressures to segregate distinctive markets for grain (AER, 2006).

In those ways, a new coalition has sought a competitive advantage for alternative agricultures. Their counter-narrative links several themes: precaution, environmental risk, socio-economic regionalism, market competition, consumer choice, rural development and *paysan* identity. Through these discursive links, a new coalition brought together diverse stakeholders: farmers, agronomists, grain traders, regional administrators, politicians, consumer organisations and some early opponents of agbiotech.

Those initiatives were followed up with a stronger emphasis on alternatives for agricultural biodiversity and rural development. The 2007 'GMO-free' conference emphasised these threats and opportunities:

'The erosion of biodiversity, independent farming, and regional quality food production arising from the exclusive control of seed by fewer and fewer companies. The challenges for rural development, biodiversity, food culture and food security arising from the global competition between food and fuel production' (GMO-free regions, 2007).

Speakers there demanded greater access to genetic diversity, traditional seeds and farmer-bred varieties, as a means to improve quality and pest resistance on the local level. They attacked European Commission policy for limiting such access, while instead favoring novel seeds from GM and hybrid techniques: 'Seeds grow from the bottom up. Laws usually come from the top down', according to Hannes Lorenzen, speaking for the Green Group in the European Parliament, as well as for PREPARE, the partnership for rural Europe network. In his view, the Commission's research policy favors research institutes, seed specialists and gene banks, towards conserving genetic diversity mainly as resources for genetic modification technology (*ibid.*).

Organic agriculture has become an arena for struggle between different agri-food models, which take the form of contending standards. When the EU Agriculture Council adopted new standards in June 2007, several issues were contentious. 'GM' labelling would not be required unless GM material exceeds 0.9%, i.e. the same as for conventional food. Labelling must specify the place of production but not the producer's name. Organic producers must quantitatively analyse their use of conventional agricultural inputs and materials. According to the Coordination Paysanne Européenne, that costly requirement will penalise small-scale producers, while disfavoring natural agri-inputs and traditional seeds. Overall the EU policy was putting organic agriculture 'in danger of contamination and industrialisation', claimed the *paysan* critics (CPE, 2007).

On economic as well as environmental criteria, quality alternatives have been counterposed to agbiotech by an NGO critique, *The EU's Biotechnology Strategy: mid-term review or mid-life crisis?* This report emphasised the socio-economic benefits of organic agriculture and agri-environmental schemes, on the one hand, by contrast with agri-food biotech, on the other. Benefits were quantitatively compared in terms of industrial competitiveness, market diversity, resource impacts and job creation. On all those parameters, agbiotech have provided less benefit, thus conflicting with the strategic objectives of the Commission's Lisbon agenda; by contrast, organic farming within a rural development policy provides a competitive alternative (FoEE, 2007). That report focused on organic agriculture because relevant statistics were more readily available than for quality alternatives in a broader sense.

In sum, alternative agricultures have been increasingly promoted in opposition to agri-industrial methods. In particular, agbiotech has been cast as both a liability and diversion of resources. This conflict arises most sharply wherever quality alternatives gain local support (Levidow and Boschert, 2008). GM crops are portrayed as a threat to the image of 'quality' products and their regional sources, as well as a diversion of resources; conversely, 'GM-free zones' brand entire regions as sources of quality.

7. Conclusion: contending bio-economies

EU policy features a master narrative of the Knowledge-Based Bio-Economy (KBBE), which marginalises the 'quality' narratives of alternative agri-food networks (AAFNs). Each narrative promotes contending paradigms for rural space. Each has different ways of defining agricultural problems, devising solutions, relating producers to consumers, adding value and commoditising resources, as shown in Table 1.

The KBBE portrays traditional agri-methods as backwards and less efficient. By contrast, AAFNs portray agbiotech as an economic liability, societal threat and illusory promise. Wherever AAFNs gain local support, sometimes backed by national or regional governments, these different narratives are in economic competition, even in political conflict.

Within an agri-industrial paradigm, the KBBE favors capital-intensive technological innovation as the reference point for relevant knowledge. Through its circular reasoning, agricultural problems arise from depleted agricultural resources, productive inefficiency, a morbidity of weak crops, etc. – whose remedy lies in agbiotech products (cf. Cooper, 2007; Birch, 2007). Their success depend upon neoliberal policies which reshape European institutions – for commoditising natural resources through laboratory techniques, restricting access through broader patents based on 'invention' claims, creating new forms of scarcity, intensifying competition for productivity, and marketising public-sector research.

These neoliberal policy changes are promoted and naturalised along with the promise of biotechnological remedies. Thus the KBBE promotes a self-fulfilling prophecy through institutional changes favoring some knowledges, especially those which can be privatised, while pre-empting alternatives. Within this paradigm, AAFNs remain an extra, marginal option – and perhaps a backwards, obsolete one.

The KBBE narrative intersects with a prior conflict over priorities for European agriculture and government subsidy. At issue has been how to enhance the competitiveness of European agriculture; whether to sustain or supersede intensive agriculture; how to enhance the quality of food products, farmland and regional development. In reforming the Common Agricultural Policy, the system for EU subsidy, new criteria aimed to accommodate diverse agricultural futures; yet CAP reform has extended conflict between the agri-industrial versus agrarian-based development paradigms. This antagonism arises from political-economic threats to quality alternatives, which undergo pressures of incorporation and marginalisation by conventional agri-food chains.

Table 1. Contending bioeconomies, their narratives and coalitions.

		KBBE: knowledge-based bio-economy	AAFNs: alternative agri-food networks
Narrative	Institutions	European Commission, Europabio, industrial farmers (COPA), European People's Party MEPs, OECD.	Coordination Paysanne Européenne (CPE), environmental NGOs, Assembly of European Regions, Alimenterra, Green MEPs.
	Paradigm	Agri-industrial development.	Agrarian-based rural development (including post-productivist too).
Issues	Agricultural-economic threats	Inefficient farm inputs and outputs keeping agriculture at a competitive disadvantage in the global economy.	Standardised commodity production, economic-technological treadmill, farmers' dependence on multinational companies.
	Agri-environmental weakness (diagnosis)	Genetic deficiencies leaving crops vulnerable, inefficient, etc.	Intensive monoculture attracting pests and diseases, while undermining local resources.
	Knowledge and resources as assets	Crop improvements – for more efficient production, for defence against environmental threats, for extraction of valuable substances, etc. – thus realising the cornucopian potential of nature.	Local resources as bio-vital strengths, e.g. biological methods of pest control, immaterial environmental characteristics, aesthetic qualities of food products, and farmers' skills in using these resources.
	Value-added sources	Proprietary knowledge from laboratory science and as main basis of added-value inputs and outputs.	Quality agri-food chains as means for primary producers (especially farmers) to gain from the value that they add.
	Markets	Global value chains of standard commodities with flexible sourcing.	Consumer-producer links through alternative networks, quality reputations and collective knowledge.
	Subsidy criteria	Withdraw subsidy from price support, thus stimulating and favouring productive efficiency, while shifting subsidy to non-agricultural activities.	Use subsidy to maintain high prices that can support rural livelihoods from diverse, quality agriculture.
	Agriculture to sustain what?	Sustaining competitiveness in global markets and agri-industrial investment.	Sustaining local community networks through quality processes and products.

MEPs: members of European Parliament.
COPA: Committee of Agricultural Organisations of the EU.

AAFNs promote high-quality, high-skill agri-production, which can also provide a basis for eco-tourism. AAFNs depend upon cooperative social networks which link food producers with consumers. These networks give 'quality' meanings to immaterial characteristics of agricultural processes and products. Here 'quality' depends upon commoditising various local resources which symbolise bio-vitality, as a commons to be developed through communities and collective knowledge. Civil society actors are mobilised around new supply chains – potentially involving 'traditional' methods, aesthetic qualities of products, socio-political commitments beyond economic interests, and local networks. AAFNs re-localise production through short supply chains, while also providing a specific territorial basis for longer-distance specialty markets.

As defined through AAFNs, quality agriculture implies an alternative type of knowledge-based bio-economy. These networks create, add and capture market value for the benefit of primary producers, especially farmers, by contrast to the technological-economic treadmill of agri-industry. Quality agriculture valorises local resources through diverse inputs, outputs and societal benefits. Agricultural labour is re-professionalised, while products are re-differentiated, in ways which build local reputations and networks. This can involve retro-innovation, e.g. re-appropriating traditional methods. By contrast, the agri-industrial system may re-professionalise labour and re-differentiate products in ways which favor laboratory-based, proprietary knowledge.

In all those ways, agri-industrial systems and AAFNs promote contending bio-economies. Beyond simply an alternative, AAFNs have been turned into a counter-narrative: 'Another agriculture is possible' – and necessary. AAFNs challenge the dominant neoliberal form of capitalism, if only in order to avoid being incorporated or marginalised. Alternatives portray agbiotech as a multiple threat of symbolic contamination, globalisation, economic competition and corporate political domination over government policy.

Through regional authorities, a Europe-wide 'GM-free' network brands regions as sources of quality agri-food products, by contrast to agri-industrial systems. This counter-narrative challenges the supposedly objective imperatives of global competition for bulk commodity production. Through this challenge, at once discursive and practical, new alliances can play an oppositional role.

In this struggle for hegemony between two forms of bioeconomy, policy coalitions operate across several arenas: narratives diagnosing agricultural problems in ways which favour specific future developments; rural development strategies for land use, employment and communities; subsidy criteria of the CAP and national implementation; markets structuring relations between production and consumption; rules on GM labelling, segregation and coexistence; civil society actors and their roles. The outcome will depend upon struggles across all those arenas, with all their EU-wide diversity.

Questions for further analysis:
- What are the dynamic relations between competing bio-economies in the agri-food sector?
- How do AAFNs stimulate and validate 'quality' characteristics of agri-food products?
- How do such products undergo pressures to be incorporated or marginalised by conventional food chains?
- How do AAFNs avoid or resist those pressures, e.g. by opposing agri-industrial systems?
- What are the prospects and limits of such opposition?

Acknowledgements

This paper draws upon a background literature survey from a research project, 'Facilitating Alternative Agro-food Networks (FAANs): Stakeholder Perspectives on Research Needs', coordinated by the Inter-University Research Center for Technology, Work and Culture IFZ (http://www.ifz.tugraz.at). It involves partners in several member states, with funding from the European Commission, during 2008-2010. It will coordinate partners in several member states, with funding from the European Commission, starting in spring 2008. Thanks for comments on this paper from Sandra Karner of the IFZ, as well as from Keane Birch, Melinda Cooper, Bernhard Gill, Steve Hinchliffe, Joost Jongerden and Bron Szerszynski. Also helpful were comments at the conference on 'Reconstruction Agro-Biotechnologies for Development', held in November 2007 in Kyoto.

References

AER/FoEE (2005). Safeguarding Sustainable European Agriculture: Coexistence, GMO free zones and the promotion of quality food produce in Europe, 17 May conference, www.a-e-r.org.

AER (2006). 2nd Berlin conference of GMO-free regions, Assembly of European Regions, www.a-e-r.org.

Almekinders, C. and Jongerden, J. (2002). On Visions and New Approaches: Case studies of organizational forms in organic plant breeding and seed production. Working Paper, Technology and Agrarian Development, Wageningen University. Available at: http://library.wur.nl/wasp/bestanden/LUWPUBRD_00321165_A502_001.pdf

Allaire, G. (2002). L'économie de la qualité, en ses secteurs, ses territoires et ses myths. *Géographie, Économie, Société* 4(2): 155-180.

Barben, D. (1998). The political economy of genetic engineering. *Organization and Environment* 11(4): 406-420.

Birch, K. (2006). The neoliberal underpinnings of the bioeconomy: the ideological discourses and practices of economic competitiveness. *Genomics, Society and Policy* 2(3): 1-15, www.gspjournal.com.

Birch, K. (2007). The virtual bioeconomy: the 'failure' of performativity and the implications for bioeconomics. *Distinktion* 14: 83-99.

Bové, J. and Dufour, F. (2001). *The World is Not For Sale: Farmers Against Junk Food.* Verso.

Burfitt, A., Collinge, C. and MacNeill, S. (2006). The discursive constitution of regional knowledge economies: a case study of the European Lisbon Agenda. Available at: http://www.eurodite.bham.ac.uk/Papers/WP1b/discourse%20paper.doc

CEC (1993). *Growth, Competitiveness, Employment: The Challenges and Ways Forward into the 21st Century.* Brussels, Commission of the European Communities.

CEC (1997). Agenda 2000: For a Stronger and Wider Union. *Bulletin of the European Communities,* supplement 5/97, Brussels, Commission of the European Communities.

CEC (2003). *Agriculture and Environment.* Brussels, Directorate-General for Agriculture. Available at: http://ec.europa.eu/agriculture/publi/fact/envir/2003_en.pdf

CEC (2004). *The Common Agricultural Policy Explained.*

CEC (2005). New *Perspectives on the Knowledge-Based Bio-Economy: conference report.* Brussels, DG-Research. Available at: http://ec.europa.eu/research/conferences/2005/kbb/report_en.html.

CEC (2006). Food, Agriculture, Fisheries and Biotechnology. Work programme. Available at: http://cordis.europa.eu/fp7/research.htm

Cooper, M. (2007). Life, autopoiesis, debt: inventing the bioeconomy. *Distinktion* 14: 25-43.

CPE (1999). Agenda 2000 negotiations: the funding of the CAP. Brussels, Coordination Paysanne Européenne, www.cpefarmers.org.

CPE (2001). To change the CAP. Brussels, Coordination Paysanne Européenne, www.cpefarmers.org.

CPE (2007). L'agriculture biologique en danger de contamination et d'industrialization.

Dalla Costa, M. and De Bortoli, D. (2005). For another agriculture and another food policy in Italy. *The Commoner* no. 10, Spring-Summer, www.thecommoner.org.

EC (1998). Directive 98/44/EC of the European Parliament and of the Council on Protection of Biotechnological Inventions. *Official Journal of the European Communities*, 30 July, L 213: 13.

EGSG (2007). *Taking European Knowledge Society Seriously*. Report of the Expert Group on Science and Governance, Brussels, European Commission, EUR 22700. Available at: http://ec.europa.eu/research/science-society/document_library/pdf_06/european-knowledge-society_en.pdf

EU (2000). *Agenda 2000: Reform of the Common Agricultural Policy* (CAP).

Euro Coop (2002). Euro Coop believes the CAP has to be deeply reformed. Available at: http://www.eurocoop.org/publications/en/position/Pac2002.asp

Europabio (2005). New bio-economy web portal launched, 15 Sept. 2005.

Evans N., Morris C. and Winter M. (2002). Conceptualizing agriculture: a critique of postproductivism as the new orthodoxy. *Progress in Human Geography* 26 (3): 313-332.

FFA (2005). Foundation Future Farming, Assembly of European Regions. Berlin Manifesto for GMO-free Regions and Biodiversity in Europe, January, www.are-regions-europe.org.

FoEE (2005). Regions demand 'power-sharing' over GMO decisions. *FoEE Biotech Mailout*, July 15-16, 2005.

FoEE (2007). The EU's Biotechnology Strategy: mid-term review or mid-life crisis? Friends of the Earth Europe. Available at: http://www.foeeurope.org/publications/2007/FoEE_biotech_MTR_midlifecrisis_March07.pdf

GMO-Free Regions (2007). 3rd International Conference on GMO-Free Regions, Biodiversity, and Rural Development. http://www.gmo-free-regions.org/.

Goodman, D. (2004). Rural Europe redux? Reflections on alternative agro-food networks and paradigm change. *Sociologia Ruralis* 44(1): 3-16.

Gottweis, H. (1998). *Governing Molecules: The Discursive Politics of Genetic Engineering in Europe and the United States*. MIT Press.

Hajer, M. (1995). *The Politics of Environmental Discourse*. Oxford, Oxford University Press.

Häring, A.M., Dabbert, S., Aurbacher, J., Bichler, B., Eichert, C., Gambelli, D.,Lampin, N., Offermann, F., Olmos, S., Tuson, J. and Zanoli, R. (2004). Organic farming and measures of European agricultural policy. *Organic Farming in Europe: Economics and Policy* 11. Hohenheim, Universität Hohenheim.

Haskins, Lord (2002). *The Future of European Rural Communities*. London, Foreign Policy Center.

Ikejima, Y. and Hisano, S. (2007). Rediscovering locality? A case of traditional vegetables in Kyoto, article in this collection, based on talk at the conference on 'Reconstruction Agro-Biotechnologies for Development', held in November 2007 in Kyoto. Available at: www.tailoringbiotechnologies.com.

Ilbery, B. and Kneafsey, M. (2000). Producer constructions of quality in regional speciality food production: a case study from South West England. *Journal of Rural Studies* 16: 217-230.

Ilbery, B., Morris, C., Buller H., Maye D. and Kneafsey M. (2005a). Product, process and place: an examination of food marketing and labelling schemes in Europe and North America. *European Urban and Regional Studies* 12(2): 116-132.

Ilbery, B., Little, J.K., Kneafsey, M. and Gilg, A. (2005b). Relocalization and alternative food networks: a comparison of two regions. End of Award Report for ESRC project R000239980.

Jongerden, J. (2006). The urban village: territorialization of sustainable development. *Tailoring Biotechnologies* 2(1): 95-104.

Jessop, B. (2005). Cultural political economy, the knowledge-based economy and the state. *The Technological Economy*, In: (Eds. A. Barry and D. Slater), London: Routledge pp. 144-166.

Kuper, R. (2007). European agriculture in the crucible of the WTO. *Capitalism Nature Socialism* 18(3): 68-80.

Lang, T. and Heasman, M. (2004). *Food Wars: The Global Battle for Mouths, Minds and Markets*. London, Earthscan.

Lawrence, F. (2004). *Not on the Label: What really goes into the food on your plate*. London, Penguin.

Levidow, L. (2005). Governing conflicts over sustainability: agricultural biotechnology in Europe. In: *Agricultural Governance: Globalization and the New Politics of Regulation* (Eds. V. Higgins and G. Lawrence), London, Routledge, pp. 98-117.

Levidow, L. (2008). Making Europe unsafe for agbiotech. In: *Handbook of Genetics & Society*. London, Routledge, in press.

Levidow, L. and Boschert, K. (2008). Coexistence or contradiction? GM crops versus alternative agricultures in Europe. *Geoforum* 39(1): 174-190.

Magnaghi, A. (2005). Local self-sustainable development: subjects of transformation. *Tailoring Biotechnologies* 1(1): 79-102.

Magnien, E. and de Nettancourt, D. (1993). What drives European biotechnological research? In: *Biotechnology Review no.1: The Management and Economic Potential of Biotechnology* (Eds. E.J. Blakelely and K.W. Willoughby), Brussels, Commission of the European Communities, pp. 47-48.

Marsden, T.K. (2004). The quest for ecological modernization: re-spacing rural development and agri-food studies. *Sociologia Ruralis* 44(2): 129-146.

Marsden, T.K. and Smith, E. (2005). Ecological entrepreneurship: sustainable development in local communities through quality food production and local branding. *Geoforum* 36: 440-451.

Marsden, T.K. and Sonnino, R. (2005). Rural food and agri-food governance in Europe: tracing the development of alternatives. In: *Agricultural Governance: Globalization and the New Politics of Regulation* (Eds. Higgins, V. and Lawrence, G.), London, Routledge, pp. 50-68.

Maye, D., Holloway, L. and Kneafsey, M. (Eds.) (2007). *Alternative Food Geographies*. Elsevier.

Murdoch, J., Marsden, T. and Banks, J. (2000). Quality, nature, and embeddedness: some theoretical considerations in the context of the food sector. *Economic Geography* 76(2): 107-125.

OECD (2005). *Proposal for a Major Project on the Bioeconomy in 2030*. Paris, Organization for Economic Cooperation and Development.

OECD (2006). *The Bioeconomy to 2030: Designing a Policy Agenda. Scoping paper*. Paris, Organization for Economic Cooperation and Development.

Penker, M. (2006). Mapping and measuring the ecological embeddedness of food supply chains. *Geoforum* 37: 368-379.

Rein, M. and Schön, D. (1991). *Frame Reflection: Toward the Resolution of Intractable Policy Controversies*. New York, Basic Books.

Renting, H., Marsden, T. and Banks, J. (2003). Understanding alternative food networks: exploring the role of short food supply chains in rural development. *Environment and Planning A* 35: 393-411.

Rigby, D. and Bown, S. (2007) Whatever happened to organic? Food, nature and the market for 'sustainable' food, *Capitalism Nature Socialism* 18(3): 81-102.

Roep, D. (2002). The added value of quality and region: The Waddengroup Foundation. In: *Living Countryside. Rural Development Process in Europe: The State of the Art* (Eds. Van der Ploeg D., Banks J., Long A.), Doetinchem, Elsevier, pp. 88-98.

Rose, N. (2001). The politics of life itself. *Theory, Culture and Society* 18(6): 1-30.

Ruivenkamp, G. (2005). Tailor-made biotechnologies: between bio-power and sub-politics. *Tailoring Biotechnologies* 1(1): 11-33.

Sarasuà, C. and Scholliers, P. (2003). Technology, and Food Production, Distribution and Consumption: a research essay and agenda for Budapest. Available at: http://www.histech.nl/tensphase2/Publications/Working/essayagr.pdf

Smith, A. (2006). Green niches in sustainable development: the case of organic food in the United Kingdom, *Environment and Planning C: Government and Policy* 24: 439-458.

Sonnino, R. and Marsden, T.K. (2006). Beyond the divide: rethinking relations between alternative and conventional food networks in Europe. *Journal of Economic Geography* 6(2): 181-199.

Van der Ploeg, J.D. (2003). *The Virtual Farmer*. Assen, Royal Van Gorcum.

Van der Ploeg, J.D., Renting, H., Brunori, G., Knickel, K., Mannion, J., Marsden, T., de Roest, K., Sevilla-Guzmán, E. and Ventura, F. (2000). Rural development: from practices and policies towards theory. *Sociologia Ruralis* 4(4): 391-408.

Whatmore, S. (1997). Global agro-food complexes and the refashioning of rural Europe. In: *Globalization, Institutions and Regional Development* (Eds. A. Amin and N. Thrift), Oxford, Oxford University Press, pp. 87-106.

Whatmore, S. and Thorne, L. (1997). Nourishing networks: alternative geographies of food. In: *Globalising Food: Agrarian Questions and Global Restructuring* (Eds. D. Goodman and M. Watts), London, Routledge, pp. 287-304.

Xenos, N. (1989). *Scarcity and Modernity*. London, Routledge.

Agriculture, food and design: new food networks for a distributed economy

Ezio Manzini

1. Introduction

Agriculture is very special production activity. And food is a very special product. Pushing agriculture and food towards traditional industrial models has led to an agro-food system that, as a whole, is not only totally unsustainable in the long run, but also holds negative implications in the short term.

Fortunately, the agro-food system has not been entirely shaped in this disastrous way. Observing it more attentively, we can detect different sub-systems in which various organisational models have been adopted and different values are generated. Of great interest among these is the emerging sub-system in which new kinds of relationships between farms and consumers are defined and tested.

This article presents the relationships between these alternative agricultural/food sub-systems and different kinds of design activities. In particular, it focuses on the (possible) design role in enhancing *new food networks* and promoting them as seeds of a *sustainable multi-local society*.

2. Food and design

The history of design is traditionally linked to the history of industry. Up to now its role in agriculture has been minimal as is also the case in food, or rather gastronomy. It has always been said that agriculture and gastronomy are a world apart from industry and so, almost by definition, a world apart from design.

Nowadays, however, there is more and more talk of a possible meeting between design and the food system. The expression, 'food design', has become something of a buzz word (though as of yet its meaning is far from clear). All this may seem to prove beyond doubt that agriculture is now industrialised and that food has become, to all intents and purposes, an industrial product like any other (Meroni, 2003; 2004).

Following this line of reasoning, design could be seen as yet another agent driving us towards a full industrialisation of this field of human activity. This is a legitimate way of thinking supported by numerous examples, but does it really have to be like this? Is industrialisation as presented so far really the only feasible proposition and all that can be seen on the design horizon? The reply we are suggesting here is 'no', and for various reasons.

It is true that design was born with industrial society and carries deep within itself the concepts, value systems and ways of doing things characteristic of the early stages of industrial development. However, industrial society has changed and continues to change, rapidly. Industry itself has changed and design with it, or even before it, being in itself one of the drivers of industrial change. It follows that if design can and must have a role in agricultural cultivation and food production, this should arise now out of a profound awareness of the crisis in the dominant economic and cultural model, and out of a recognition of the possible role of design as co-promoter of alternative agricultural and food systems that can become promise real steps in the direction of sustainability.

This does not mean that design must deny its nature of industrial actor (i.e. of being one of the main drivers in the industrial culture definition). It means that it can and must collaborate to redefine the very concept of industry itself. Especially, so far as we are concerned here, it means collaborating in the consolidation of an agriculture, food industry and distribution system capable of moving in the opposite direction to what has been the prevailing trend until now.

3. Beauty and agriculture

'That's a beautiful olive grove'. That's what they still often say in Tuscany, where I live, when they want to speak well of an olive grove. Beauty is still considered by many to be the most concise way of expressing the quality a field should have – a beauty that obviously also includes its productivity, but does not stop there. A beautiful olive grove must be productive but it must also be looked after just like, or even more than, a garden.

A beautiful olive grove is the result of a diverse range of activities ranging from pruning and caring for the trees themselves, to tending the meadowland around them, to the constant clearance of the irrigation canals and the upkeep of the dry stone walls that hold the terraced hillside. The frequency with which these different activities are repeated varies from annual tasks, like pruning, to lifelong labour that will effect generations to come, as in the maintenance of dry stone walls.

A beautiful olive grove marries individual interest and collective advantage. The cultivator does something for himself, but he also carries out a fundamental social task in managing two common goods of great importance to the whole community: the hydro-geological system and the quality of landscape. In so doing he produces socialised economic value, since the landscape he helps to maintain, in the case of Tuscany and similar places, is one of the main driving forces behind the tourist economy.

A beautiful olive grove produces good olive oil, and, moreover, an oil that not only looks good, smells good, tastes good and is nutritionally good, for the oil this field produces is also a good social operator. As the product of a process shared by the whole local community, it becomes a topic for conversation and, as such, contributes to social regeneration (Malaspina and Vugliano, 2005).

A beautiful olive grove and its world of supporting values, as pictured here, is an inheritance that reaches us from long ago. In many parts of the world these values and customs may now be seen as 'cultural fossils', the remains of a bygone world. In other places they may be seen as limitations on development, part of what must be left behind, if we are finally to enter the modern age.

In these notes we shall try to show that we can, or maybe must, think differently. Our beautiful olive grove with its value system and supporting customs can be seen in a completely different way: not as a cultural fossil, not as a limitation on development, but on the contrary, as 'a seed for the future' – a thing from the past, that is, but one that could develop on new ground, giving rise to new possible futures.

Before discussing these issues further, I would like to add another introductory consideration. The arguments supporting the prospect outlined have been drawn mainly from experiences in the north of the world, from social and cultural contexts where, with rare exceptions such as the Tuscan hillside mentioned, this way of thinking and doing things has by now been almost totally overrun by the new ideas on productive efficiency and destructive pervasiveness of market culture. It is clear that in these contexts the prospect indicated, although feasible, is objectively speaking difficult to achieve since it involves reviving discontinued traditions that are on the way to becoming extinct. On the other hand, it should be stressed that this same prospect, if acknowledged in time, is far more likely to be successful in those areas of the planet, mainly the South and East of the world, where such values and customs as we are describing are still solid and potentially vital.

4. Out-dated industrial models

Let's leave the Tuscan hillside with our olive grove and look down, towards the valley bottom, and out, to the world as a whole. What we see dominating is an agro-food system that works like a huge paradoxical machine. A perverse system that fails to resolve the problem of hunger, yet at the same time has made obesity one of the greatest plagues of our time. Furthermore, to achieve these results, it acts as a powerful waste-layer, consuming resources, impoverishing land and reducing diversity, both genetic and cultural.

How is it possible that such a paradoxical situation can have developed and be considered acceptable by (almost) everybody? That's a long story (see Capatti *et al.*, 1998; Capatti and Montanari, 2002). The agro-food system we know today is the application, in agriculture, of ideas and organisational methods that we could nowadays call 'out-dated industrial', but which seemed for many years to be successful formulas. Such ideas and organisational models have led us to see fields as industrial areas, plants and animals as machines, and food as goods to be standardised and trivialised ... at all costs: at the cost of degrading the ecosystem, erasing age-old patterns of social organisation, wisdom and expertise, ultimately to the detriment of the health of those very consumers who were to be the beneficiaries.

Faced with the emerging problematical issues inherent in this way of conducting things, the prevailing attitude in the past, which is still widespread, is that all this is a necessary evil, the sacrifice to lay on the altar of growth (in the so-called developing regions of the world), or of economic survival (in areas of long-standing industrialised agriculture). However, nowadays things are changing. The manifest visibility of environmental problems; the diffusion of epidemics in breeding farms and human illnesses associated with bad eating habits; aversion to genetically modified organisms; evidence that, in spite of the quest for efficiency, said to be sought at all costs, a large part of humanity is still suffering hunger ... taken together, these phenomena have gone beyond the point where the crisis in our industrial agro-food model can still remain hidden.

Stimulated by the crisis in the dominant model, other new idea ideas are emerging about how a sustainable, industrial agro-food system might look; new ways of thinking, that is, about the sense of land cultivation and food production, about how it might no longer respond just to the logic of economic productivity, but be recognised as one of the most profound expressions of human action, related to the individual, society and nature. This is a new way of seeing things that implies a new idea of industry, economy and society: a society in which cultivating a field means first of all taking care of 'mother earth', looking after the most precious good that humanity has at its disposal now and for future generations. This is a society in which the production, preparation and consumption of food is considered to be at the same time a response to a necessity, a quest for pleasure and a form of social rapport – profound activities that go well beyond simply nourishing our biological machinery, and that should be conceived and actuated as part of a more general framework. This wider framework we today know by the name of 'transition towards sustainability' (Manzini and Jegou, 2003).

It is not among the aims of these notes, nor in the capacity of the writer, to draw the complex picture of how these emerging ideas may turn into a new general development model, or indeed how a fully sustainable agro-food model might work. On the other hand, since the transition towards sustainability is a social learning process, no-one can really claim to be capable of so doing.

The emerging ideas proposed here are a contribution to this vast, collective-learning process, undoubtedly an incomplete contribution, but maybe a useful one in indicating some of the characteristics of the evolution in progress, and particularly in underlining those most relevant to what we are interested in, i.e. to what, hopefully, design can do.

5. Agro-food system articulation

The contemporary agro-food system can be described as a stratified reality, a macro-system in which different agricultural philosophies and food cultures exist side by side. Here we shall focus on five: *the traditional, classic, experiential, and advanced agro-business systems,* and *social experimentations.* This stratified reality is the context wherein design operates and selects its own options.

- *The traditional system.* In different ways from region to region, we still find forms of organisation and traditional lore which reach us from the depths of rural history underlying the contemporary agro-food system. This stratification of the system is what remains of pre-industrial agriculture, of its learning, its organisations and its local and seasonal food circuits. As we have been able to observe, this underlying stratum can be presented from region to region either as a still vital, living tradition or as what remains of a disappearing system.
  This subsystem as it is presented today does not make demands on design. If and when it does, as we shall see, that means that it is already turning into a different form, which we shall discuss later. This underlying stratum is threatened and often overwhelmed by what we can call the dominant agro-industrial system.
- *Classic agro-business.* This is the agro-food system organised by archaic industrialisation formulas as discussed previously. It leads to mass production and consumption by agricultural firms and breeders which we call industrialised because they are mechanised, 'chemicalised' and, more recently, 'bio-technologised'. This is the dominant component of the current food and agricultural sector in industrialised countries and, considering the major dynamics in action, tends also to be so in those not yet industrialised.
  The classic agro-industrial system places equally classic demands on design: agricultural machinery, product packaging, apparatus and equipment for food preparation and communication strategies for an undifferentiating undifferentiated and undifferentiating public (hooked by low prices and a profusion of alluring advertising).
- *Experiential agro-business.* This is the component of the agro-food system which is most highly influenced by the most forceful logic of the moment, and whose primary objective is to research and promote the particularities of experiences that products and services bring (or should bring). This research has found in food, and the places where food is produced, a privileged application field (as is obvious, given the peculiarly sensory and experiential nature of food and its typical places of production).

With the emergence of this new and growing component of the agro-food system over the last few decades, new demands have emerged for design related to product identity and place of origin, to the conception of new sale and restaurant services, even to the planning ex-novo of new food products as applications of design of/for experience (this is the application field of *food design* in its strict sense). Given the importance of this issue, we shall return to it later.

- *Advanced agro-business.* This is the aspect of the contemporary agro-food system that lays its bets on technological solutions to the growing environmental and social problems. It tries to respond industrially to the huge demand for controlled, organic food products. Advanced agro-business is the expression in agriculture of the most interesting shapes that industry is taking. It entails the extensive application of organic and biodynamic cultivation methods, and the use of advanced *minimal food processing* systems. DOP (*produced by guaranteed production process*) and IGP (*of guaranteed geographical origin*) labels can be seen in the same light, as a legal representation of the idea of advanced agro-business.

  It requires considerable design capacity to see the food industry as an advanced industry with these characteristics. Obviously this is true on a technical and organisational level, but it is also true on a cultural and communicational plane. This gives rise to an as yet embryonic demand on design: an industrial design for advanced industry orientated towards such a 'sensitive' production area as food. Again we shall come back to this later.

- *Social experimentation.* This is the latest and most dynamic layer of the agro-food system, and one whose future is as yet unclear. It is the mover of some of the major dynamics, such as the spread of networks, the demand for 'natural' foodstuffs and, more generally, the quest for sustainable solutions. It is these macro-tendencies as a whole that give rise to the social experimentation we are referring to. They are experiments that come from both the demand side (such as collective purchasing groups) and from the supply side (the Slow food organisation, organic product networks). We also refer to them as 'creative communities' because they are mostly the outcome of individual and collective self-organisations inventing new ways of resolving a problem or opening up a new opportunity. Except for a few special cases, these social experiments have not yet expressed a clear, deliberate demand for design. However, in my opinion, it is precisely on such projects that design should focus its attention, in order to play its potentially constructive role in promoting a sustainable agro-food system. The reasons for this conviction will be the subject of the following paragraphs.

I would now like to describe in greater detail the role design can play in the two most recent layers of the agro-food system: experimental agro-business and social experimentation.

6. Experiential agro-business and food design

Experiential agro-business is the way in which the emerging service and experience economy (Pine and Gilmore, 1999; Jordan, 2000) is taking shape within the conventional (industrial) agro-food system.

One prerogative of this phenomenon is the importance attributed to the diversity of products and places of production. This implies an inversion of tendency with respect to the standardisation policy offered and imposed by the classic industrial model, which is undoubtedly positive. However, this potentially positive shift has been more than counterbalanced by a series of negative implications that can be grouped and expressed as 'the spectacularisation of food and agriculture'.

In practice the process runs like this: the service and experience economy in its present form has to be constantly refueled with 'fresh' cultural and social resources able to trigger a strong emotional response, in other words, able to create a spectacle. These fresh resources are often drawn from the pool of knowledge, customs and characteristic local places that the traditional system (or rather, what remains of the traditional system) intrinsically possesses. In itself, this would not be a bad thing, if using these resources provided an opportunity for their regeneration. However, this is not the case: their use for the spectacular tends rather to turn them into empty images behind which nothing of what they really were remains.

In the absence of any profound reflection on the identity of places, communities or their products (and in the absence of any sensitivity towards issues relating to the sustainable use of physical and social resources), the experience and service economy leads us to treat food, community and local identity as though they themselves were products to be promoted and consumed. The result is that typical local products are transformed into commercial brands, and places of production and producer communities turned into theme parks and the characters that populate them. Furthermore, any genuine public interest is reduced to its most hedonistic dimension (as in most television programs on the subject, which are conceived in such a way that they can be seen as a sort of food pornography).

These dynamics, as interesting at their outset as they are disheartening in their practical consequences, are counterbalanced by the extraordinary activities of Slow Food. I shall come back to this in the following paragraphs. First, however, we must briefly consider how design has confronted today's dominant tendencies: what has it done and what could/should it do?

We have already mentioned how the experiential agro-business has placed clear, high demands on design, demands to which designers have generally responded by adapting their own ideas and habits to the new necessities. In other words they have adapted themselves to the behavior and thinking of the service and experience society. In so doing, updating their 'classic' competences (product and communication design) and adjusting their more innovative ones (like strategic and service design), designers have started to play a significant, active role in the emerging agro-food system. Having said this, we should add, to my regret, that this significant, active role has not so far led them to any profound reflection on the meaning of what they have been doing. Except in exceptional cases (which certainly exist), designers have gone uncritically into the service and experience economy, themselves joining the main forces driving towards this spectacularisation.

To get out of this role governed by the thinking currently dominant in economics and the media, designers should ask themselves some more profound questions about the sense of the experiences they are proposing. It is a difficult reflection, but one facilitated by the existence of an extraordinary phenomenon for comparison, that of Slow Food. This is an organisation that has successfully shown us all that it is possible to link sensory experience with the safeguarding and valorisation of characteristic products, together with the knowing and organisational forms they spring from. In this way, it is playing an extraordinarily important role (first in Italy, but now on a worldwide level) in the safeguarding and regeneration of such a precious common good, as is the cultural variety of local food production.

The 'presidi dello Slow Food' are diffuse, local organisations spread throughout Italy that aim to protect specific local products. Looking at Slow Food as a 'designer', what we can see in my opinion is the most positive example of strategic design, of service design and of experience design applied to the world of agriculture and food today (and this even though, to my knowledge, until a few years ago, none of its promoters had had any significant relationship with the designer community). Slow Food teaches us that it is possible to carry out a design activity that goes beyond the spectacular consumption of what remains of a precious, historical heritage of knowing, flavors, places and social customs. Slow Food promotes a *post-spectacular design* (Thackara, 2005), able to promote identity and generate significant experiences, without entailing their transformation into empty images and rapid consumption; able to make of this activity an occasion for regenerating our traditional heritage, matching it to the most advanced technological and organisational possibilities (*advanced agro-business* as mentioned earlier) and able to turn it into a seed for a sustainable future (Manzini and Vezzoli, 2002; Manzini and Jegou, 2003; Meroni, 2005).

In order to explore the implied possibilities it is useful to take another step forward and consider the theme of social experimentation and its possible implications for design.

7. Social experimentation and multi-local society

The issue has already been introduced in a previous paragraph: the spread of network systems, the widespread demand for 'natural' foodstuffs, and the quest for sustainable solutions have given rise to new ways of thinking and doing. This is happening both on the side of demand and of supply.

Let us consider in particular the implications of the spread of network organisations so much talked about in recent years. This phenomenon has led to a huge increase in connectivity (i.e. in the number of meaningful interactions concretely possible). In turn, the high level of connectivity achieved has served as an enabling platform for new forms of organisation where the network is not only a technical infrastructure, but also becomes a powerful, new organisation model that breaks vertical hierarchies and generates horizontal, un-intermediated, potentially peer-to-peer solutions.

All this enables us to imagine a new family of organisations, at the same time decentralised and open to wider systems, an organisational model, that is, which leads us to redesign from scratch consolidated ways of doing things, classically based on low-connectivity systems. Clearly the radical adoption of network models is not in itself a solution to the social and environmental problems we are faced with today (even Al Qaeda and certain pedophile organisations are based on the intelligent application of network organisation models employing the horizontal communication technology available today). Nevertheless, these organisation models, and the technology that makes them possible, present interesting and promising opportunities of potential value to the development of new approaches in the agro-food system.

The spread of the Internet in particular has promoted certain relevant and potentially generative issues, ideas that are also capable of generating new ideas in operational areas far apart from those that produced them: notably, the *network economy*, *open source systems*, and *peer-to-peer organisations* (Stalder and Hirsh, 2002; Cottam and Leadbeater, 2004). Can all this be translated in some way to the agro-food system? What could we understand from the expression 'food-network?' This question still does not have a clear, detailed answer. However, some partial answers have already emerged in the social experiments referred to.

Let us consider activities like fairtrade purchasing groups, organic markets in cities (*farmer markets*), new producer/consumer relationships (such as 'adopt a tree' or 'vegetable season tickets') ... but let us also note how the success of fairtrade shows in a concrete way that direct, fair relations are possible between producers and consumers even when far apart.

Let us link these activities, which are mainly centered on the issue of virtuous un-intermediating, with those related to the valorisation of local products (EMUDE, 2004). Once again, Slow Food activities come to mind: from the diffusion in Italy of the local organisations of the 'presidi dello Slow Food', local organisations aiming to protect specific local products, to the extraordinary initiative of *Terra Madre,* whereby thousands of small agricultural, animal husbandry and fishing communities all over the world have been identified and networked together, united by their possession of specific production and food know-how.

If we try to see these and other similar promising cases as a whole, a new vision of the agro-food system emerges (maybe even a new vision of the world!). What appears is the image of a multi-local system, of a 'world' endowed with an extensive variety of places and communities – communities with their own individual identities, but open and well-disposed towards contact; local communities with a high connective potential, in peer contact with other local communities, with whoever and whenever useful, just as in peer-to-peer organisations on the Internet.

We see a multi-local system in a network economy where the number of knots and links available is more important than the knots themselves, and where basic knowledge, like knowing about food and its production, is a common good accessible to communities, within the limits of the sustainable use of any common good. In short, we are envisaging a multi-local system able to orientate the development of advanced agro-business, steering it more clearly towards sustainability (Manzini, 2004a,b; Distributed Economy Labs, 2005).

8. Strategic design and new food networks

If the vision outlined is to become feasible and the multi-local agro-food system is to become a reality for the majority, then communities of producers, consumers and producer-consumers need to consolidate and 'make themselves visible'. They must be able to display their products and skills, their needs and wishes and their willingness to do something towards satisfying them. There must be a platform, an infrastructure, that gives them the real possibility of making contacts, of presenting their offers, of building relationships that are not only economic, but are also neighborly and, where appropriate, expressive of solidarity.

In my opinion, the agro-food system lends itself to being reorganised in this way. The experiences we have discussed, though minority experiences, tell us that this is possible. Furthermore, food and land are two fundamental elements in the lives of everybody. It is possible to develop interest and movement around them on a wider scale. When talking of them we can talk about necessity and pleasure, about past and future. Knowledge and skills that risked being forgotten can be revived around them. Food can help us rediscover the quality of local and seasonal products, but it can also lead to the enjoyment of products from far away that are produced by a familiar, friendly community and so, as the Slow Food slogan says, are 'good to eat and good to think'.

All this can develop from the bottom up, as forms of social self-organisation, but designers too can take part in this virtuous process. They can bring their specific skills to assist in community building, improving visibility, making communication channels more fluid, implementing enabling platforms that facilitate the diffusion and effectiveness of the activities of these communities. From here, bringing their skills in the field of *experience design* into play, they can contribute to the promotion of food networks where aesthetic and sensory qualities *also* circulate, but freed of the tendency towards spectacularisation: indeed aesthetics and sensory perception are fundamental dimensions in any human relations, and all the more so if the object of these relations is something as profound and important as food.

To conclude, in the metaphor of agriculture, designers can collaborate in working the ground in which seeds of both new and ancient cultures can really germinate and grow into the plants of a sustainable food and agriculture future: a future where there are beautiful olive groves, with all the wealth of meaning implied in that expression.

References

Capatti, A., de Bernardi, A. and Varni, A. (Eds.) (1998). *L'Alimentazione*. Annali 13, Storia d'Italia, Torino: Giulio Einaudi editore

Capatti, A. and Montanari, M. (2002). *La Cucina Italiana. Storia di una Cultura*. Roma-Bari, Editori Laterza

Cottam, H. and Leadbeater, C. (2004). *Open Welfare: Designs on the Public Good*. London, Design Council.

Distributed Economy Labs (2005). *Developing and Managing Regional Value-Networks*. Landskrona, DELabs.

EMUDE - Emerging user demands for sustainable solutions (2004). *Social Innovation as a Driver for Technological and System Innovation*. research working papers.

Jordan, P. (2000). *Designing Pleasurable Products*. London, Taylor and Francis.

Malaspina, R. and Vugliano, S. (2005). *Paesaggi Coltivati*. Interreg III MEDOC, Paesaggi Mediterranei, research working paper.

Manzini, E. (2004a). Towards a cosmopolitan localism. In: *Spark! Design and Locality* (Eds. J. Verwijnen and H. Karkku), Helsinki, University of Arts and Design.

Manzini, E. (2004b). A cosmopolitan localism. In: *Medesign_Forme del Mediterraneo*, (Eds. Fagnoni, R., Gambaro, P. and Vannicola, C.), Firenze, Alinea Editrice.

Manzini, E. and Jegou, F. (2003). *Sustainable Everyday. Scenarios of Urban Life*. Milano, Edizioni Ambiente.

Manzini, E. and Vezzoli, C. (2002). *Product-Service Systems and Sustainability. Opportunities for Sustainable Solutions*. Paris, UNEP Publisher.

Meroni, A. (2003). Il design dei sistemi alimentary. In: *Design Multiverso. Appunti di Fenomenologia del Design* (Eds. Bertola, P. and Manzini, E.), Milano, Edizioni Polidesign.

Meroni, A. (2004). *Design e Innovazione di Sistema Nel Settore Alimentare. Una Ricerca Progettuale*. Milano, Edizioni Polidesign.

Meroni, A. (2005). Strategic Design for the Food Sector: The Food-System Innovation, paper presented at the Agrindustrial Design, Universty of Economics, Izmir, 27-29 April 2005.

Pine, J.B. and Gilmore, J.H. (1999). *The Experience Economy. Work is Theatre and Every Business a Stage*. Boston, Harvard Business School Press.

Stalder, F. and Hirsh, J. (2002). Open Source Intelligence. *First Monday* 7(6).

Thackara, J. (2005). *In the Bubble*. Boston, MIT Press.

Quality agriculture and the issue of technology: a short note on reconstruction

Joost Jongerden and Guido Ruivenkamp

1. Introduction

Over the last quarter century or so, there has been a growing interest in alternatives to the dominant mode of modern, industrialised agriculture, with its large-scale, mechanised structures and global, capitalist superstructure. This is exemplified by the development in Europe and North America of local networks ('regional initiatives') for the production and distribution of high quality (e.g. 'organic') food products. In general, these alternatives are represented in public and scientific discourses as a consumer-issue. This is not only the consequence of the attempts of regional initiatives to reinvent markets on the basis of sustainable production, quality products and fair trade: it may well be the result also, and more fundamentally perhaps, of a consumerist rather than a civic conception of the public and scientific discourse on food and agriculture. By and large, this focus on consumer concerns passed over in virtual silence broader public anxieties concerning the entrenchment of global corporate power and radical instrumentalisation (Davison *et al.*, 1997). An important characteristic of the regional initiatives is the break they represent with neo-liberal politics and the related representation of the world as a global (super)market.

Referring to 'regional agriculture' or 'quality agriculture', analyses of regional initiatives in agriculture tend to focus mainly on the issues of distribution and marketing, thereby presenting the initiatives in terms of a consumerist issue. Here, we aim to look at such initiatives from the perspective of technology. This approach leads to and develops from the proposition that *the development of alternatives in agriculture is contingent upon the development of alternatives in technology*. Regional initiatives and the developing alternatives in agriculture (and, by inference, any full analysis of them), that is, must assume the fundamental import of a technological approach in its broadest sense – otherwise, the conservative implications of a consumerist frame of reference will seriously hamper the movement and probably doom it to failure. Empirical substantiation for this argument will be based mainly on a case in the Netherlands, which provides an excellent illustration of how the reorganisation of agricultural production is mirrored in the development of technology, and *vice versa*. First, by way of background and relevant to the debate on alternatives, certain agricultural policies and the modernisation of agriculture are discussed. The main conclusion of this article is not simply that technology matters, but that the development of alternatives in agriculture forces us to think about the development of alternatives in technology. Without a radical shift in the development of technology, the development of alternatives in agriculture will be greatly impeded, pervasively undermined and ultimately blocked.

2. Backgrounds

Over the last decennia, agricultural policies in the Netherlands at the level of the individual farm have been formulated within the triangle of specialisation[146], increase in scale[147] and intensification[148]; while policy for the agricultural sector as a whole has focused on the integration of farms in agro-industrial production chains. In terms of trade and markets, policy had been based on protectionism, mainly formulated within the framework of a common European agricultural policy. This protectionism has been marked by trade barriers (to parties external to the European Community), and fixed prices (later in combination with quotas) for internal parties (farmers, agro-industry). Since the 1980s, this policy of protectionism has been replaced by a neo-liberal policy of global competition. The reduction of prices resulting from this strategic shift has been used to re-emphasise the necessity of further enhancement of the specialisation/scale-increase/intensification triad, and a closer, tighter integration of farms into the agro-industrial production chain. The transformation of farming practices in accordance with the policies mentioned was referred to as *the modernisation of agriculture*. Farmers who were unable or unwilling to adapt their farming practices to the demands of this modernisation were considered 'backward', or just discounted as 'bad farmers'. Characterised as '*wijkers*' (lit. 'those who retreat'), the natural fate of these farmers was simply to disappear. This was considered to be not only inevitable but necessary, since the disappearance of farmers who did not organise their enterprises (sic) in accordance with the modernity project contributed to the (desired) increase in scale of others; if they insisted and somehow succeeded in holding on to their farms, these *wijkers* were considered nothing less than an impendent to progress.

The policy of modernisation contributed to an increase in production. For example, the potential production of wheat under ideal circumstances (the theoretical production limit) in the 1950s of 6,000 kilos per hectare had doubled to 12,000 kilos per hectare by the 1990s. However, this increase was co-produced by the application of high levels of inputs, such as chemical fertiliser and pesticides, which has contributed to major environmental problems. One summer day in the 1980s, for example, the wind blew away the entire upper layer of the peaty soils in a part of Drenthe, a province in the northern part of the Netherlands, a local calamity which was a direct result of the destruction of the soil's structure, caused by intensive input potato production. In addition, the one-sided focus on yields had turned farmers into producers of bulk agricultural products (or of agricultural components to be processed into end products by a processing industry), products and components that were large in mass but low in quality. In this process, the lack of quality of agricultural products (such as taste) was compensated for with the supplementation of additives. Farmers had become dependent on and subject to the demands of the processing industry, and then later those of retail conglomerates, whose power increased.

In the 1970s, an increasing number of farmers started to question the prevailing agricultural policies. This resulted, among things, in the contra-productivity thesis. In short, this thesis holds that the increase in agricultural productivity has not been as high as is often assumed, because of the use and waste

---

[146] The mixed farm gradually disappeared and became replaced by farms specialised in either arable or livestock production. In arable farming, winter wheat, potatoes and sugar beet became the predominant crops (Wiskerke, 2000: 81).

[147] Increase in farm size: for example, the number of farms decreased from 410,000 in 1950 to 111,000 in 1996 (Wiskerke, 2000: 81), and then to slightly less than 80,000 in 2006 (LEI, 2006: 120); the surface-area of land under cultivation reduced somewhat over this period (1950 to 2006), from approximately 2,300,000 ha to 1,899,000 ha (LEI, 2000: 98; LEI, 2006: 125). The median farm size thus quadrupled, from 5.7 ha in 1950 to 23.7 ha in 2006.

[148] Recorded as an increase of yield, per ha and per labour unit: for example, in arable farming, an increase in winter wheat production of some 250% was realised between 1950 (3,900 kg per ha) and 1996 (9,600 kg per ha) (Wiskerke, 2000:81).

of resources that are not operationalised in the standard policy models. For example, the increase in productivity in dairy production in the Netherlands was only made possible by the importation of fodder (soya, etc.) from the South: the land used for the production of this fodder ought to be included in models calculating the amount of land needed to support one cow, but is not. Equally, the costs of pollution in dairy and arable farming are not included in the pricing of agricultural products, enabling these products to seem cheaper than they really are. Furthermore, the associated categorisation of farmers came to be called into question, and agricultural policy exposed as problem rather than solution. The Working Group for a Better Dairy Policy (*Werkgroep Beter Zuivelbeleid*) used an old parable to illustrate their position:

> 'Once, a person ordered a suit at a tailor's. When the person came to try on the suit, the collar did not fit, but the tailor said, 'Sir, you are not standing on your feet properly. You need to lean backwards.' When the person did that, the trousers did not fit, but the tailor said, 'You need to push your belly forward.' Finally, one of the trouser-legs turned out to be longer than the other, but the tailor convinced the man that he needed to walk a bit askew. When the man went for a walk in the park, he moved and stumbled his way forward exactly as the tailor had prescribed. People saw the man walking and said to each other, 'It must have been a very good tailor to make a suit for such an unfortunate man.'

In the course of the 1970s, Dutch farmers started to raise a number of issues. In addition to environmental issues (and also gender issues, the invisibility and under-appreciation of female farmers), attention was focused on and the negative effects of EU agricultural policies for farmers in the south of the country, in particular the dumping of food, which was disrupting local production capacity and bringing misery to local farmers. Plans for the establishment of mega-farms met with fierce criticism and were successfully resisted. It was argued that these mega-farms implied a deskilling of the workforce and reduce craftsmanship, turning farmers into the mere executors of externally conceived prescriptions. Also, these mega-farms would increase the rat race of producing more agricultural products for lower prices, putting more pressure on farmers' incomes. The farmers demanded a fair return, and, moreover, not only for themselves.

In 1977 the farmer-based Working Group for a Better Dairy Policy decided to support a strike of dairy-workers demanding wage increases to cover price-inflation. This practical position followed a radical discursive shift. The interests of the farmers and the workers had conventionally been assumed to be opposed, with the farmers as the employers, the owners of the land and the farming businesses, and the workers the employed. The Working Group, however, argued that the farmers' ownership was only virtual – the farming enterprises were in fact controlled by the company that processed, marketed and distributed the dairy produce, which although starting out as a local farmers' cooperative had long since been transformed into multi-national company with little to no affinity with the cooperative ideal (notwithstanding the fact that its cooperative status continued). The Working Group for a Better Dairy Policy argued that farmers and workers had a common interest – fair prices/wages to be paid by the company – and considered the struggle of the workers as parallel to, and part of, their own struggle.

Not only farmers but also students organised themselves, at the Agricultural University in Wageningen. Students there contested and denied the classical idea that science and technology were politically neutral, and challenged the position of the university in the so-called *Green Front*, a name attached to the close collaboration and cooperation between the ministry of agriculture, conventional farmers' associations and knowledge institutes, which applied and advanced the idea of agricultural 'modernisation'. The students argued that knowledge production was difficult or impossible to disentangle from choices with political dimensions, and that such choices needed to be made explicit. The organisations for research and knowledge production established by these students associated themselves with the new social movements in agriculture. In spite of the accusation leveled at them of being communist umbrella organisations, these students' organisations for research and knowledge production succeeded in triggering debates about agricultural policies (e.g. Boerengroep, 1973, 1988; Kwekkeboom and Kleiboer, 1989) and paradigms of development and agrarian sciences (e.g. Studium Generale, Boerengroep, Imperialisme Kollectief and WSO, 1974; Imperialisme Kollektief, 1988; Imperialisme Kollektief & Studium Generale, 1985; Beeker *et al.*, 1995), and also in carrying out more applied forms of research for farmers seeking alternatives.

3. From resistance to reconstruction

The members of the new social movements[149] in agriculture were in many cases practicing farmers, who were concerned not only with pressuring policy-makers and politicians, but also with the future of their own farm. For them, the question was how they would be able to practice alternatives in agricultural and rural development, i.e. how, notwithstanding the modern policy measures and strategic approach of government, they could develop a farming practice reconciled with their social and political values and conviction, a farming practice that also contributed to the maintenance and development of the countryside. Some of these farmers converted to organic farming practices, but others, inspired by examples from abroad, in particular Italy and France, considered the idea of producing what were termed 'regional products', high value agricultural products for consumer markets that were produced in a clearly demarcated locality (the region) and processed in that region from raw materials produced in that region in an 'environmentally friendly' way. The latter implied a reduction in the use of pesticides and herbicides, but not necessarily an absolute ban even though some farmers considered the environmental friendly production method a stepping stone towards organic agriculture (Jongerden and Ruivenkamp, 1996).

Beyond the immediate features of regional products, the efforts of this approach towards reorganising production could be described in terms of four interrelated forms of social struggle. First, there was a social struggle over the *production of added value* and by whom (or under whose hegemony) this added value should be produced – the agricultural producers, the processing industry, or the retailers? Second, there was a struggle over the *power of decision over labour*: should this be in the hands of Brussels, the processing industry or retailers, or the citizens, i.e. people in their different roles as farmers, consumers or environmentalists (etc.)? Third was a *struggle over markets*, i.e. unraveling the farm from the web of global markets and international capital in which it had been interwoven. Fourth, there was a *struggle over technology* development, particularly in respect of the biased effects of technology for the scale level of farms. This all led to the common political project for the *re-establishment of the farm as a junction in local food production and consumption*.

A striking case in point exemplifying these issues was the *Zeeuwse Vlegel*. This initiative had its roots in both types of the new social movements that had emerged in agriculture was established at more or less the same time as the Union for Arable Farmers in the Netherlands (*Nederlandse Akkerbouw Vakbond*)

---

[149] The term 'new social movements' refers to a diversity of movements which emerged in various societies after the 1960s, and are considered to be different (in composition and outlook) from the old social movements of class struggle (labour movement). For a treatment of the issue see Offe (1985), Laclau and Mouffe (1985) and Buechler (1999).

was founded (1993), in which *Zeeuwse Vlegel* farmers participated prominently and which organised farmers to agitate against neo-liberal policies. Another important institutional background of the *Zeeuwse Vlegel* were the so called 'wheat-study clubs' that flourished in the 1980s. In these study clubs, farmers discussed their farming practices and compared farm results, a form of peer-group learning.

One group of young farmers, combining their common practices and conviction in the Zeeland young farmers' regional organisation, the NAK (*Nederlands Agrarisch Jongeren Kontakt*, together with a regional environmental organisation, the ZMF (*Zeeuwse Milieu Federatie*), launched a plan for the production of a regional type of bread in 1988. Farmers, the NAK and the ZMF were convinced that is was better to develop an environmentally sound farming practice than continuing emphasising and amplifying the contrasts and contradictions between agriculture and environment; that it was preferable to develop a plan to produce, process and market this wheat with an eco-friendly ethos, establishing closer contacts between producers and consumers. In 1990, four farmers started to experiment with 'wheat-for-bread' production, and in 1991 the organisation adopted the name '*Zeeuwse Vlegel*'.

The name *Zeeuwse Vlegel* had a multiple meanings. *Zeeuws* referred to a locality, i.e. Zeeland, a province of the Netherlands situated in the south-west of the country. With a surface area of about 2,930 km^2, of which almost 1,140 km^2 is water, the province basically consists of a number of islands and a strip of land bordering Belgium. A *vlegel* is a flail, a wooden instrument to thresh grains. In days gone by, the ears of a plant would be spread out over a threshing floor, while several people separated the grains from the ears by hitting the ears with the flail. This had to be done in a coordinated matter, in some kind of rhythm, otherwise the flails would collide with each other, and the threshing fail. A *vlegel* is also, however, a rascal, a daredevil challenging his (or her) environment. The *Zeeuwse Vlegel*, or *Vlegel*, project to process grain – wheat – into consumer-products was a daring and challenging initiative, since it implied a rejection of the modernisation paradigm, and therefore doomed to fail according to its critics.

The *Zeeuwse Vlegel* started in the late 1980s with four farmers aiming to cultivate wheat on a couple of hectares in an environmental sound way. It grew rapidly, to 22 farmers cultivating 130 hectares in 1994. The initiative could easily increase the cultivated area of grain, since farmers showed considerable interest in shifting towards an environmental sound cultivation practice; knowing that this would probably keep pace with an increased sale of their bread. The sale of bread was, indeed, an important enabling (and limiting) factor. The amount of *Zeeuse Vlegel* bread sold over recent years has been totaling about 20,000 loaves a month, from about 20 ha wheat a year. The *Zeeuwse Vlegel* currently has 15 farmer-members and is approaching its 20th anniversary.

In the initial period, the *Zeeuwse Vlegel* had to deal with many production, processing and marketing barriers, barriers associated with the bulk orientation of the agricultural sector. To give just one example, wheat varieties cultivated in the Netherlands perform well in terms of kilo per hectare, but are in most cases of poor baking quality. The emphasis on weight is reflected in the storage system: no selection is made on the basis of quality. In order to process flour of baking quality, the wheat produced in the Netherlands is enriched with wheat from France or other countries whose wheat has good baking qualities, or else with specific additives. This enrichment actually shows the poverty of Dutch agriculture: it is able to produce quantity, but lacks quality. The *Zeeuwse Vlegel* posed a challenge to the social organisation of the bulk production system, and its success showed that it is possible to develop a production system that is fair and viable, both ecologically sound and economically robust. This had several consequences for the organisation of production. The *Zeeuwse Vlegel* needed baking wheat varieties – and preferably a mix of specific varieties – attuned to the specific natural (soil) and social (small-scale production) resources of the region. By cultivating, processing and blending several varieties, a flour of a relatively constant quality could be produced throughout the year. To this end, therefore, the *Zeeuwse Vlegel* involved the cultivation of some eight different winter and spring wheat varieties. Because baking quality is a potentiality of a variety, a characteristic needing to be realised, differences could and did occur (as a consequence, for example, of variations in farming practices, or soil), and batches from different farms needed to be stored separately, in order to determine the baking quality of each batch/farm. This was rather unusual in the Netherlands, where all wheat was implicitly stored as a bulk product. Moreover, the farmers' initiative extended itself to other actors, including millers, contracted to process the wheat into flour, bakeries, to produce the bread according to the *Zeeuwse Vlegel* specifications, and consumers, to communicate the story behind the bread: environmentally sound and based on a philosophy of fair trade, fair price.

4. The issue of technology

A major need for the *Zeeuwse Vlegel* was to obtain access to bread wheat varieties of a good baking quality that were resistant to various fungal diseases and robust in their response to environmental variation. The *Zeeuwse Vlegel* had difficulty identifying varieties meeting these criteria from the varieties placed on the Dutch Variety List. The Variety List is a descriptive list of permitted varieties, a list indicating some specific characteristics and which operates as a regulatory register proscribing unlisted varieties – with the notable exception of varieties that have been approved in other countries in the European Union (the European Community when the *Zeeuwse Vlegel* was developed). EU-regulations state that a variety included in the list of one member state may be cultivated and traded throughout the Union. The varieties included in the Dutch Variety List were high yielding non-baking types, attuned to the practice of animal feed production, with its concomitant assumption of wheat as a bulk product, as well as to the practices of the milling industry of enrichment, by combining high quality wheat from abroad with the low quality Dutch varieties, or by adding specific additives to the flour processed from homegrown wheat.

In view of the strong bias towards varieties attuned to large-scale agriculture, the *Zeeuwse Vlegel* decided to search for varieties that were outside the Dutch list but allowed and cultivated in other EU (EC) member states. An agricultural advisor of the extension service in Zealand, who had a keen interest in wheat and was familiar with varieties in other countries, recommended particular varieties cultivated in Belgium and France. By 1990, the *Zeeuwse Vlegel* had identified a number of promising foreign varieties and proceeded to contract the Zeeland Experimental Station (a research body linked to the ministry o agriculture) to test these varieties. Testing was to refer to criteria related to the specific cultivation practice of the varieties – indicating varieties with a reasonable degree of pest resistance, weed suppression, robustness, etc. – and also to determine baking quality. Later, the *Zeeuwse Vlegel* decided to contract the station annually for such evaluations (Boef and Jongerden, 2000).

In 1998, the *Zeeuwse Vlegel* cultivated eight varieties on a total area of 70 ha. Of these varieties, two proved to be of particular importance: the winter variety Renan[150], a variety bred in France and

---

[150] Renan had been bred by INRA (Institut National de la Recherche Agronomique) in Rennes, France. It is a very hardy variety which is today being rediscovered both by organic farmers and those working in genomics, because it carries an original chromosomal region for resistance to fusariosis. See: http://www.international.inra.fr/press/the_genetic_resources_centre_for_straw_cereals

cultivated predominantly in France, Belgium and Germany, and the spring variety Sunnan, bred and mainly cultivated in. Furthermore, dough made from a mixture of varieties with Sunnan gave a yellowy color to the bread, which came to be considered one of the fine distinguishing marks of the *Zeeuwse Vlegel* bread. Unfortunately, the availability of Sunnan was under threat since it had been withdrawn from the (international) market in 1991, and was no longer placed on the national Variety List. Sunnan is a spring wheat variety developed by Weibull AB, a Swedish breeding company. It had first become available on the Dutch market in 1986. However, testing results for Sunnan for recommendation on the List of Varieties turned out to be rather poor. The *Sunnan* yield performance appeared low – at only 93 percent of the average of that of recommended varieties – and Sunnan was not marked very well in the List. For the *Zeeuwse Vlegel* farmers this marking was not very important, as the good baking quality and high level of disease resistance of Sunnan were considered more important than simple yield performance. For the wider sales prospects of *Sunnan* however, the low marking was fatal. In 1991, Cebeco-Handelsraad, an agro-conglomerate and Weibull agent, withdrew the variety, specifically due to its limited sales in the Netherlands, and Weibull abandoned its breeder's right. From 1991, Sunnan was no longer available on the market.

The withdrawal of Sunnan from the market did not mean that it would be unobtainable and unusable by the *Zeeuwse Vlegel* farmers. Farmers could still save seeds, make use of their farmer's privilege, and produce Sunnan on their farm. However, under the seed law, farmers were not allowed to sell or exchange Sunnan seed. This implied that *Zeeuwse Vlegel* could not save the seed at the level of the collective. Instead, each member-farmer of the *Zeeuwse Vlegel* had to save Sunnan seed individually if he wanted to continue to be able to produce the variety. With the presence in Sunnan of those characteristics which were highly attuned to the specific farming practice of the *Zeeuwse Vlegel* initiative, its farmer-members wanted the variety to be available without such restriction and searched for a solution.

Since the breeder had abandoned its breeder's right, Sunnan became a public or 'free' variety. A possible solution would be for *Zeeuwse Vlegel* to submit an application to become an official variety maintainer[151] and to request that Sunnan be reinstated onto the Variety List in the Netherlands. This request was granted by the committee responsible for the compiling of the List of Varieties. The *Zeeuwse Vlegel* contacted Weibull to obtain a sample of original seed, and compared the sample with one maintained and used in Zeeland by themselves. It came to the remarkable conclusion that the Zeeland sample was the more uniform, and the *Zeeuwse Vlegel* therefore decided to use its own sample for seed production. The *Zeeuwse Vlegel* farmers had clearly done rather well in variety maintenance.

Zeeuwse Vlegel made history by becoming an official *Sunnan* variety maintainer. Its action resulted in a first time *reappearance of a removed variety* on the List. Moreover, this reappearance came at the request of a farmer group. The case was also unique in that a *farmer organisation* became *an official variety maintainer*. The reappearance of the removed Sunnan variety on the Variety List and appropriation by the *Zeeuwse Vlegel* of maintainer status for a variety illustrates that a re-appropriation of technological artifacts such as specific varieties is also possible within modern industrialised agriculture; it also shows that the process of (agricultural) modernisation is not inevitable and irreversible. Instead, the act of challenging social-technical arrangements may create new opportunities. The *Zeeuwse Vlegel* now

---

[151] A maintainer of a variety is a person or organisation responsible for maintaining the variety and ensuring that it remains true to type throughout its full life-span. See also: http://www.oecd.org/dataoecd/0/63/33999126.PDF

produces *Sunnan* seed, and not only for cultivation by its members but also for sale to organic farmers, who value the variety's characteristics of robustness and good baking quality[152].

The *Zeeuwse Vlegel* was not the first organisation to be confronted with problems in the socio-technical construction of variety protection and regulation. Zelder, a grain breeding company specialising in wheat, had had some negative experiences when trying for a new and more sustainable approach to crop improvement. At the end of the 1970s, the company developed a variety named 'Tumult' (an appropriate name, as soon became clear). Tumult was a multi-line wheat variety, i.e. a variety composed of different lines, or components, that are *indistinguishable* from each other in terms of agronomic characteristics. Each bag of seeds contains a mix of lines, the only difference between the components being that they have different resistances to disease, the basic idea of the multi-line being to bring into a variety a mix of resistances, in order to prevent a disease from breaking the resistance of, and thus destroying, a whole variety. Tumult was composed of five components, and resistant against the fungal disease yellow rust. The Committee for Plant Breeders Rights in the Netherlands, however, concluded that Tumult could not be granted breeders' rights. Being composed of five different plant types with distinguishable differences in disease resistance, the 'variety' was deemed *insufficiently uniform*. There was also a problem with the naming of the components. Zelder had named them 'Tumult 1', 'Tumult 2', etc., which was rejected since varieties had to have phantasm names, without numbers, in order to be the subject of applications for breeders' rights. Zelder responded by naming the components *'Knal'*, *'Boem'*, *'Pang'*, *'Plof'* and *'Dom'* ('Bang', 'Boom', 'Thud', 'Splat', and 'Stupid'). The option to apply separately for breeders' rights for each of the five components was not realistic. Different only in respect of a single resistance gene, they were barely distinguishable from each other, and hardly distinctive plant varieties. In addition, testing the lines individually for their cultivation value (by the committee for the composition of the Variety List) would cause problems. The resistance to yellow rust of individual lines was below standard, but when tested as a multiline the variety resistance was superior. Eventually, the breeder was able to convince the registration authority to test and approve the multiline variety as one entity. Unfortunately, however, commercially the multiline turned out to be a failure. Maintaining the lines and producing the seed separately proved too expensive (Louwaars 1997). After Tumult, Zelder did develop one more multiline, named 'Grand Prix' (composed of Grand Ferrari, Grand Matra and Grand Romeo), but the breeder had difficulties in maintaining the *stability* of the multiline. After a few generations, the ratio between the components had changed from 20 percent per line, to one ranging between 37 percent and 7 percent. In the end, Zelder gave up on this resistance strategy.

The Tumult multiline variety example clearly demonstrates that seeds as technical artifacts are developed within specific contexts, from specific techno-institutional arrangements. They show us that the construction of technological artifacts – and seeds may be considered a technological artifact – is socially constructed, and that the development trajectories of these technological arte-facts, and choices between trajectories, are (also) the result of political arrangements.

The *Zeeuwse Vlegel* was not only interested in maintaining and storing available varieties that had proved to be of crucial importance. It was interested also in identifying existing varieties of which it had no knowledge, but which may be of potential use for the organisation, and in stimulating the development of new varieties appropriate for its purposes. The main selection criteria for identifying and developing

---

[152] History repeated itself. Renan, the other variety highly appreciated by the *Zeeuwse Vlegel*, was also withdrawn from the List of Varieties and the market. The *Zeeuwse Vlegel* wanted to keep the variety available and successfully applied to become an official maintainer of Renan.

such varieties were quality (baking quality) and robustness (resistance against diseases). Two initiatives were taken: first, an attempt to engage in a collaboration with the Center of Genetic Resources in the Netherlands for the identification of sessions which could be of interest for the *Zeeuwse Vlegel*; and second, an attempt to become involved in breeding in the Netherlands.

In 1996, efforts were made to establish linkages between the *Zeeuwse Vlegel* and the Center for Genetic Resources in the Netherlands, CGN, or the Genebank. These efforts made within the framework of the so-called 'Sustainable Development Agreement' (SDA), in which biodiversity was a key issue. The SDA principle of participation of interests groups made it possible to include a pioneer project in which the *Zeeuwse Vlegel* and the Genebank could explore the possibilities of utilising genetic resources *in situ*.[153] An evaluation by the project Breeding for Diversity[154], funded by a Dutch NGO called Eco-operation, the coordinating agency for the SDA, concluded that there were three main barriers to establishing links between conservation (Genebank) and utilisation (farmers). First, there was a *limited access* to the material stored in the Genebank. The characterisation of the material in the Genebank did not include data on, for example, baking quality. Evaluation of the material on other characteristics (such as baking quality), which might be required by farmers (like those of the *Zeeuwse Vlegel* initiative), was rather expensive. Such costs operated as a prohibitory factor, making utilisation of the Genebank resources difficult and thus effectively denying access. Second, there was a *limited availability* of material in the collection of the Genebank. The Genebank only provided bona fide users with small seed sample collections. These samples were sufficient for breeders, the main user group, but not farmers, who needed larger samples. Third, there was a *limited usefulness* of material in the Genebank. Most of the Genebank material could not be directly used in agriculture. Breeding and further selection were needed to make the material useful, which was not in the mandate of the Genebank.

With funding from the SDA, the *Zeeuwse Vlegel* was paid for a trial evaluating thirteen winter wheat and ten spring wheat varieties from the Genebank for baking quality, aimed at the identification of suitable baking wheat varieties. All of the twenty-three varieties proved unsuitable. In view of this negative result, other attempts to identify suitable varieties for baking in the collection were abandoned. The material collected and stored in the Genebank proved to be of little value for direct use. Although a Center for Genetic Resources, the Genebank was under-utilised and lacked the political mandate to develop the resources stored: it was simply equipped with the facilities to store and evaluate the diversity in its collection (Boef and Jongerden, 2000).

---

[153] The SDA itself basically consisted of three bilateral Sustainable Development Treaties signed in March, 1994 between the government of Netherlands and those of Costa Rica, Bhutan and Benin. The governments of the four countries involved desired to promote the implementation of the 1992 Declaration on the Environment and Development and Agenda 21, adopted by UNCED in Rio de Janeiro, Brazil, with 'the goal of establishing a new and equitable global partnership through the creation of new levels of cooperation among States, key sectors of societies and people' (for the complete declaration, follow the link at http://www.un.org/documents/ga/conf151/aconf15126-1annex1.htm). In the 1994 bilateral treaties, the countries agreed to 'establish long-term cooperation between their countries based on equality and reciprocity as well as consultation and mutual assistance to effectively and efficiently pursue all aspects of sustainable development, thereby promoting the participation of all interest groups in their respective societies'. Included in each of these treaties was a stipulation whereby the cosignatory to the agreement with the Netherlands could nominate a Dutch project to be supported by the Dutch government. Bhutan selected the *Vlegel*.

[154] The Breeding for Diversity project has been implemented by the authors in cooperation with the Regional Food Network composed of the *Zeeuwse Vlegel* Initiative, the Lois Bolk Institute (representing organic farmers), the Van der Have seed enterprise and the Wageningen University Science Shop.

One may suggest that this minimal operation of the Genebank only represented a first stage. Initially, that is, the idea was just to establish a bank for storage, like a library or museum, and the use to which it would be put could be left for later, as a separate (secondary) issue. This apolitical perspective does not really account for the facts, however. Why, for example, was information on wheat yield and disease resistance included, but that on baking quality not? Are we really to believe that the inclusion of (quantity-related) information useful to agrobusiness and the non-inclusion of (quality-related) information not useful to agrobusiness was mere coincidence? In fact, just like the functioning of the Variety List, the functioning of the Genebank, with its limited access, availability and usefulness (evidenced by the *Zeeuwse Vlegel*-related trial), showed that stored materials – just like the placed varieties – are, indeed, socially constructed artifacts. Furthermore, they are socially constructed in such a way as to be hardly useful at all for alternative agricultural trajectories, which implies that they also ideologically affirm the existing reality, i.e. the functioning of modern industrialised agriculture. In this sense, the failed attempt to link the Genebank to the *Zeeuwse Vlegel* initiative described above illustrates well the difficulty of linking institutions developed within the modernity project to alternatives. The very nature of the hegemonic is to define and repel alternatives: it does this in many ways, perhaps the most pervasive of which is through the assumed ideology governing the operation of its sanctioned bodies[155].

Another effort to create room for maneuver for regional and organic farming initiatives in the Netherlands concerned the efforts made to develop a participatory breeding approach for quality agriculture. Rethinking conventional breeding strategies means, above all, recognising the key roles of farmers and their knowledge and social organisation(s) in the management and maintenance of agrobiodiversity. Recognising these roles is the basis of the approach to agricultural research known as 'participatory plant breeding' (PPB).

Participatory plant breeding is a crop improvement strategy that assumes and promotes the active involvement of farmers. Historically, participatory plant breeding 'principally aims at more effectively addressing the needs of farmers in marginal areas in developing countries' (Almekinders and Elings, 2001). Participatory approaches have typically been developed and implemented in areas where farmers rely on their own varieties and relations with institutional breeding are weak, i.e. farmers do not, or only to a very limited extent, make use of breeders' varieties. These are predominantly regions in the southern hemisphere, where commercial plant breeding has not ubiquitously achieved a hegemonic position. By stimulating farmers to participate in the research agenda of the commercial plant breeder, the breeder hopes to be able to increase the relevancy of the farmers' work and to penetrate production systems it was previously unable to. Relevancy in this context refers to the end product, an institutional variety, actually being adopted by the farmer in his/her production system, while penetration implies that the farmer is willing to return to the institution to obtain seeds. This concept of participation does not imply that the farmer is a genuine or equal partner of the breeder, since farmers are in most cases not involved in agenda setting. Their participation is merely instrumental, to ensure the adoption of institutional varieties by farmers working under harsh conditions.

In the Netherlands and with the case of quality agriculture, the attempt to design a participatory approach was of another dimension. It was not oriented towards an extension of institutional breeding, but a re-enhancement of breeding activities by farmers (although limited to selection). This implied a break with

---

[155] It is pertinent to note here that although the case cited concerns a historical narrative, some of the difficulties encountered by the *Zeeuwse Vlegel* initiative during the 1990s, the problems related to the Genebank (analysed as three types of barriers), and the deeper issues thereby revealed, remain unresolved.

a structural trend in breeding over the last decades, or even centuries. In modern history, institutional varieties (both private and public) had become the dominant varieties, replacing and displacing farmers' varieties, and terminating the role of farmers in crop-improvement (at least, in grains). The plan *to re-involve farmers* in breeding and to allow them to select in the breeding programs of institutional breeders the varieties they considered to be best suited to their farming practices, implied a reconsideration of the role of farmers in breeding – a restructuring, in fact, of the whole plant-breeding system. It implied a bottom-up agenda setting by farmers, who would thus penetrate commercial breeding programs. The breeding companies were not very keen on such a form of participation, since it would bring too much transparency to their breeding strategies and expose for scrutiny the value of their material, which is considered a corporate secret. Neither, of course, did the companies have an economic interest in this type of PPB, since the size of the independent-farmer agriculture sector was too small to form a market of any significance. The moves towards PPB came to nothing. All in all, an attempt to break down the walls of institutional breeding, and re-value the role of farmers as breeders, failed.

5. Conclusions

This article has explored some issues around the argument that challenging the kinds of technology incorporated in institutional arrangements is crucial for the development of alternatives in agriculture. It has been emphasised that alternatives in agriculture need alternatives in technology development. Evidence presented – concerning the Variety List, the Genebank and participatory plant breeding – has shown how transformation from an industrial production system to a quality and socially embedded production system requires the development of alternatives in technology and a radical review of the institutions involved in technology development. The case discussed, the *Zeeuwse Vlegel* initiative, focused on arable farming and crop improvement, showing that the transformation to a new production system – regional/quality agriculture, here – needs a change in technological choice. Such change is only rendered possible by the emergence of alternative production systems, production systems based on the development of new social relations in production, which must then deal with in-built ideological assumptions supporting and supported by the currently dominant system (the large-scale, mechanised production mode of modern, industrial agriculture). Even though the focus has been on crop improvement, we would claim the argument itself to be of much wider significance, clarifying this with the statement that improved seeds are not simply a biological product but also a social product.

The politics and policies of agricultural production and development are reflected and embedded in technology. Varieties can be the seeds of industrial production but also of socio-economically and ecologically sound production; they can be the seeds of global flows of agricultural components or the seeds of locally embedded systems. This implies that the social struggle for alternatives in agriculture also is a social struggle over the development of technology. And this makes the issue of technology too important to leave it to so-called experts and policy-makers. By way of conclusion, let us cite two important philosophers of our time, Andrew Feenberg and Antonio Negri. In Questioning Technology, a critical theory of technology-issues, Feenberg concludes that 'technology is not a fate one most chose for or against, but a challenge to political and social creativity' (Feenberg, 1999: 225). Negri, in his turn, argues that 'it falls to society to claim (bio)technology as a field on which to fight for freedom, and not as an extension of the subjugation, exploitation, and marketing of life' (Negri and Dufourmantelle, 2004: 83-4). These citations cover precisely the concerns of the idea of a reconstruction of technology, and put the challenge to develop new techno-research agendas squarely on record.

References

Almekinders, C. and Elings, A. (2001). Collaboration of farmers en breeders: Participatory crop improvement in perspective. *Euphytica* 122 (3): 425-438.

Beeker, J., van Ginkel, G., Mol, E. and de Wolf, J. (1995). *Postmarxisme als alternatief: een aanzet tot een debat over postmarxistische ontwikkelingstheorieën.* Wageningen, Imperialisme Kollektief.

Boerengroep (1973). *Over de landbouwvoorlichting: materiaal voor een kritiek op de voorlichtingskunde en agrarische sociologie.* Wageningen, De Boerengroep.

Boerengroep (1988). *De witte motor in revisie: discussiebrochure over de toekomst van de zuivelindustrie.* Wageningen, De Boerengroep.

Buechler, S.M. (1999). *Social Movements in Advanced Capitalism.* Oxford University Press.

Davison, A., Barns, I. and Schibeci, R. (1997). Problematic Publics: A Critical Review of Surveys of Public Attitudes to Biotechnology. *Science, Technology & Human Values* 22 (3): 317-348.

De Boef, W.S. and Jongerden, J. (2000). The Netherlands: A genebank, a farmer organization and wheat diversity. In: *Tales of the Unpredictable. Learning about Institutional Frameworks that Support Farmer Management of Agrobiodiversity.* PhD Thesis, W.S. de Boef. Wageningen: Wageningen University.

Feenberg, A. (1999). *Questioning Technology.* London, New York, Routledge.

Imperialisme Kollektief (1988). *Beunhazen en projektontwikkelaars: Landbouwuniversiteit en Derde Wereld.* Wageningen, Imperialisme Kollektief.

Imperialisme Kollektief & Studium Generale (1985). Technology and agricultural underdevelopment in the third world: lectures of the congress organized by the Imperialisme Kollektief and Studium Generale of the Wageningen Agricultural University, October 24, 25, 26 in 1985 at the International Agrarian Center, SG Paper 93. Wageningen, Studium Generale.

Jongerden, J. and Ruivenkamp, G. (1996). *Patronen van Verscheidenheid.* Wageningen, Wetenschapswinkel.

Kwekkeboom, M. and Kleiboer, B. (1989). *Het EG Zuivelbeleid ter Discussie.* Wageningen, De Boerengroep.

Laclau, E. and Mouffe, C. (1985). *Hegemony and Socialist Strategy: Towards a Radical and Democratic Politics.* London, Verso.

LEI. (2000). Landbouw Economische Berichten. Den Haag, Landbouw Economisch Instituut.

LEI. (2006). Landbouw Economische Berichten. Den Haag, Landbouw Economisch Instituut.

Louwaars, N.P. (1997). Regulatory Aspects of Breeding for Field Resistance in Crops. In: *Biotechnology and Development Monitor*, No. 33. Amsterdam.

Negri, A. and Dufourmantelle, A. (2004). *Antoinio Negri in conversation with Anne Dufourmantelle, Negri on Negri.* London, New York, Routlegde.

Offe, C. (1985). New Social Movements: Challenging the Boundaries of Institutional Politics. *Social Research* 52(4): 817-869.

Studium Generale, Boerengroep, Imperialisme Kollektief & WSO. (1974). *Studium generale kongres. Technische vooruitgang: waarom en voor wie?* Wageningen, Studium Generale.

Wiskerke, J.S.C. (2000). The Netherlands: farmers' renewed interest in genetic diversity. In: *Encouraging diversity, the conservation and development of plant genetic diversity* (Eds. C. Almenkinders and W. de Boef). London, Intermediate Technology Publications.

Communic(e)ating: communication and the social embedding of food

Guido Nicolosi

1. Introduction

The public debate on environmental conditions is to a large extent founded on judgments, appraisals, forecasts and desires that concern two key dimensions of the life of Man: *technique* and *territory*.

The disagreeable sensation pervading some of those following this debate is that it proves marred by a serious error of perspective. I refer, in particular, to the consolidated tendency to think of technique and territory as neutral and objective entities; and to the consequential choice of leaving them to be dealt with by only those with specific technical-scientific expertise (engineers, geologists, etc.).

Territory and technique, in truth, are socially rooted entities and no serious reasoning on them should neglect a deep reflection on the socio-historical conditions that determine their definition. Territory is socially rooted to the extent in which it is defined in a continuous and mutual exchange between ecological environment and socio-cultural conditions of human groups that inhabit it. It is precisely technique that is the main form of mediation implied in such exchange. But far from neutral, technique is itself produced by Man. By means of technique, Man projects himself, his culture, to the outside, in time, creating an objective and *meaningful* world, in which he lives.

Such a relationship is not determinist, neither unidirectional: by means of technique Man also expresses the way in which the natural world conditions and constrains him (Figure 1). The relationship with the world needs mediating[156] and this can be done with material objects (forks, umbrellas, etc.), but also with immaterial or symbolic objects (beliefs, codes, etc.). But the same material objects incorporate

---

[156] That there is a direct relationship of human beings with nature is a much empirically doubted fact (Vygotskij, 1974).

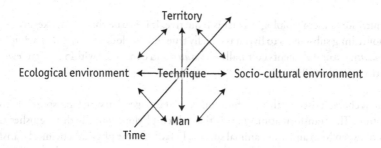

Figure 1. Model of eco-cultural evolution.

symbolic systems and cultural practices and by mediating Man's relationship with the world carry out an important communicative function[157].

But technique is not only the medium of Man's relationship with the non-human, the non-social. Technical action, namely 'a kind of acting that creates artifacts', is also one of the fundamental conditions that determine the very social nature of Man. Indeed, there are three modalities of technical acting: (a) for the purpose of *employment*; (b) directed toward the *modification* of what is found; (c) and *productive* (Popitz, 1990: 95). But to employ, modify and produce implies, next to a subject-object relationship, also a subject-subject relationship:

> '...in this acting there are also determined social conditions of the human species at the same time. The intention of employment unavoidably raises the issue of property rights, modification brings a determined form of social exercise of power (therefore not only 'power' over objects), and production leads to the diversification of activities, and therefore always a form of division of labour. Since men act technically, their living together is determined by *property*, *power* and the *division of labour*. (*ibid.*: 95-96)

In this sense, technique is 'a form of social relationship and production'. Rather than the social consequences of technique, we should discuss 'socio-technical systems', namely: '*networks of human and technological elements operating jointly in order to achieve a determined purpose*' (Gallino, 1998: 18-19; but also see Latour, 1995).

The case of food is, in my view, one of the most emblematic. It is an entity that is at the crossroads between territory and technique highlights well the need we have to update the 'toolbox' for the appraisal of environmental and development problems. This is what, here, we resolve to show. In particular, in the first paragraph, we will discuss how food is placed between nature and culture and how it is embedded in a socio-cultural matrix; in the second paragraph, we will try to show how the existence of such matrix is rendered evident by the influence played by communication in alimentary processes; and, finally, in the two successive paragraphs, we will analyse how in modernity, despite processes of uprooting nutrition from the context of appurtenance, human beings continue to feel the need to 're-embed' food in a meaningful universe. Moreover, how communication continues, for this reason, to play a decisive role that any project of redefining alimentary technology must take into consideration.

2. Food: nature and culture

To consider nutrition a merely biological fact is reductive. If it is true that Man, like any other animal, must ingest nourishing substances to live, it is equally true that he does so only after having transformed them into *aliments*: natural elements culturally *elaborated* and consumed within the framework of social codified practices (Poulain, 2002: 11).

The food chain is characterised by three or four fundamental stages: harvest, (transport), transformation and consumption. The transformation stage is that which phylogenetically distinguishes the greatest discontinuity between Man and other animal species. In fact, if some physical-chemical transformations are also done by other living species, humans are unique in practicing cooking and combining foods. For

---

[157] Naturally, the meanings communicated must then be interpreted (Geertz, 1987) in the context of practices exposed to changes of socio-historical conditions in time, in a ceaseless activity of production and *re*-production of the world in which the distinction between structure and subject and between theory and practice tends to vanish (Bourdieu, 1980).

this reason, for Man feeding means using natural products from a specific territory and transforming them through culinary techniques. That is, if the alimentary act unites Man with the animal, the culinary act (which is a techno-cultural act) belongs to an order of experience that is exclusive to the human species (Perlés, 1979: 5). Already in this, Man shows his specificity: to crossbreed nature and culture with techno-transformative action[158]. In this sense, I propose to consider the human condition as 'physiologically' *bio-technical* and bio-cultural (and therefore bio-social, bio-political, etc.): from his very beginnings, Man has always lived in a bio-technical environment, in which nature and culture are continually interbred.

There is a complex relationship between culture and nutrition. Survival is guaranteed by satisfying alimentary needs; but, in the case of Man, this is achieved in different and at times hostile natural environments. This is possible (beyond that of a specific omnivorous physiology) because Man is a social animal endowed with culture and symbolic ability to construct more favorable environmental conditions. For this reason, in the case of the human animal, many anthropologists rightly speak of bio-cultural evolution. Now, we know well enough the laws of biological evolution. But in the case of culture things are infinitely more complex and it is difficult to be able to speak of cultural adaptation as *fitness*. It is necessary to be wary of socio-biological readings or those that explain cultural development as a function or aligned to fundamental biological needs (in the field of alimentation, see Harris, 1990). The two developments *can* be aligned, but are often independent. Indeed, some developments or cultural options can also be counter-productive from the biological adaptation point of view (De Garine, 1979).

Man does not only feed on food, but also on culture and must satisfy the two levels that compose him: the material (corporeal) and immaterial (symbolic). And it may happen that symbolic-cultural requirements provoke disastrous endemic deficiencies of a nutritional kind[159]. If we add the fact that various cultures respond to alimentary needs in different ways, we can easily see that in the case of cultural processes it can be misleading to speak of 'evolution'[160]. It would be more correct to admit a good dose of symbolic *arbitrariness*.

But it would be wrong for this reason to remove the cultural dimension from the context of what must be scientifically analysed. At times, genes adapt more readily than human cultures to changes. Any kind of new intervention aimed at the introduction of new practices, technologies or food production must be able to deal with this aspect.

---

[158] Already these short and banal considerations might induce reflection by those who from 'integralist' positions oppose any kind of technological intervention on nature (and food).

[159] It is no accident that the phase of weaning (passage from 'natural' to 'cultural' nutrition) is the riskiest moment for babies' survival.

[160] For example, with the advent of the Neolithic age the agricultural revolution had the effect of increasing enormously the amount of food available and the possibility of storing foods, but at the expense of a slow but significant narrowing of the qualitative spectrum of alimentary consumption (Gaulin, 1979: 48). In effect, agriculture marks a paradox: an animal that bases its evolutionary successes on the condition of omnivore, 'chooses' to practice a relative but important alimentary specialisation. Such a choice is evidently cultural and to some extent can be read as a veritable 'regression': 'Agricultural societies, by partially reducing the fluctuation in resources, or at least, the irregularity of alimentary cycles, introduced the risk of crises with catastrophic consequences' (Fischler, 1979: 197). Nevertheless, what from the point of view of mere biological adaptation might be considered dysfunctional behavior, from the historical point of view has had exceptional implications: the affirmation of the earliest kinds of private property by means of production, construction of cities, foundation of empires, until arriving at the modern industrial era.

3. Food and communication: an inseparable binomial

Before the advent of the so-called alimentary modernity, there were substantially two fundamental junctures of the history of human feeding (Arsuaga, 2004: 10-11): the first marks the introduction, approximately 2.5 million years ago (*Australopithecus* and *Homo habilis*), of significant amounts of products of animal origin[161]. The second, approximately 10,000 years ago (Neolithic), marks the passage from a form of feeding based on the harvest of spontaneous products in nature to their organised and planned production by means of agriculture and breeding. Both moments have represented fundamental stages in the evolution of the forms of social organisation of Man.

The introduction of the products of animal origin meant the passage from the condition of gathering to that of hunting (after a probable stage of necrophagy). Such a change most likely contributed to make feeding become a collective and sexually differentiated action. In particular, since Man was not endowed with suitable 'natural instruments' (tusks, claws, etc.), there was the need to develop an action of collective hunting with an advanced level of coordination[162]. This implied, besides lengthening the alimentary process (between need and its satisfaction), also a meaningful development of the system of communication between the players involved (Perlés, 1979: 6).

But as already anticipated, it was the affirmation of the practice of cooking food by fire (culinary action) that marked the real phylogenetic rupture between Man and animal. Though difficult to date exactly, it most likely happened in the Paleolithic. From the socio-anthropological point of view, it marks a decisive change: the reinforcing of the processes of differentiation of tasks and the consequential end of the alimentary autarchy of the individual. The gap between the need for food and its satisfaction, between hunger and consumption, was further widened. The sequential chain of the necessary operations became relatively long and stimulated the elaboration of a complex *project* that implies the refining of the human characteristics: oral communication founded on a shared system of signs and previsional ability.

Moreover, the practice according to which feeding became first of all a social action founded on the division of labour, exchange and reciprocity was consolidated. Most likely, from this moment on the consuming of meals was to be done in a non-individualised way, in a shared place and time, becoming, therefore, a fundamental practice of community integration. Probably in this particular moment of the hominisation process food takes on a privileged function of *medium* of communication, acquiring in such a way an exceptional symbolic charge[163] (*ibid.*: p. 9).

Many authors have found this charge stimulating for their own reflections. Lévi-Strauss, for example, in the context of his structuralist theory of culture as communication (2002, 2004). He highlighted[164] how through cooking food Man uses the binary opposition between the non-elaborated and elaborated

---

[161] At the same time, and the two things are probably correlated, rudimentary stone tools made their first appearance.

[162] Also the harvest of vegetable products may not be exclusively individual, but it implies a decidedly reduced coordination and division of labour.

[163] For example, meat, the product of hunting becomes a male food par excellence. The distribution of the venison among the members of the group, moreover follows a logic of social importance based on roles, hierarchies, exchange and allegiance relationships.

[164] Extending Jakobson's theory of the acquisition of the sounds of language to the distinction between nature and culture.

to represent the transforming function of culture over nature (against the transforming action of nature). As asserted by Deborah Lupton:

'According to Lévi-Strauss, culture is the complex of those practices that distinguish human beings and make them unique. Alimentary practices exemplify this opposition, above all in concert with other dichotomies like that between the raw and cooked, between food and non-food. In fact, for Lévi-Strauss, cooked food is a cultural transformation of raw food, a way to modify and demarcate nature. Therefore, cultures are defined by the transposition into daily life of this transformation' (Lupton, 1999: 19).

In truth, the raw falls mid-way between nature and culture (see Figure 2), to the extent in which albeit not elaborated like cooked food, it can be served as a course of a meal (fruit, vegetables, etc.). As is known, moreover, for Lévi-Strauss a society's cuisine can be analysed, exactly like a language, according to structures of opposition and correlation and based on specific *gustemes*.

Mary Douglas has often stated that the alimentary dimension is able to go beyond the border between the sphere of the natural and that of the cultural. Needs of the body and needs of society are, in food, inextricably interwoven, so much so that 'it is culture that creates, among men the *communication system* that defines the edible, the toxic, satiety' (my italics) (Douglas, 1979: 146). According to the English anthropologist, food practices are rigidly grounded and codified within precise cultural systems; a *gastronomy* is always a rich system that comprises a cosmological taxonomy and a broad ensemble of rules that express representations and beliefs of the subjects that belong to a certain culture (or a group within it) (Douglas, 1993).

Following this formulation, with reference to the alimentary problems of development, Douglas emphasises how in communities (often rural ones that feed following the close ties with seasonal correspondences) it is utopian to think of introducing important changes in food practices without deeply affecting their fragile inner equilibrium and what binds them to the environment. In general, the overly abrupt introduction of new foods may even provoke serious forms of widespread anxiety. Equally, the forced realisation of a plan of agricultural development foreign to local ethno-gastronomic traditions can have the same negative effect as the compulsory introduction of a foreign language.

'In order to understand and evaluate a local system, it is necessary to set it in the context of the net of interconnections that bind it to other family institutions and in the framework of the articulation of the family with the broader social institutions of the community.

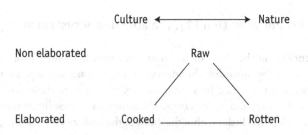

Figure 2. Lévi-Strauss' culinary triangle (1965).

Analysing the global food crisis, some insist on the need to rehabilitate traditionally neglected alimentary resources. Perhaps it would be simpler to improve a staple food rather than manage the effects of the introduction of a new produce. But sometimes these attempts to improve the quality of a traditional food clash with the resistance of the local population. Some new foods are readily introduced into the traditional system, whereas others (whose taste is not imperceptibly different) are categorically refused. There is, therefore, from a nutritional point of view, an immediate interest to focus on the cultural aspects of alimentary systems' (Douglas, 1979: 150).

The close network of relationships between alimentary practices and social institutions, often family[165], are evident not only in the case of rural communities or traditional societies. Suffice it to recall the existing connection, in western contemporary societies, between alimentary modernity and the processes of female emancipation and the consequent radical transformation of familiar structures. In this sense, we reiterate, for Douglas food represents a real *system of communication* able to express social and symbolic relations that exist within the family, groups and communities (Douglas, 1985).

For Igor de Garine, the food-communication nexus is also central. It is verified empirically that human groups living in very similar environmental conditions develop alimentary behaviors and give value, from nutritional and symbolic points of view, also highly varied foods. This fact demonstrates that feeding represents one of the main communicative registers that allow asserting the inner cohesion of a society (De Garine, 1979). Alimentary behaviors, in such a perspective, carry out a *demarcative* function aimed at boosting cultural heterogeneity. Moreover, according to De Garine, valued foods serve to explicate the *differential gaps* between groups and categories that cohabit in each culture. The interpretation given by the French anthropologist is highly evocative. The differential gaps are not found at a biological level, but at cultural one that performs a function similar to that carried out in the field of sexuality by the taboo of incest: by differentiating alimentary behaviors, the inner cohesion of the original groups, as well as the social heterogeneity between individuals, categories, social groups and cultures is strengthened at the same time. This makes communication and exchange possible, rather the main 'instruments' by means of which human societies are formed and persist.

In conclusion, we have briefly shown how communication has great importance in the definition of alimentary processes. This importance, that provides the main confirmation of their embedding in a social-cultural matrix, has always had a twofold valence: on one hand, nutrition (in all its stages: harvest, transformation, circulation, etc.) always implies a social exchange of a communicative kind; on the other, food may be interpreted as a symbolic system of communication in itself. Both levels have been radically overturned by the advent of alimentary modernity.

4. Modernity and communication (1): the alimentary system as expert system

Various views, depending on the more or less nostalgic attitude of their proponents, paint a past characterised respectively by splendid abundance or frightening scarcity. In truth, pre-modern alimentation was always characterised by a cyclical *alternation* of both these conditions (in addition to periodic states of normality) and *Homo sapiens* has shown an extraordinary physiological ability at adapting: to eat too much or too little (within the limits of the biologically possible). Closely correlated

---

[165] Feeding must be considered in the wider context of family systems of care of the body (Douglas, 1985, p. 197).

to this condition (the potential risk of hunger), Man has historically developed a marked propensity toward alimentary *differentiation* (Montanari, 2003).

The affirmation of alimentary modernity has been a slow and gradual process. It has undergone a violent and radical acceleration with the industrial revolution and the affirmation of the capitalist and globalised economies. The main effects of alimentary modernity have had an impact precisely on differentiation and alternation[166]: both have undergone an inexorable specialist reduction[167].

Alimentary modernity and the affirmation of the long globalised chains of production create a process of *uprooting* of the alimentary act from the local and specific 'eco-bio-cultural' ('techno-territorial') context. Through the *taylorisation* of production, transformation and distribution[168] food 'takes off' from the territory; and it become 'indifferent' to the characterisations that in the past[169], by permeating it, rendered it unique. In this way, an *abstract* alimentary system is created.

Namely, this is the concrete application to food of those radical transformations that Anthony Giddens has described with reference to modernity in general (1994): *space-time distanciation, disembedding* and *reflexivity.*

For Giddens, modernity has lifted out social relationships from the 'local contexts of interaction', assuring at the same time their restructuring 'across indefinite spans of time and space' (Giddens, 1994: 32). The most important examples of such uprooting processes are the so-called *expert systems[170].* In my view, the modern alimentary system must be studied and understood, in its features and social consequences, precisely as an expert system.

The disembedding of the expert system from the local context of interaction implies that 'trust' becomes essential to its workings. That because social expectations are created in space-time distanced contexts. As Giddens often states, such trust is at the same time an article of faith and what Simmel called 'weak inductive knowledge'. Because of this dual matrix the confidence entrusted to expert systems is, at the same time, a quasi-magical component (of symbolic type) and a pragmatic one (tied to observation and experience). We have confidence in the expert systems because they are constructed on specialist *knowledge* that is inaccessible and mysterious to us, but which some elected few possess (professionals, technicians). These are considered the clergymen of the churches (with their own technical and legal apparatuses and, above all, with their own methods of monopolistic certification). But we also have faith because we know by experience that such systems generally work and satisfy our expectations. This

---

[166] Indeed, that qualitative and quantitative variable cyclic nature characteristic of pre-modern feeding and linked to ecological and cultural constraints: productive seasons, work/rest alternation, festivity, rituals tied to agricultural work, phases of shortage or abundance, religious precepts, etc.

[167] Industrialisation and growing modernisation of agriculture have provoked a two-fold integrated process: specialisation and homogenisation. Every territory is by now part of broad international systems of agricultural and food production. The local/regional dimension loses importance and much of the foods *in loco* are in truth produced outside. The potential increment of the alimentary repertoire effectively becomes transformed into its opposite: homogeneity of the products distributed in the supermarkets in all geographic areas and drastic reduction of the intraspecific variety of foods (Fischler, 1979: 200).

[168] To this should be added the processes of rationalising alimentary consumption through: a) the substantial participation of nutritional science; b) the taylorist-fordist organisation of the food offer (see the *Mcdonaldization* of society by Ritzer, 1993).

[169] Thanks to the earth, climate, seasons and material and symbolic practices.

[170] Giddens defines them systems of technical realisation or professional competence that organise wide areas in material and social ambient in which we live today (Giddens, 1994: 37).

dual nature can sometimes provoke important dyscrasia (at times airplanes crash or cows go mad) that fracture the confidence that is generally granted to expert systems, creating a climate of anxiety.

The expert system is prevalently disembodied, founded on functions, roles, flows, etc. Uprooted from local contexts of human interaction, it works as a professional structure that sets aside sympathetic relations. Nevertheless, the same Giddens recognises that trust given is not only tied to abstract emblems (anonymous); an important part of this trust is conveyed by particular *'facework commitments'171*. In this sense, a fundamental component of the expert systems are the so-called *access points*: points in which the abstract systems open up to the external world, guaranteeing contact between its representatives and the profane. They are very important to ensure the maintenance of trust, precisely so that common people *physically* meet the operators of the system. These *openings* to the profane world represent the interface of the system; in fact, they are literally, spaces of shared interaction. For this reason, the access point constitute, for certain aspects, the most delicate point of the system. The place in which great efforts must be concentrated in order to limit the onset of 'friction' that may throw the entire system into crisis. A crisis that takes on the guise of 'crisis of confidence'.

Now, is important, for our scope, to remember that much of this effort is *communicative*. All the force or weakness of the expert system depends on the workings of the communicative processes that drive it. And, equally important, to a large extent the success of such processes depends on the quality of the non-verbal dimension of the communication. Giddens writes:

> 'Even if everyone knows that the true repository of trust is the abstract system more so than the individuals that in determined contexts 'represent' it, the access points remind that its operators are persons in flesh and blood (and therefore potentially fallible). The facework commitments tend to depend notably on what one may call the *conduct* of the representatives or the operators of the system' (Giddens, 1994: 90).

Naturally, the type of *conduct* varies according to the type of system in question. Giddens refers to the importance of the *nonchalance* that the operators, whose activity is carried out in contexts of manifest danger, must learn to show. He cites the case of the airlines, but the reasoning works readily if applied to the alimentary system as well. Giddens, in particular, uses the theatrical metaphor of Goffman (1959) in order to recall that the management of the communication in the access points consists in the 'mastering the faint border between the stage and backstage', namely:

> 'The clear distinction between stage and backstage strengthens the role of the behavior as a means of reducing the impact of human incapacity and fallibility. Patients would not nurture the same unconditioned trust in the medical staff if they knew well the errors that are committed in the outpatients' clinics and in surgical wards. Another reason concerns the chance spaces that nevertheless persist in the mechanisms of the abstract systems. There is neither ability, however highly practiced, nor form of expert knowledge, however great, to exclude every possible element of risk or chance' (Giddens, 1994: 91).

The border between 'stage' and 'backstage' of an interaction is faint since the relationship between these two dimensions is not necessarily linked to the physical positioning of the actors. What counts in

---

171 Facework commitments are defined by Giddens as relationships of trust supported by or expressed in social bonds instituted in circumstances of simultaneous presence (Giddens, 1994: 85).

determining way are the *models of informative flows* that regulate a certain social situation. To be on stage or backstage depends directly on the *quality* of the (social) information to which one has access:

> 'In effect, the analysis of the definition of the situation may be entirely detached from the problem of the direct physical presence to concentrate our interest solely on the access to the information. If by mistake the intercom in the kitchen is left on and a customer listens to waiters' jokes from backstage, the definition of the situation will be modified even if changes in the physical place or the physical positioning of the participants did not happen. Analogously, in general, the social situations and the behaviors within society can be modified by the introduction of new media' (Meyrowitz, 1995: 60).

This aspect is determining for understanding the role and influence that mass media and the ICT (Information and Communication Technologies) have had in provoking the current crisis of confidence in the expert alimentary system. Electronic and digital media are able to influence the 'definition of the situation' circumventing the material limits imposed by physical structures and geographical and temporal distances; and to break down the barriers constructed to maintain the separation between stage and backstage. To what extent did the images of the staggering cows broadcast by the media during the 'mad cow crisis' influence the propagation of the food crisis? In our view, very greatly indeed. They managed to circumvent the many official reassurances that, in those troubled times, were issued by various official operators of the system (veterinary surgeons, ministers, retailers, breeders, etc.).

This aspect shows the determining role that the third founding characteristic of modernity plays: *reflexivity*, that is, 'the regularised use of knowledge about circumstances of social life as a constitutive element in its organisation and transformation' (Giddens, 1991: 20). And the media performs a key function in diffusing, accelerating and reinforcing modern reflexivity. Here reflexivity is understood not so much as an enlightened project of cumulative growth of knowledge at the expense of the arbitrariness of tradition. It is seen, on the contrary, as a systematic affirmation of a radical principle of methodological 'doubting'. With reference to food, the example of the media coverage of the so-called 'mad cow' or BSE (Bovine Spongiform Encephalopathy) crisis was emblematic.

And it is for this reason that the importance of communicative processes must be stressed. In fact, what we are witnessing, at least since the BSE scandal, is a ferocious *communication war* with still uncertain results. The protagonists of this communication war are various: environmentalist and consumer protection groups, the assorted *stakeholders* of the powerful food industry, the public powers, and in the middle, the media.

In particular, the food industry and the public powers have been creating a communicative strategy aimed at reducing anxieties heightened by the media representation of the crisis in food safety. They have had to touch on what many sociologists and anthropologists have been saying for a long time: eating unknown manufactured products, without a past and social roots can mean losing the last sense of oneself; and eating 'against nature' (Genetically Modified Organisms) can mean symbolically de-naturalising oneself. Therefore, in order to face the current of modern alimentary anxieties, two paths have been undertaken to restore identity to food: labeling (traceability, guarantee of origin, etc.) and advertising narration. Both realise a new artificial form of *'communicative embedding'*[172].

---

[172] At the Department of Analysis of Political, Social and Institutional Processes of the University of Catania, we have been making an empirical diachronic investigation into the narrative strategies adopted by food advertising (press) that seems to confirm this (see Nicolosi, 2007).

It is interesting to notice, in fact, that advertising strategies carry out a function in many aspects analogous to that carried out by traceability. In both cases, in fact, this concerns, through *narration*, restoring an identity to 'objects' that have become 'opaque' and unrecognisable over time sine they are culturally and socially uprooted. Naturally, they involve different forms of narration. Traceability seeks to tell an 'objective' history, a biographical history of a determined food. Publicity, through *discourse* and *branding*, often tells a mythical or fantastic story (stories), so that the consumer can newly incorporate meanings with the food.

5. Modernity and communication (2): individualised alimentary consumptions

Human identity is a bi-dimensional entity founded both on material (corporeal) and immaterial (symbolic) aspects. We often remove this fact in order to re-discover it in the liminal stages of life (birth, death, sickness), in which the incarnated essence of Man is perceived dramatically. In such a sense, it is certainly true that the human being is largely and has always been what he eats. That means that the consumption of food has always represented an exceptional and fundamental action of symbolic-identity communication.

There are two authors who, in my view, have given prominence to such a 'function': Pierre Bourdieu (1979), and Mary Douglas (1984; 1993; 1999). Albeit having deeply different theories, premises and implications in the background, both have skillfully shown that the choices of consumption (above all food) carry out a *discriminating* function. By consuming, people express their own choices of belonging (alliance) or extraneity (conflict); they communicate to the world their own preferences with respect to: with whom and against whom to be, which world to live in or dream of and which to refuse. Indeed the *refusal* expresses even better than choice the cultural semantics of consumption. In this light, consumer goods are a veritable complex system of information and communication and food shows this aspect emblematically. It is registered in a kind of language endowed with a matrix that is both numerical and analogical. Numerical in that it is based on a conventional code shared and constructed on discreet entities that are easily recognisable:

> 'Let us now take another signifying system: food. We shall find there without difficulty Saussure's distinction. The alimentary language is made of (1) rules of exclusion (alimentary taboos); (2) signifying oppositions of units, the type of which remains to be determined (for instance the *type savory/sweet*); (3) rules of association, either simultaneous (at the level of a dish) or successive (at the level of a menu); (4) rituals of use which function, perhaps, as a kind of alimentary *rhetoric*. As for alimentary 'speech', which is very rich, it comprises all the personal (or family) variations of preparation and association [...] The *menu*, for instance, illustrates very well this relationship between the Language and Speech [...]' (Barthes, 1992: 29).

Analogical in that food ingestion contributes to mould in a non-verbal way the cultural representation of our body:

> 'The action of incorporating food can be considered the apotheosis of the sign left by consumption choices on the body, in an almost permanent way both from the exterior point of view as well as interior: commonly, one believes that skin tone, body weight, the robustness of bones, the condition of hair and nails, digestion, are all directly conditioned by diet' (Lupton, 1999: 41)

Sociology has always discussed the characteristics of the determining factors conditioning the act of consumption. Yet we may say, perhaps a little crudely, that there is a certain agreement on the fact that parallel to the individualisation of the paths of life, also the choices of alimentary consumption, in western modern societies, are increasingly conditioned by 'autonomous' and 'independent' actions of individual will. The sociological implications tied to this phenomenon accompanying the diachronic and synchronic passage from traditional societies to (late)modern ones are enormous and some sociologists have skillfully described it.

Giddens, for instance, describing the individualised reflectivity of the 'Trajectory of the Self', understood as a self-planned identity based also on the planning of the body. In contemporary society, the body is the mirror of the soul (its main communicative medium, one might say) and for such reason, Giddens says, reflexivity extends to the body. Here body is not understood merely as a passive object, but as part of a system of action (*embodied self*) founded on the need to continuously realise individual *choices* oriented at the definition of a particular *lifestyle*:

> 'Observation of bodily processes – 'How am I breathing?' – is intrinsic to the continuous reflexive attention which the agent is called on to pay on her behaviour. Awareness of the body is basic to 'grasping the fullness of the moment', and entails the conscious monitoring of sensory input from the environment, as well as the major bodily organs and body dispositions as a whole. Body awareness also includes awareness of requirements of exercise and diet. Rainwater points out that people speak of 'going on a diet' – but we are all on a diet! Our diet is what we eat; at many junctures of the day we take decisions about whether or not to eat and drink, and exactly what to eat and drink' (Giddens, 1991: 77).

Naturally, self-reflection was not created in contemporary society, but has its roots in complex processes of the past dating back at least to the 12th century. But what is new in contemporaneity is the mass spread of the perception of the body/self as a project and the access to the instruments to giving concrete enactment of this project. In this, the mass-media has carried out a decisive role.

For Pasi Falk (1994), instead, the emotional feature that grounds the social bond in the traditional society is based on the sharing of ritual forms of collective meals (*eating community*); as the affective bond between mother and son is founded on breast-feeding. And it is for this reason that, in simple societies, eating is a rigidly codified activity. One eats together as an act of *sharing*, welding the principles of social *cohesion* and *reciprocity*. This symbolic-communicative dimension of a communitarian nature, for Falk, is lost in modernity, since the alimentary action undergoes an inexorable process of individualisation: in the choices (tastes) and in the practices (eating alone). For Falk, however, the modern body is a 'closed' body and the incorporation of food is the prime moment in which *control* (individualistic) in the definition and communication of the self is achieved (Falk, 1994: 27). The surface of the body, in fact, has assumed the most important communicative function (simil-linguistic) of contemporaneity; and to carry out an integrating function of an oral kind, Falk maintains, there remains only *conversation*. As when, emblematically, it remains the only thing able to ensure the collective dimension in the practice of 'dining out'; the choices on the 'what to eat' being freely and individually decided from the menu.

6. Conclusions

The reach of the 'great transformation' which the alimentary system underwent with the advent of modernity is effectively shown by the semantic change that the expression food (un)safety has progressively been subjected to in the western world.

If referred to the pre-modern era, the meaning of the formula may unequivocally be traced back to the threat, always present, of *hunger*. In the contemporary western world, because of the developments described up to here, the expression has radically changed its semantic characterisation in order to become a synonym of anxiety with respect to the risks of an always potential contamination (physical and symbolical) of food.

More generally, thanks to the transformations which alimentation has undergone, an obsessive attention (individual and collective) seems to have gained ground regarding food. I have sought to describe this phenomenon elsewhere (Nicolosi, 2006, 2007) by means of a new analytical-conceptual proposal based on the analysis of the narrative forms applied to food. The idea is to define (late)modern society as an 'orthorexic society'; that is, a society founded on an alimentary hyper-reflexivity in its various meanings: dietetic (fitness), ethical (critical consumption), aesthetic (food-design), symbolic (slow food), psycho-pathological (eating disorders), anxiogenic (alimentary fears), etc. A hyper-reflexivity embodied in an alarming communicative fragmentation[173]: a veritable narrative Babel (publicity, nutritional science, *television chefs*, etc.) where, however, hegemonic discursive forms also emerge. In short, the orthorexic society is a society obsessed by food, by the 'right' and 'wrong' foods.

Food, once again, is therefore a central symbolic and communicative issue, whose analysis can help us understand the most intimate features of an age[174], also ours. And to reflect on *'Reconstructing agro-biotechnologies for development'* means to never lose sight that the culture of a society, be it traditional or modern, is defined (even) by the way in which that society feeds or produces the food it needs. Suffice it to recall that the term *culture*, of Latin origin, is etymologically linked to the verb *colĕre*, namely to cultivate, and to the practice of working the land (*agri culturae*); and that, for example, the agricultural revolution has marked for Man an exceptional cultural watershed that has its principle expression in the great symbolic valorisation attributed to such a practice by the Greeks and Romans:

> 'Roman culture – like the Greek – did not have a great appreciation for uncultivated nature. In the system of values elaborated by Greek and Latin intellectuals, it had little room: it was, indeed, the real antithesis of *civilisation* – a notion in turn also etymologically connected to that of the *city*, namely an artificial order created by man in order to differ and separate himself from nature. From the productive point of view, that culture had cut out its ideal space in a tidily organised countryside around the cities: what the Latins called *ager*, the ensemble of cultivated lands, rigorously distinguished from the *saltus*, virgin nature, not-human, not-civil, not-productive... the forest was synonymous with marginality, exclusion; ... Clear priority to the practices of cultivation:

---

[173] Fischler (1979) speaks of the era of *gastro-anomy* in which a mosaic or a 'cacophony' of dissonant alimentary criteria prevails: moral, scientific, dietetic, identitary discourses, etc.

[174] 'Dans ce qu'il consomme, l'homme se révèle, mais aussi dans la façon dont il consomme'. In this way, eloquently, Catherine Perlès (1979: 4) opens her interesting text on 'L'acte alimentaire dans l'historie de l'homme'; and it precisely true, that what we eat, but perhaps still more the way we eat, can explain who we are.

agriculture and orchard farming were the core of the Greek and Roman economy and culture (at least if we refer to the dominant models). Wheat, vines, olives were the strong points: a triad of productive and cultural values that those civilisations had adopted as a symbol of their own identity... Entirely different were the modes of production and cultural values of the 'Barbarians' – as the Greeks and Romans called them. The Celtic and Germanic populations, for centuries accustomed to wandering the great forests of Central and Northern Europe, had developed a strong predilection for the exploitation of untouched nature and uncultivated spaces. Hunting, fishing, harvesting wild fruit, wild breeding in the forests (above all pigs, but also horses and cattle) were central activities and characterised their way of living' (Montanari, 2003: 12-13).

In western societies we have forgotten this radication. Equally, technology embodies an ideology that denies the complex socio-technical matrix that generates it. Techno-science hides its delicate function of mediation of the relationship between Man and Environment, proclaiming itself Queen of modernity. This technology, in the age of globalisation, refuses the feedback with a specific territory and sets itself up as a unidirectional and universal flow.

The scientific and economic world is, obviously, the most tenacious supporter of such ideology. Indeed, it is afraid that recognising the social and cultural roots of the forms of human organisation and production may reveal the deception of the presumed neutrality of the development processes and help the affirmation of a principle of responsibility. Rather, the idea of a 'governance' of the techno-scientific development that takes account of the appraisal of the ends that it proposes (Gallino, 1998).

It would surely be an error if techno-scientific development in itself were denied. To the extent in which, as we have said, the nature of Man is bio-technical by definition. As Jonas would say (2002: 13), the *Homo faber* was born as an auxiliary and important part of *Homo sapiens*. The problem is to hinder the definitive and absolute replacement of the latter by the former.

In this sense, research becomes urgent that, uniting scientific and socio-humanistic knowledge, produces a new *cultural theory* of the development of alimentary (bio)technologies. Drawing on an old idea of Mary Douglas, the importance of this research is comparable to that begun in the field of the *new energy sources*. We need one as much as the other, but it must be understood how to render both compatible with the cultural and environmental requirements of Man.

References

Arsuaga, J.L. (2004). *A cena dai Neanderthal*. Milano, Mondadori.

Barthes, R. (1992). *Elementi di semiologia*. Torino, Einaudi.

Bourdieu, P. (1979). *La distinction: critique sociale du jugement*. Paris, Les Éditions de Minuit.

Bourdieu, P. (1980). *Le sens pratique*. Paris: Les Éditions de Minuit.

Douglas, M. (1979). Les structures du culinaire. *Communications* 31: 145-170.

Douglas, M. (1984). *Il mondo delle cose. Oggetti, valori, consumi*. Bologna, Il Mulino

Douglas, M. (1985). *Antropologia e simbolismo. Religione, cibo e denaro nella vita sociale*. Bologna, Il Mulino.

Douglas, M. (1993). *Purezza e pericolo*. Bologna, Il Mulino.

Douglas, M. (1999). *Questioni di gusto*. Bologna, Il Mulino

De Garine, I. (1979). Culture e nutrition. *Communications* 31: 70-92.

Fischler, C. (1979). Gastro-nomie et gastro-anomie. *Communications* 31: 189-210.

Falk, P. (1994). *The consuming body.* London, Sage.

Gallino, L. (1998). Critica della ragione tecnologica. Valutazione, governo, responsabilità dei sistemi sociotecnici. In: *La tecnologia per il XXI secolo,* (Eds. Ceri, P. and Borgna, P.), Torino, Einaudi, pp. 5-24.

Gaulin, S.J.C. (1979). Choix des aliments et Évolution. *Communications* 31: 33-52.

Geertz, C. (1987). *Interpretazione di culture.* Bologna, Il Mulino.

Giddens, A. (1991). *Modernity and self-identity: self and society in the late modern age.* Cambridge, Polity Press.

Giddens, A. (1994). *Le conseguenze della modernità.* Bologna, Il Mulino.

Goffman, E. (1959). *The presentation of self in everyday life.* New York, Anchor.

Harris, M. (1990). *Buono da mangiare. Enigmi del gusto e consuetudini alimentari.* Torino, Einaudi.

Jonas, H. (2002). *Il principio responsabilità.* Torino, Einaudi.

Latour, B. (1995). *Non siamo mai stati moderni. Saggio di antropologia simmetrica.* Milano, Elèuthera.

Lévi-Strauss, C. (1965). Le triangle culinarie. *L'arc* 26: 19-29.

Lévi-Strauss, C. (2002). *Antropologia strutturale.* Milano, Il Saggiatore.

Lévi-Strauss, C. (2004). *Il crudo e il cotto.* Milano, Il Saggiatore.

Lupton, D. (1999). *L'anima nel piatto.* Bologna, Il Mulino.

Meyrowitz, J. (1995). *Oltre il senso del luogo,* Bologna, Baskerville.

Montanari, M. (2003). *La fame e l'abbondanza. Storia dell'alimentazione in Europa.* Bari-Roma, Laterza.

Nicolosi, G. (2006). Biotechnologies, alimentary fears and the orthorexic society. *Tailoring Biotechnologies* 2 (3): 37-56.

Nicolosi, G. (2007). *Lost food. Comunicazione e cibo nella società ortoressica.* Catania, Ed.it.

Perlès, C. (1979). Les origines de la cuisine. L'acte alimentaire dans l'historie de l'homme. *Communications* 31: 4-14.

Popitz, H. (1990). *Fenomenologia del potere.* Bologna, Il Mulino.

Poulain, J.-P. (2002). *Sociologies de l'alimentation.* Paris, Puf.

Ritzer, G. (1993). *The Mcdonaldization of society.* London, Sage.

Vygotskij, L.S. (1974). *Storia dello sviluppo delle funzioni psichiche superiori.* Firenze, Giunti.

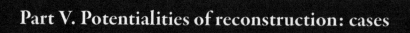

Part V. Potentialities of reconstruction: cases

Risk, rights, and regulation: the politics of agricultural biotechnology in South Africa

William A. Munro

1. Makhathini: the world and a very small place in Africa

The Makhathini Flats is an area of about 13,000 ha that straddles the Pongola river (downstream of the Jozini dam) in north-east KwaZulu-Natal province, South Africa. It falls largely within the Umkhanyakude district, which is one of the poorest districts in the province (Witt *et al.*, 2006). It is also the main area of smallholder cotton farming in South Africa, with some 3,000 farmers farming a total of 2,500 to 10,000 ha of cotton per year, depending on projected prices and access to producer credit[175]. There are over 40 farmers' organisations in the area, ranging in size from 15 to 300 (Gouse *et al.*, 2002). Almost all the cotton cropping in Makhathini is rainfed, and the area is highly susceptible to drought[176]. Cotton cropping is by far the dominant agricultural activity in the region, with about 80% of land going to cotton. This is largely because of the local ecology and the absence of producer credit for other crops (Gouse *et al.*, 2002). Most farmers occupy between one hectare and three hectares of land but some farm much larger cotton plots if they can get access to the land (Ismael *et al.*, 2001). At the same time, smallholders produce a very small proportion – about 5% in 2000/1 – of South Africa's cotton crop, with the remainder being produced by some 300 large-scale commercial farmers, most of whom are white (Gouse *et al.*, 2002). South Africa's cotton industry is small; it generally produces less than half the cotton needed to meet domestic needs, and imports the rest. Consequently, cotton prices are set very much at the international level, and local producers are very vulnerable to international price fluctuations.

In the larger scheme of things, then, one might consider Makhathini cotton to be profoundly important for local people but of not much interest beyond. And yet the Makhathini Flats has an international profile. In 2003, several Makhathini farmers traveled to the UK in order to address the Commonwealth Business Council (before heading to Denmark and Germany) (GmWatch July 8, 2004). That same year, one Makhathini farmer, T.J. Buthelezi, who was chairman of the Ubongwa Farmers' Union, testified before the US House Committee on Science[177]. In turn, farmers and journalists from numerous countries have visited T.J. Buthelezi's farm on the Makhathini Flats. What made these farmers interesting in these fora far from the floodplains of the Pongola river was that they represented the first smallholder farmers in Africa to adopt genetically engineered crops – specifically the Monsanto company's Bollgard Bt cotton seed – and they were there (on Monsanto's dime) to extol the virtues of agricultural biotechnology in improving their lives and livelihoods[178].

---

[175] Witt *et al.* (2006) note a dramatic falloff in number of farmers planting cotton at all in the past few years because of drought.

[176] The significance of this is that the yield differential between Bt and non-Bt cotton, while significant on irrigated crops, is far less so on rainfed crops (Joubert *et al.*, 2001). In Makhathini Flats, one study found that there is no significant difference between Bt and non-Bt cotton yields due to high between-farm variability.

[177] http: www.house.gov/science/hearings/research03/jun12/buthelezi.htm retrieved 11/12/2006.

[178] Freidberg and Horowitz (2004: 4) describe the appearance of the Ubongwa Farmers' Union at the WSSD in Johannesburg in 2002, to demand their 'freedom to farm... the freedom to grow any crop of their choice, the freedom to access the best available technology, and the freedom to improve agricultural productivity'.

Their arguments fit into a broader narrative of 'hope' that agricultural biotechnology brings to Africa's hungry which has been aggressively advanced by proponents of the technology on a number of fronts: by the industry itself (especially through its public relations arm, the Biotechnology Information Organisation); by its civil society allies, such as the industry-funded South African lobbying organisation, AfricaBio; by African research institutions, such as Florence Wambugu's Africa Harvest Biotechnology Foundation International in Kenya, and their international allies such as the International Service for the Acquisition of Agri-Biotechnology Applications (ISAAA) and the National Science Foundation in the United States; international aid-and-development agencies such as the United States Agency for International Development (USAID) and the British Department for International Development (DfID); and international foundations such as the Rockefeller Foundation and the Gates Foundation[179]. This 'narrative of hope' vigorously advances the argument that transgenic technologies offer the *only* feasible solution to agrarian poverty and hunger in the developing world, to the crisis of African agricultural productivity, and even to the loss of biodiversity associated with industrial agriculture and climate change. As such, it constitutes a key discursive strategy in what we elsewhere have called the 'Biotechnology Project', a concerted effort to consolidate a particular global agricultural order based on corporate control of agriculture, strong property rights, transgenic technologies, and supportive regulatory frameworks (Schurman and Munro, 2007).

Certainly, at first glance, the Biotechnology Project is well under way in South Africa. South Africa is the first African country to release agricultural GMOs commercially. In fact, in the early 2000s it led both India and Brazil in this respect (Scoones, 2005: 4; Monsanto, 2006). Bt cotton was the first GM crop to be approved and was first commercially deployed in 1997 (Monsanto's Bollgard, commercialised by Delta and Pine Land Corporation – only a year after the first commercial plantings in the United States). This was followed the next year by Bt yellow maize (insect protected and licenced to Durban-based Pannar) and in 2000 by Bt white maize. Monsanto's Roundup Ready cotton and soya were released in 2001. In 2004, Syngenta's Bt11 maize was approved for commercial production (over an appeal by anti-GM activists). Of the entire South African cotton crop, more than 85% is today planted with transgenic seed (Monsanto, 2006), principally the Monsanto company's Bt product marketed through the Delta and Pine Land Co. Estimates of the total maize crop that are planted with transgenic seed vary between 6% and 20% (2004)[180]. Most notably, however, all three GM crops that have been commercialised are the products of multi-national seed companies (Monsanto, Delta & Pineland, Pioneer Hi-Bred, Syngenta). In particular, Monsanto has set out to consolidate its power in the seed sector. In 1999 Monsanto bought two South African seed companies, Sensako and Carnia, and merged them under the DeKalb brand, giving the company extensive access to the soybean, wheat, barley and sunflower seed markets as well. In 2005, it acquired Seminis, the global vegetable company, and thus gained an entry point into the South African vegetable seed market through Seminis South Africa, which holds nearly 60 vegetable and melon seed varieties (SAFeAGE, 2005). Subsequently, Monsanto also acquired Delta and PineLand, as well as Syngenta's global cotton operations. Thus, these multinational companies hold an enormous amount of power in the South African seed sector. Indeed, they hold patents on 100%

---

[179] For the metaphor of 'hope,' see especially Boudreaux (2006) and Monsanto (2006). For an especially far-reaching discussion of the benefits of Bt Cotton for African smallholders, see the interview with ISAAA's Clive James in ABIC 2004, no.10 (9/04).

[180] The low figure is Monsanto South Africa's estimate of its own GM maize; the high figure is the ISAAA's estimate of the entire 2004 maize crop (see SAFeAGE, 2005).

of the GM crops marketed in South Africa[181]. As Table 1 indicates, foreign companies dominate all categories of GM activity in South Africa, and this position gives them enormous power to control intellectual property rights and to significantly shape the public research domain. For instance, a joint project by the Agricultural Research Council (ARC) and Michigan State University to develop Bt potatoes has been constrained at the research stage by proprietary gene technologies owned by Syngenta (Bt-cry11a1) and Monsanto (35s promoter and NTP-II selectable marker gene) (Brenner 2004: 42). At the same time, it is worth noting that, as elsewhere, these companies have been reluctant to exercise that power directly, preferring to enter royalty-free licencing arrangements with research organisations as part of their 'corporate social responsibility' profile (cf. Smith and Rajan, 2007)[182]. In general, then, corporate power and private property rights seem to be in the driving seat of agricultural technological development and transgenic technologies are privileged. The epistemologies, interests, and concerns of Makhathini farmers are marginalised; they are the *vehicle* and not the *agents* of the narrative of 'hope.'

But the narrative of hope is not the only narrative about the impacts of GM crops on the environment and livelihoods of African farmers (and farmers elsewhere); it is not entirely hegemonic. Part of the reason that proponents of genetic engineering have gone to such trouble to foreground Makhathini and the narrative of 'hope' is to counter an oppositional narrative that stresses the *risks* and uncertainties associated with GMOs. To critics, these risks range from concerns about genetic erosion to uncontrollable genetic species-jumping, to unfettered corporate control and commodification of nature. In South Africa itself, the anti-GM organisation, SAFeAGE (South African Freeze Alliance against Genetic Engineering) was established in 2000 in order to call for a five-year freeze on the deployment of GMOs because of uncertainty concerning their environmental and health impacts. Activists in South Africa, as elsewhere in the world, have called for labeling of food products on the argument that consumers have a right to know what they are consuming and ought to have the effective choice to reject it. An intense international diplomatic furor exploded in 2002, when Zambia refused to accept World Food Program aid that was genetically modified on the argument that GM 'contamination' of Zambian land or agricultural products might close off access to crucial European markets (in the light of the EU moratorium) or adversely affect biodiversity in ways that Zambian farmers could not manage; in effect, the Zambian government argued, acceptance of genetically modified food aid entailed an unacceptably high risk of actually *increasing* food insecurity and hunger.

These dueling narratives are rooted in a global politics of genetic engineering that has its heart in Europe – the epicentre of global resistance – and a nervous system defined by US corporate interests in trade and intellectual property right protection. Both these narratives claim to be pro-poor. Thus, they convey complex normative propositions and contradictory knowledge claims about risks and rights across vast global spaces via transnational networks of scientists, policymakers, activists, business professionals, and the like. Equally clearly, they are narratives about power in the governance of technology[183]. Thus the farmers of Makhathini had been propelled, as both an icon and a lightning rod, into a complicated and

---

[181] Note, however, the on-going development, and recent successful glasshouse trials of maize engineered for resistance to rust streak virus – the 'first African produced GM seed' – under patents held by Pannar and the University of Cape Town (author interview July 30, 2007).

[182] As Brenner (2004) notes, Syngenta negotiated a royalty-free commercial licence with MSU to make its Bt technology available to developing country partners for Bt potato research under particular terms and conditions. Wright and Pardey (2005) argue that multinational corporations tend to be lenient with researchers who use their patented technologies up until they reach the point of commercialisation. It is at that point that multinationals tend to exercise their patent rights.

[183] For an interesting and useful critique of the institutional power framework associated with the mainstream discourse on biotechnology for the poor, see Hisano (2005).

Table 1. GM Approvals under the GMO Act, 1997 (1997-2005). Adapted from National Department of Agriculture: Genetically Modified Organisms Act 1997: Annual Report 2004/2005.

Type of approval	Crop	Number app/crop	Company	Number app/comp
General release	3	8	2	8
	Cotton	4	Monsanto	7
	Maize	3	Syngenta	1
	Soybean	1		
Commodity clearance	3	10	4	10
	Maize	8	AgrEvo	3
	Oilseed Rape	1	Monsanto	4
	Soybean	1	Pioneer Hi-Bred	1
			Syngenta	2
Trial release (since Dec 1999)	11	40	18	40
	Canola	1	AgrEvo	2
	Cotton	15	ARC	2
	Groundnut	1	ARC-GCI	1
	HIV vaccine	1	Aventis	1
	Maize	13	Calgene	1
	Potato	2	Cato Research	1
	Soybean	2	CSIR	1
	Sugarcane	2	Dow AgroScience	1
	TB vaccine	1	First Potato Dynamics	1
	Vaccine	1	Monsanto	12
	Wheat	1	MSD	1
			Novartis (Syngenta)	1
			Pioneer Hi-Bred	2
			SASEX	1
			Stoneville	2
			Syngenta	8
			Triclinium	1
			University of Natal	1
Contained use	9	10	7	10
	C. Glutanicum	1	AECI Bioproducts	1
	E.coli VNII	1	ARC	4
	E.coli XL1 Blue	1	Capespan	1
	Granulovirus	1	Dow Agrosciences	1
	Maize	2	Natal Bioproducts	1
	Pathogen (*Zylophilus ampelinus*)	1	Pioneer Hi-Bred	1
	Pathogen	1	SA Bioproducts	1
	Potato	1		
	Sweet Potato	1		

intensely conflictual global political field defined by competing claims of risks, rights, and regulation. In this field, they have particular iconic utility for the Biotechnology Project in several ways. First, they enable the industry to stress the 'scale neutrality' and portability of the technology – it may have been developed and designed for large-scale monocropping agriculture in the USA, but it is equally suitable for small-scale farmers in Africa (Monsanto, 2006). As I suggest below, this argument has significant ramifications for the politics of regulation and the practice of risk assessment, as well as the authoritative status of portable scientific 'facts'. Second, because cotton is a non-food crop, it enables the industry to efface the critiques of genetic engineering that focus on food products, health, and consumption (a major focus of conflicts over GMOs globally). This weakens the appeal for a precautionary approach to the technology. Third, since South African cotton is not exported, it enables the industry to avoid the argument that producing GM crops can destroy export markets if consumers in those markets (notably Europe) refuse to buy them[184]. Finally, the rapid adoption of Bt cotton by Makhathini farmers – from some 8% to over 90% in the first three years – indicates that farmers want the technology. Surely they have a right to use it.

Moreover, the very act of associating GMOs directly with the Makhathini Flats assigns them specific local meanings. The capacity to shape those meanings has significant ramifications for the development trajectory of agricultural biotechnology. As several scholars have pointed out (Scoones, 2005; Freidberg and Horowitz, 2004), conflicts over agricultural biotechnology are inexorably both local struggles and global struggles; they bring together the fate of farmers' livelihoods, on the one hand, and the logics of the global economy, on the other. They are multi-level conflicts that involve a range of transnational social and professional networks as well as transnational discourses – of modernity, of development, of science, of rights, and of risk. Indeed, agricultural biotechnology is both a strand in the complex relationships between globalisation and livelihoods, and a useful prism through which to consider those relationships. Questions therefore arise about what the political field of biotechnology means for the farmers of Makhathini (and other resource-poor African farmers), and what the farmers of the Makhathini (and other resource-poor farmers) mean for this political field. How do local flows of power and conflict shape the trajectory of the Biotechnology Project?

This paper approaches this question by focusing on the politics of agricultural biotechnology regulation in South Africa, for it is here that the dueling discourses of rights and risks have converged most intensely to catalyse an escalating and increasingly polarised conflict over the meaning of this technology for South African society, biodiversity, and environment. To make sense of these struggles, and their effect on the Biotechnology Project, the paper first examines the *logics* of regulation to show how they shape the politics of rights and risks in two particular ways. First, it argues that the location of agricultural biotechnology regulation by the state within the framework of science and technology policy pushes conflicts over regulation into the realm of 'boundary making,' which involves attempts by various interested parties (regulators, experts, activists, citizens, farmers) to draw the epistemic and normative boundaries of authoritative knowledge either broadly or narrowly. Second, it shows that these logics are rooted in a meta-narrative of economic modernisation and development that makes the South African state sympathetic to the Biotechnology Project, and pushes it to 'bring regulators within the engine of

---

[184] As Wright and Pardey (2005) note, the regulatory politics become considerably more complicated when GE products are exported. The South African National Biotechnology Strategy also notes that in those products for which there are export markets (and also perhaps in food products), there is a serious need to maintain a strict separation of GM and non-GM value chains (p.28). Yet this requires an extensive infrastructure of separate storage facilities, inspection, documentation, etc. which is not easy to provide reliably in a case where the agricultural support infrastructure is already weak, such as among black smallholding areas in South Africa.

industry' (to paraphrase Robert Bud, 1995). The paper then sketches the politics of boundary making in the construction of South Africa's regulatory framework for agricultural biotechnology. It shows how a small group of scientific experts was able to 'capture' the regulatory process, defining authoritative knowledge narrowly and marginalising social constituencies such as farmers. As such, the regulatory framework for biotechnology sustains the Biotechnology Project.

Nevertheless, the paper concludes, in part by re-invoking the Makhathini farmers, the political conflicts generated around agricultural biotechnology have opened up spaces for new actors and directions in agricultural development, and perhaps reduced the control of the biotechnology industry by creating new opportunities for 'independent participatory action' (Levidow, 2007) by marginalised groups (such as the Makhathini farmers). In sum, biotechnology is there to stay, but it may not represent the Biotechnology Project that anyone envisaged.

3. Of science, sovereignty, and struggle

Science and technology are profoundly transformative, not only of nature but of social institutions, culture, etc. The policy frameworks that shape their development trajectories reverberate right through society. One need only look at the Green Revolution, for instance, and its impact on India's social and political economy – characterised on the one hand by increased agricultural exports which boosted India's position in the world economy, and on the other hand by increased farmer suicides. Governments are, of course, aware of these effects as is indicated by the fraught politics and diplomacy associated with Zambia's decision to reject genetically engineered food aid from the World Food Program in 2002[185]. For the Zambian government, one serious political consideration in reaching its decision was the recognition that it lacked the regulatory capacity to manage either the process or the potential effects of disseminating GM maize to the hungry; as such, its internal sovereignty was threatened as well as its external sovereignty, and *political* risk was as much a consideration as environmental or economic risk in deciding to reject the food aid.

Certainly, the social and institutional transformations associated with technology policy are frequently a source of considerable social struggle. They also establish a significant link – and sometimes tension – between the nature and authority of expertise, on the one hand, and democratic accountability, on the other (when and why, for instance, should experts – as opposed to farmers or representatives of the public – be assigned the exclusive authority to evaluate the environmental risks of genetic engineering?). This link allows us a useful point of entry into the analysis of the normative politics of technology promotion and regulation. In the first place, struggles over the boundaries of authoritative scientific knowledge are also *national* struggles, ie. struggles about the parameters of state power and how the public good is defined/pursued. Consequently, as Gottweis (1998: 11) has noted, there are 'complicated mutually constitutive and mutually reinforcing relationships between the state and science and technology'. Following this argument, Sheila Jasanoff (2005) argues in her comparative analysis of regulation in the United States, Germany and Britain, that in each case one should recognise biotechnology policy as a 'nation-building'exercise, or perhaps more accurately as an exercise in new ways of imagining the nation. It is therefore useful to start with policy frameworks (cf. Paarlberg, 2000).

---

[185] See also Mnyulwa and Mugwagwa (2003) on the wake-up call that the crisis constituted for African governments to focus more attention on devising their regulatory frameworks for biotechnology.

In order to understand policy frameworks, it is useful to analyse technologies as located in discursive fields of policy-making, and constructed as opportunities, as problems, as dangers, etc. Hebert Gottweis (1998) has pointed to the importance of understanding the construction of authority in defining sound science and authoritative knowledge through what Thomas Gieryn (1983) has called 'boundary work'. As Gottweis (1998: 25) puts it: 'The distinction between science and non-science is a critical element in boundary work, but on a more general level boundary work encompasses those acts and processes that create, maintain, and break down boundaries between knowledge units, such as disciplines or subdisciplines. Thus, boundary work incorporates the processes whereby legitimacy and cognitive authority are accorded to knowledge units. In this reading, science is a space that acquires its authority from and through episodic negotiations of its flexible and contextually contingent borders and territories'. In other words, boundary work defines those actors, methods, stocks of knowledge, values, and types of organisation that are included in the production of authoritative knowledge. And the question that arises, then, is when and how do those boundaries become more stable or more contingent; and how does the opening or closing of boundaries affect the politics of technology and development?

4. South Africa and the biotechnology project: the meta-narrative of development and the logics of regulation

For the South African government, the starting point for developing an agricultural biotechnology policy is the role of science and technology in creating wealth and improving the quality of life in contemporary South Africa (South Africa, 2001: 4). The government sees technology as crucial to meeting these objectives. But technology development is also seen as a central strategy to help South Africa position itself as competitive in the global economy and perhaps even to enable the country to '*leapfrog* into the future' (South Africa, 2001: 8). Thus, biotechnology policy is part of a larger national and nationalising project – a project that is both geo-political and geo-developmental in ways that stretch well beyond the realm of agriculture, but may have profoundly important implications for how agriculture develops in the country. It is a project about how South Africa positions itself in the world, in Africa, and in modernity. This project defines the logics of South Africa's biotechnology regulation frameworks.

In order to fully appreciate these logics, it is useful to note that the South African government developed its regulatory framework during the mid-1990s, a time of two transitional moments in the construction of a post-apartheid development regime. One transition, straightforwardly enough, was the transition from apartheid-era development thinking, which was widely recognised as insular, inward-looking, hidebound, and uncoordinated. Given the priorities of the apartheid regime, and the increasing pressure it had experienced during the 1980s, this was not surprising. The government had tended to focus its interest on the chemical and military sectors. It had encouraged little venture capital or private-sector development in the life sciences, and had devised no coordinated policy for research and development (Ofir, 1994). While there were many life scientists pursuing active agricultural research agendas – notably on plant tissue culture (second generation biotechnology) – in universities and government research institutes, these endeavours tended to be relatively localised, isolated, and poorly funded[186]. The political transition, however, gave the post-apartheid government both the opportunity and an imperative to devise a new, more inclusive, national development strategy (in the venerable tradition

---

[186] The Council for Scientific and Industrial Research (CSIR) was the public agency charged with promoting scientific R&D. While it did show enthusiastic support for researchers' own projects, it lacked the resources and the capacity to coordinate or advance any systematic agenda. Indeed, by the mid-1980s, the leading role in biotechnology had been taken over by a private-sector explosives company, AECI (interview with South African scientist, July 2007).

of post-liberation governments). It was in this context that the government produced the White Paper on Science and Technology (WPST) in 1996 (South Africa, 1996), proudly presented as the first ever effort to establish a comprehensive framework for scientific and technological development as the keystone to economic modernisation.

The second transition was the politically more awkward transition in national development strategy away from the redistributivist Reconstruction and Development Programme (RDP), promulgated in 1994, to the quite stringently neoliberal Growth, Employment and Redistribution strategy (GEAR), adopted in 1996 (see Habib and Padayachee, 2000; Bond, 2000; Padayachee, 2005). The awkwardness of this shift in strategic and ideological vision was strongly reflected both in the WPST, and in its sectoral elaboration, the 2001 National Biotechnology Strategy, which moved back and forth between a stress on science for the public welfare and science for global competitiveness, before settling on the latter. The strategy laid out in these documents rested on three core pillars. One was a conception of modernity founded on a 'knowledge-based economy'; the second was a geo-economic vision based on the political-economic imperatives of global competitiveness; the third was a belief in the virtues of privately funded and privately pursued innovation[187].

The concept of a 'knowledge-based economy' has been theorised in a variety of ways, but it is generally understood to describe a post-industrial economic system in which the organisation of production, the character of productive relations, and indeed the logics of accumulation, are qualitatively distinct from those that exist under industrial capitalism (cf. Touraine, 1971; Castells, 1996). In this system, leadership is based on Information and Communication Technology rather than on industrial capacity, and the centrality of capital/labour relations in economic organisation has been displaced by the importance of knowledge and education. Indeed, capital ownership is understood to be less salient as a source of economic power than intellectual property. On this account, in the knowledge-based economy, 'technological change is the primary source of economic growth, [which] means that economic and S&T policies have to recognise as central concerns the two processes – innovation and technology diffusion – which are the agents driving that technological change' (South Africa, 1996: ch.3 (1)). Furthermore, since innovation is understood as the motor of economic growth, the core economic activity is knowledge production. But knowledge production in itself is not necessarily innovative. To be innovative, it must be understood as a more multivalent, output-oriented, process in which new knowledge is created, acquired, diffused, and put into practice to provide economic and social goods[188].

In the innovation-based knowledge economy, the organisation of production is also considered to have distinct characteristics: production takes place not in a specific location that produces a packaged final item, but in cooperative information-sharing networks of actors, organisations, and institutions that are both multidisciplinary and move across public and private realms through 'webs of dialogue'

---

[187] As the WPST put it: 'there is general agreement that the collective organisations in the private sector are fundamental in the financing and implementation of innovation in society. The prime engine for economic growth is located in a country's firms – a nation's competitiveness depends upon the ability of its firms to innovate, and thereby provide sustainable output and employment growth' (South Africa, 1996).

[188] The WPST makes a distinction between *creativity*, which is the generation and articulation of new ideas, and *innovation*, which is 'the application in practice of new ideas, which in many cases involves the introduction of inventions into the marketplace'. Thus it argues that 'a national system of innovation can only be judged as healthy if the knowledge, technologies, products and processes produced by the national system of science, engineering and technology have been converted into increased wealth, by industry and business, and into an improved quality of life for all members of society' (South Africa, 1996).

(see Sell, 2003). The key outcome in this economic system is innovation itself (measured and realised in intellectual property rights). The two key features of this that are important here are (1) the key organisational imperative of the production system is to develop the most efficient set of networks for producing innovation (a kind of postindustrial version of postsfordist production systems); and (2) the key struggles are not the class struggles of the industrial age but struggles over the inclusion, exclusion, or weighting of particular types of knowledge in producing innovation.

South Africa does not, of course, have a postindustrial economy. Yet the adoption of the 'knowledge-based economy' as the pervasive reference point for its Science and Technology strategy signaled the government's ambitions and expectations to be a globally competitive post-industrial political-economic player. Certainly, at the point of liberation in 1994, policy makers believed that the fundamental infrastructure – skills, R&D ability, institutional capacity – were in place to seize the future, and that venture capital would be forthcoming as South Africa opened its economy to the world (van Renssen, 2005; Burton and Cowan, 2002). This liberatory moment was, then, a moment of psychological relief and political optimism. No doubt it was also psychologically buoyed by the warm welcome extended to post-apartheid South African participation in multilateral organisations and institutions; and South Africa moved rapidly to establish itself as a global player in these settings (see Gupta and Falkner, 2006). But it was also a moment of considerable nervousness and urgency among policy-makers and think-tank analysts fueled by the fear of being left behind in the 'fast race' to competitive success in the global economy as other emerging markets (Brazil, India, China, Argentina) took opportunities to get in on the technological revolution (see South Africa 2001)[189]. As the WPST (South Africa, 1996: ch.1(2)) put it, 'In the world-wide race for competitiveness the finishing line keeps moving away. The ability to maximise the use of information is now considered to be the single most important factor in deciding the competitiveness of countries as well as their ability to empower their citizens through enhanced access to information'[190]. Across a broad spectrum of policy areas, South Africa's policy-framing documents reflect a powerful urge to keep that finishing line in sight.

In addition, South Africa's technology and development policy frameworks are influenced by its continental geopolitical ambitions to be a driver of the 'African renaissance' (see Lodge, 2002). In the same way that South African capital has pushed northwards into the African continent over the past decade, so the government believes that South Africa should be a technology leader on the continent, spreading South African know-how and products around the continent. Biotechnology is a part of this vision, and the government believes that it has a competitive advantage in this respect. It has a long tradition of public agricultural research, several lively research institutes that are already internationally networked, and considerably more research capacity than other African countries (except, perhaps, Nigeria, Egypt, and Kenya with whose research establishments South African scientists were developing increasingly strong links). As Ofir (1994) notes, 'It is expected that South Africa will play a major role

---

[189] This sense of international competitiveness was also perhaps intensified by the UNDP's emphasis on technology development, and the publication in 2003 of its *Technology Achievement Index*, in which South Africa's position appeared quite lowly.

[190] The narrative of a knowledge economy as the centre of modern geo-economic competition is indeed a global narrative. Thus, David Byrne, European Commissioner for Health and Consumer Protection, in an address to the National Press Club in Washington in October 2001: 'In line with the European Union's strategic goal of becoming the most competitive and dynamic knowledge-based economy in the world, the European Commission is placing particular emphasis on the potential of life sciences and biotechnology'. In addition, it was the notion that if you did not put your development eggs in this basket you would fall behind other countries in ways that could never be made up that kept British Prime Minister Tony Blair firmly committed to biotechnology development against a rising tide of political opposition in the late 1992 and early 2000s. See Munro and Schurman (2008).

in the development of human resources in biotechnology elsewhere in Africa, directed towards the empowerment of African scientists on the continent to serve the needs of their own countries. Already postgraduate students from different African countries are starting to participate in research projects' (see also Woodward *et al.*, 1999; South Africa, 2005). In addition, South Africa has stronger regulatory capacity than most African countries. It is one of very few to have statutory plant breeder's rights, and is a member of UPOV 1978. There is a widespread expectation in South Africa and in the region that South Africa's regulatory framework for biotechnology development and deployment will provide the model for other African countries still formulating their own policies.

For all its pre-occupations with global competitiveness and its geopolitical ambitions, however, the South African government was not eager to place the state at the centre of its science, technology and development strategy. As the WPST (South Africa, 1996: ch7 (4)(4)) put it: 'Government can play an enabling role in the encouragement of an innovation culture within the private sector. It is then the responsibility of the private sector to recognise the importance of innovation spending from the point of view of competitiveness... The White Paper therefore proposes that organised business develop a policy for building innovation capacity and supporting innovation within the country's enterprises'. This vision of science development policy called for tripartite arrangements between the state, industry, and public research (mainly universities, but also government research institutions such as the Agricultural Research Council (ARC) and the Centre for Social and Scientific Research (CSIR)) that constituted a kind of postindustrial corporatism in which the role of the state is to enable entrepreneurial activity[191]. In the global context of neoliberal development ideology, it reflected the significant strategic shift from the post World War 2 policy approach that had called on states to 'push science' to one that called on states to 'enable innovation' (see Thackray, 1998). As such, it also reflected quite precisely the World Bank's 1997 vision for positioning and defining the role of the state in development (World Bank, 1997). As a strategic, practical, and ideological matter, the government's development policy framework shifted from a *direct* focus on public welfare to an *environmental* focus on public welfare.

It was in the context of this broad meta-narrative of technology and development – the knowledge-based economy, global competitiveness, and private-sector leadership – that South Africa became the first developing country to adopt the concept of a 'national innovation system' as its "intellectual framework for [Science and Technology] policy..." in which a set of functioning institutions, organisations, individuals and policies interact in pursuit of a common set of social and economic goals' (GCIS, 2007: 479). In particular, this would involve the incorporation of government, industry, and university researchers into cooperative networks of knowledge production (see Van Renssen, 2005). In doing so, the government drew directly on the defining work on realising frameworks for innovation that had already been done within the OECD and elaborated in considerable detail by the European Union[192].

---

[191] WPST: '[I]nnovation is not a trivial process. It is costly, often high risk, and generally operates on a longer time frame in years than the traditional quarterly periods or annual periods over which a company's performance is evaluated. Investors in innovation programmes require vision, strategic thinking and confidence. Very often they require to be driven by external competitive pressures and threats. Amongst other factors, to innovate successfully often means that a nation must be prepared to probe the unknown and to take calculated risks. Government has an important role to play in the sharing of this risk and in its attempts to provide an enabling environment for innovation' (South Afrika, 1996).

[192] In a sense, South Africa positioned itself in a quintessential semiperipheral role: as a receiver of technology, ideas, procedures and institutional rules that it can in turn employ in subimperial ways, it both asserts its sovereignty and caves to the intellectual categories of outsiders. This is most clearly observable in the fight over generics in the pharmaceutical industry. One might also note here the UNCTAD 2007 LDC Report, which sets out to define the role of LDCs as 'creative imitators'. South Africa is well positioned in this vision to play a subregional role.

The concept of a system of innovation is organisational. It is based on the recognition that effective innovation cannot occur without a range of actors and organisations performing specific roles, including research, realisation, funding, property rights protection, infrastructural support, and regulation. Thus, a system of innovation comprises 'a network of organisations within an economic system that are directly involved in the creation, diffusion, and use of scientific and technological knowledge, as well as the organisations responsible for the coordination and support of these processes' (Dantas, 2005). In short, a system of innovation is a design for governing technology and development, in a context where the lines of governance can readily become blurred and conflictual (see especially Freidberg and Horowitz, 2004).

Biotechnology development is an important component of South Africa's national system of innovation. The National Biotechnology Strategy called for the establishment of a number of Biotechnology Regional Innovation Centres (BRICs), which would promote and coordinate active Research and Development programs to be carried out by university and industry consortia in ways that would draw both investment and expertise (thereby building the South African skills base). Four BRICs have been constructed so far, with one (Pietermaritzburg) devoted to agricultural research. The government also established an Innovation Fund to support product development 'at the concept stage'; but the political fallout from using public funds as venture capital and the pressure to show quick returns on investment (in a field where returns on R&D can be very long term indeed) made this a poor instrument for stimulating innovation (Van Renssen, 2005). Certainly, given national pressures and needs, there were real resource constraints, and the proportion of South Africa's national budget directed towards science R&D continued to be very much lower than it is in developed countries (Burton and Cowan, 2002). For this reason, the South African government focused very heavily on marketable output and the importance of close collaboration between academic researchers and industry – what the National Biotechnology Strategy called the 'relationship between the academic domain and the commercial or market-facing domain' (South Africa, 2001: 14; see also Aerni, 2001). R&D funding should be raised through contracts, which would perforce be with international companies until an 'entrepreneurial culture' emerged in South Africa. The national system of innovation was intended to promote these outcomes.

The notion of a network-based development framework implies a governance structure that is inclusive, non-hierarchical, and functional: the trick simply is to identify 'relevant stakeholders' in the particular issue-area according to the specific roles that they play in the organisational network[193]. But of course these networks are not infinitely fluid, and they do not bind all actors or organisations with equal intensity. Networks and sub-networks overlap and intersect. They are bounded and hierarchical (though hierarchies may shift) and they are structured to promote particular practices among actors and organisations. Not least, the core principles of the development meta-narrative – knowledge-based modernisation, global competitiveness, private-sector-driven innovation – privilege particular nodes, discourses, norms, and participants in the system. These ends are pursued both through the formal rules – for instance, property protection laws and risk assessments – and the informal norms and procedures – for instance, patterns of consultation or inclusion on decision-making committees, or support for publication in peer-reviewed international journals – that constrain actors' activities. As such, the system of innovation is not only a governance structure but also a political field subject to

---

[193] South African policy analyses, documents and decisions routinely list the 'relevant stakeholders' in a particular issue-area; cf. the GMO Act.

intense 'boundary work' as different constituencies jockey to establish or shape the 'rules of the game' for coordination and cooperation in the system.

Such boundary work determines how exclusively or inclusively the system's rules and norms are drawn with respect to modes of knowledge, public voices, social constituencies, economic interests, etc. There are divergent tendencies. Given the emphasis on private-sector initiative, there is an exclusionary tendency to establish more effective controls over knowledge. This can be done formally by strengthening the Intellectual Property Rights regime through patent laws, and it can be done informally by privileging specific kinds of scientific/technical knowledge in a way that cedes a great deal of boundary-making power to the scientific and research establishment, whether it is domestic or international, whether based in academia or industry. And it tends to strengthen the social and knowledge networks of this 'epistemic community.'

On the other hand, there is an inclusive tendency that stresses multi-disciplinarity, different modes of knowledge, and public voice[194]. This tendency gains support from South Africa's democratic transition. While this is still a work in progress, South Africans are proud and protective of their new democratic credentials, and of the association of those credentials with a commitment to human rights, both domestically and internationally. In particular, they are proud of their constitution – hailed as the most liberal in the world – and the 'voice' that it guarantees its citizens. And, in the aftermath of the struggle for liberation, very many South African citizens believe quite passionately that their ability to exercise that voice is a crucial (perhaps defining) part of their democracy. Particularly relevant to the discussion here is that the National Environmental Management Act (NEMA) of 1998 made extensive provision for public participation and public information (at all points in the process) in setting the conditions for activities that might have an impact on the environment (South Africa, 1998). This was followed up by the passage of the Promotion of Access to Information Act in 2001, which was aimed to increase the transparency of government activities, and the Promotion of Administrative Justice Act of 2001 which required government agencies to explain their actions and decisions, upon request from civil society organisations.

It is clear that real tensions can arise in a range of policy domains between South Africa's outward-looking development approach, with its orientation towards technology and private intellectual property rights, on the one hand, and its commitment to democratic inclusion and voice, on the other. The South African government has tried to finesse these tensions in somewhat awkward, and occasionally politically risky, ways – a classic case is the government's tendency to equivocate in its conflicts with pharmaceutical companies over the cost of HIV/AIDS drugs and the commercial deployment of generics; another is the conflicts over cultural rights and intellectual property rights for indigenous peoples (see e.g. Chan, 2004). In particular, the government tried to maintain flexibility in its IPR regime. On the one hand, it encourages South African innovators to establish intellectual property rights on their own initiative (South Africa, nd). It has also declared its intention to hew to international property rights norms. On the other hand, it declined to establish a national property rights framework. Instead, it left the

---

[194] This tendency is superbly captured in David Byrne's speech to the National Press Club: 'We stand accused that [labelling] is a major imposition, is unjustified and unenforceable. I reject these accusations. Europe is perfectly entitled to impose the labelling rules proposed. Our consumers are demanding this. They are entitled to choice and full information. To those who say that these labels are not science-based let me say this. Labelling is not an issue for science alone. It is an issue for consumer information' (2001: 3). In this framing, regulation is understood not simply as technical but as social. See also Marris (2000) and Jasanoff (2005) on the two different traditions in the USA and Europe, leading USA to rely very heavily on science-based regulation (i.e. technical emphasis).

development of property rights protocols to individual institutions, and outsourced examination of property rights claims to designated countries that already have that capacity, though this made it very expensive – often prohibitively so – for South African researchers to seek patent rights. This has created the strange situation in which the premier parastatal agricultural research institution, ARC, does not have an institutional IPR policy, though individual universities such as the Universities of Stellenbosch and Cape Town do (Brenner, 2004: 35; interview with research scientist, July 2007). In addition, the government was extremely hesitant to ratify the Cartagena Protocol of the Convention on Biodiversity, which advances the precautionary principle; South Africa only signed in November 2003, two months after the Protocol went into effect, and four years after South Africa's own GMO Act went into effect.

The point, then, is that under these circumstances, technology development – including biotechnology – becomes a potent conduit for the configuration both of sovereignty (i.e. the capacity of the state to determine the paths and strategies of national development) and of accountability (i.e. the capacity of citizens to monitor these choices and to hold the government accountable for them). What is at stake, in a sense, is the 'thickness' of democracy. In the case of a biotechnology development strategy for agriculture, everyone agrees that it should proceed in a way that maximises its social benefits and minimises its socio-economic and environmental risks (Aerni, 2001: 7-8). But it is very difficult to find agreement on what those risks actually are, how they are constituted, how they should be calculated, and how they should be contained[195]. And it is this area of uncertainty and negotiation that provides the political space for the boundary work of defining the knowledge and expertise regimes of biotechnology policy.

4. Boundary wars and biotechnology

The history of (third generation) agricultural biotechnology in South Africa goes back to 1989, when the Department of Agriculture received its first application for a GM field trial of Bt cotton[196]. The government authorised the trial under the 1983 Agricultural Pest Act and ceded the mandate for conducting scientific risk assessment and establishing field containment requirements to the South African Committee on Genetic Experimentation (SAGENE). SAGENE has played a critical role in doing the 'boundary work' that has come to define the realm of expert knowledge production and evaluation in the field of biotechnology. Established in the late 1970s as the national advisory body on biotechnology research and development (Aerni, 2001: 11), SAGENE was led by life scientists based in universities and research organisations who were pursuing their own individual research interests and projects but shared an enthusiasm for genetic and molecular research. Government science policy offered little support because the apartheid government's research interests focused on very specific outcomes in applied research, especially first generation plant biology. There was little coordination among research establishments, or between the government, academia, and industry. Thus early biotechnology development was driven largely by the personal initiative and ambition of these individuals, who recognised that the international field of genetic and molecular research was changing rapidly and dramatically under the 'gene revolution', and were concerned about the danger of being left behind professionally. These scientists established SAGENE mainly to develop sound science

---

[195] For a nice rendition of the operative normative distinctions, see the WPST (South Africa, 1996: ch.3(2)): 'South Africa has a proud record of quality basic research and it is important to sustain this. In particular, the bridge that such research provides to the international scientific environment must be preserved. *It is the quality of our science rather than the number of international agreements we are party to which will ensure this*' (italics added). In other words, international connections are important, but these are best secured through 'good science' than through 'good citizenship' in the international community.

[196] This was probably the Delta and Pine Land application to bring in Bollgard seed – though Monsanto 2006 puts the date for that at 1993.

research protocols around experimentation, safety, etc.[197] In doing so, they gave institutional form to a specialised 'epistemic community' that was cogent, but also exclusive and prescriptive.

In the early 1990s, with the advent of 'new' biotechnologies as well as the shifting in policy frameworks for science and technology associated with the end of apartheid, the research groups associated with SAGENE moved almost inexorably to the centre of the regulatory process. They also built extensive international research networks, both with academic colleagues and with funding organisations, especially in the United States[198]. As they drew in international research funding and institutional collaborations (see Woodward *et al.*, 1999), they also strengthened local networks of science research and raised the profile of the local scientific community. Consequently, they built themselves a strong position to speak with authority on issues of science development and regulatory policy, and indeed to place their own normative framework of authoritative knowledge, which stressed 'good science' (laboratory-based scientific method) at the heart of policy debates[199]. Institutionally, this authority was expressed most directly through SAGENE, which acquired new terms of reference that included the provision of advice to any Minister, statutory or government body on any form of legislation or controls pertaining to the importation and/or release into the environment of GMOs. Indeed, SAGENE provided the guidelines for field tests of transgenic crops and was responsible for the evaluation of risk assessments (food, feed and environmental impact) of all applications requesting authorisation to conduct GMO activities (South Africa, 1997; Woodward *et al.*, 1999). As GMWatch noted, 'As the new South African government, which was ushered in on the 27th April 1994, had no particular knowledge or expertise in these areas, regulatory matters were left very much in the hands of SAGENE'[200].

1997 turned out to be a big year for biotechnology in South Africa. The first commercial GM crops were planted, the Genetically Modified Organisms (GMO) Act was passed, and the first anti-GM organisation appeared. The GMO Act is the core legislation on biosafety and deals with trade, production and R&D of GMOs. Its key objective is 'to provide for measures to promote the responsible development, production, use and application of genetically modified organisms' (South Africa, 1997). As a regulatory instrument, it is in fact a very weak Act.

The Act places regulation of GMOs within the national Department of Agriculture, under the administrative authority of the Act's Registrar. It makes provision for two regulatory bodies, the Executive Council and the Advisory Committee. The Executive Council consists of officials from six government departments: Agriculture, Health, Environmental Affairs and Tourism, Labour, Trade and Industry, Science and Technology. It is the ultimate decision-making body on GMO activities under the Act. The Advisory Committee replaces SAGENE as the regulatory agent. Appointed by the Minister of Agriculture, it consists of ten scientists who are experts in the field of GMOs, and it conducts safety reviews and risk assessments (on food, feed, and environmental impact) that are submitted with an

---

[197] According to a senior scientist who was a founding member of SAGENE, they modeled their approach very closely on the approach laid out by the National Institutes of Health in the United States (interview, July 2007).

[198] Anecdotal evidence suggests that one important conduit for these networks and collaborations ran through expatriate South Africans, some of whom had left during the insurrectionary period and were working in North American research universities.

[199] As Gottweis (1998: 47) notes, though scientific work cannot be disarticulated from the institutional context in which it takes place, and is promoted by organisations themselves embedded in the international capitalist order, 'It is in the nature of scientific disciplines that over the course of time such complex intertextualities tend to move into the background, while scientific practice seems to be mainly determined by the inner logic and workings of everyday laboratory life'. There is a vast amount of evidence showing that scientists hate nothing more than technology opponents who 'don't get the science', or 'get it wrong'.

[200] (www.gmwatch.org/print-profile.asp?PrId=282 accessed November 6, 2006).

application. The Committee makes a recommendation to the Executive Council. In fact, the group of scientists who were networked around SAGENE have remained at the centre of the process. Indeed, several of them played a key role not only in writing the National Biotechnology Strategy, but also in drafting the GMO Act. In this context, it is not surprising that one of the key scientists involved in the process described this regulatory system as superior to both the US and the European models (GMO page on scientist's website).

The process of constructing this regulatory approach involved 'boundary work' that had serious ramifications for the politics of biotechnology. In the first place, the 'circle of expertise' was drawn quite tightly and strongly privileged research scientists and plant technologists (whether from the academy, the government, international consultants and foundations, or the commercial companies themselves). As a result, the scientific *methods* that characterised the framework were highly regarded, but the *process* was recognised as non-transparent and exclusive. Social scientists were scarcely included. The Advisory Committee made no allowance for public participation, which was limited to a notice and comment procedure linked to permit applications for environmental release (and there were many opportunities to evade the permit process). Further, modeled on the US regulatory approach, the Act did not empower regulatory agents under the Act to carry out risk assessments themselves but only evaluate those provided by applicants. The Act relied substantially on a 'substantial equivalence' approach, (i.e. made no effective distinction between 'new' and 'old' biotechnology techniques in devising risk assessments), and applied only to living GMOs, not to products derived from GMOs. It did not take into account the precautionary principle that was then being negotiated in the Convention for Biodiversity[201] – it is perhaps in this context that the full meaning of the WPST's line that 'it is the quality of our science rather than the number of international agreements we are party to which will ensure' South Africa's good standing in international knowledge production is best appreciated. Although the Act recognised them as crucial stakeholders, both farmers and consumers were frozen out. Perhaps more significantly, this tight drawing of the boundaries of expertise had the effect of separating the realm of biotechnology policy effectively from the realm of development policy, where social scientists played a significant role[202]. As such, it normed towards a narrow conception of relevant and authoritative knowledge in regulatory decision-making, and enabled the headlong approval of GMOs to go ahead at breakneck speed with no consideration to the capacity of the state to provide – or even to design – appropriate resources, infrastructure, extension, education, etc. to the farmers who would end up putting GMOs into the environment. This point was particularly significant since the Act enshrined the principle of 'end-user' liability for any harm caused by GMOs, rather than the 'polluter pays' principle that was written into the National Environmental Management Act.

It was in this context that a group of environmental activists, led by a biologist formerly at the University of Cape Town, founded the first anti-GM organisation, BioWatch, in 1997 'to publicise, monitor, and research issues of genetic engineering and [to] promote biological diversity and sustainable livelihoods'. In particular, alarmed at what they perceived to be a wholesale, inchoate, and non-transparent process of approval for permit applications, they declared their aspiration 'to prevent biological diversity from

---

[201] See Mayet (2006) retrieved November 11, 2006.

[202] The powerful irony about this situation – which demonstrates, perhaps, the effectiveness of the scientific epistemic community in boundary making – is that the very first regional seminar on biotechnology policy formulation and implementation, organised by the Intermediary Biotechnology Service in 1995, had stressed the importance of orienting biotechnology towards end-users (smallscale farmers) and integrating social sciences in policy planning. In particular, seminar participants – from foundations, National Science and Technology Centres, industry, etc. – had generally agreed on the importance of including farmers in the construction of biotechnology policy frameworks. See Van Roozendaal (1995).

being privatised for corporate gain'. Three years later, another organisation, the South African Freeze Alliance on Genetic Engineering (SAFeAGE) was established, modeling its demands on the European Union moratorium on GMO deployments, and demanding a five-year freeze on all GMO activity, including importation of GM products, the planting of GM crops, and the processing of GM products into food consumables without explicit labeling. Broadly, these organisations were motivated by a belief in three related sets of citizenship rights: the right to a safe food system; the right to a healthy (and sustainable) environment, and the right of consumers to choose (including not only the right of people to know what they are eating, but also the right of farmers to select the seed varieties that they wish to plant). In short, they sought a different, and broader, notion of sovereignty over seeds – one in which public voice plays a much larger role in assessing not only the risks of the technology but also its value, based at least in part on informed citizen choice. They believed that these normative commitments should be considered inseparable from knowledge claims regarding technological innovation, and should be incorporated into the regulatory system.

These organisations worked hard to generate a public awareness – and suspicion – of GMOs. But they took different strategic tacks. Biowatch, concerned principally about the dangers of corporate control and threats to biodiversity, focused attention on the adequacy of regulation. Recognising that the GMO Act was a weak and readily 'captured' mechanism, it sought to re-cast the normative terms of risk assessment by broadening the array of regulatory instruments that could and should be brought to bear on GMO regulation. It pursued this strategy most directly in two legal actions that it brought against the National Department of Agriculture (NDA). The first was a suit brought in August 2002 requiring the Registrar of the GMO Act to provide Biowatch with comprehensive data on risk assessments, in terms of a right to information secured under Section 32(1) of the Constitution. The second was an appeal that Biowatch lodged in April 2004 against the Registrar's decision to grant Syngenta a permit for the import, field trials, and general release of Bt11 maize. Biowatch won the first of these actions, but in a rather bizarre twist the court required Biowatch to pay the costs of Monsanto, which had intervened in the case as an interested party (and in particular wanted to protect information that it had released to the government for risk assessment purposes but regarded as proprietary). The costs to Biowatch were especially high because Monsanto's intervention had required the organisation to commission an expensive affidavit to address Monsanto's concerns[203]. Biowatch lost the second case, partly because it was unable to muster the scientific expertise to undermine Syngenta's documentation[204].

With these actions, Biowatch had three main objectives. One was to learn as much as possible about the actual process of evaluation under the Act in order to devise effective counter-strategies. The second was to pry open the process for public scrutiny. The third was to try to inject the National Environmental Management Act, which requires both a 'risk-averse and cautious' approach and participatory environmental governance, into the politics of regulation. The Syngenta appeal demonstrated the importance of this third aim for Biowatch's 'boundary work' in two particular ways. First, the Registrar was fast-tracking the permit process by allowing simultaneous field trials and general release. This process not only undermined the very point of field trials but was incompatible with a 'risk-averse and cautious' approach to environmental management. Second, to short-circuit the field trials phase, Syngenta's risk assessment could not include significant local data. Instead, it relied on studies carried out in the United States and on data for a similar, though not identical, product. This approach to risk

---

[203] In an interesting development, the national labour organisation, Cosatu, and a range of religious organisations petitioned the court to overturn that ruling.

[204] The organisation noted that it is difficult to find expertise in South Africa that is independent of industry.

assessment, Biowatch argued, ran counter to the participatory governance requirements of NEMA, and took inadequate account of the potential impact on *local* biodiversity. As such, it raised questions about the portability of scientific 'facts' versus the role of contextual knowledge in managing and evaluating technology transfer (Biowatch SA, 2004)[205].

While BioWatch attacked the regulatory system, SAFeAGE pursued a broader public strategy that focused on public awareness, food safety, and consumer choice. Stressing the 'precautionary principle', it called for a comprehensive five-year on all GMO activity in order to allow 'more time to properly assess the use of genetic engineering and its health, safety and environmental implications if we are to use it in our food and farming', and it urged the government to ratify the Cartagena Protocol. Constructing itself self-consciously as a network and not as a campaign organisation, the SAFeAGE leadership worked mainly to orchestrate coalitions of civil society organisations around specific activities, such as media events, information sessions, supermarket campaigns, labeling campaigns, petition drives, etc. It managed to secure the support of important constituencies, such as the Catholic Bishops Conference[206] and the labour federation, Cosatu, and enjoyed generally supportive coverage by the media. Working closely with another small NGO, the African Centre for Biosafety (ACB), (whch was run by an environmental lawyer who had previously worked for Greenpeace International, and acted as a clearing house for information related to genetic engineering), SAFeAGE made a great deal of information public in an effort to secure the precautionary principle as a public demand and a regulatory norm.

This strategy failed, however, to generate much *active* public opposition to GMOs, even though polls in 2002 and 2003/2004 indicated considerable public suspicion of GMOs (though also low public awareness) (www.africabio.com). While it was easy to get Cosatu to support it rhetorically, or to get 30 NGOs to sign on to a petition, the organisation relied heavily on shifting coalitions of consumer and producer groups and the sometimes fleeting convergence of group interests on particular issues. The approach turned out to be too diffuse and fragmented since it depended on local organisations being prepared to take up particular concerns and organise local actions[207]. In 2005, SAFeAGE set out to re-invent itself with a more focused, campaign-based, approach that targeted three areas: a labeling campaign, a GM-free food list, and the establishment of GM-Free Zones (today also a key strategy in Europe, as the European Commission develops its co-existence strategy).

While these organisations pushed the boundaries of expertise and authoritative knowledge, trying to both engage the normative framework of policymaking and to inject an alternative normative framework into the policymaking realm, the boundary-makers pushed back. In 2000, a pro-industry lobbying organisation, AfricaBio was formed, not only with industry support, but also incorporating many of the participants in the scientific network associated with SAGENE and the GMO advisory committee. For instance, one of the key figures in AfricaBio is a leading scientist at a leading South African university, was chair of SAGENE, a member of the GMO Act Advisory Committee, and was recently appointed to the government's Innovation Council. AfricaBio presented itself as a nonpolitical advocate for the technology that would both provide accurate information (and, perhaps more to the point, counter 'misinformation') to consumers, media, and decision-makers (based on 'good science'), and bring its

---

[205] At stake here is the politics of what Giddens calls disembedded 'expert systems' as a feature of modern social organisation.

[206] In 2001, the Catholic Bishops Conference issued a statement opposing GMOs titled 'Genetically Modified Food: The Impending Disaster'.

[207] In KwaZulu-Natal, for instance, local groups wanted to organise around rBGH, and its effects on milk, but the issue did not generate the same passion in other parts of the country.

members into the international biotechnology networks. The organisation was extensively networked with international organisations, notably ISAAA and organisations such as the US Grains Council with which it partnered to carry out demonstration trials of Bt white corn for small-scale farmers. It has worked tirelessly and effectively to promote the technology and the over-riding authority of good science. For instance, in a public hearing before Parliament's agriculture and land affairs committee in January 2006, the Executive Director of AfricaBio argued vigorously that South Africa should tighten the regulatory focus and streamline the process because 'there is quite an extensive amount of data that has already been collected globally', and it was important to 'not waste our resources trying to re-invent the wheel by doing it all over again ourselves'. While it was it was essential to provide safety checks and balances through regulation and legislation, she noted, it was also essential that they remain easy to use, by both scientists and farmers, and that its cost be minimised to ensure maximum benefits from the technology[208].

How do we evaluate this boundary war? Debates about GMOs in South Africa are highly polarised and vocal. They have had an impact on the regulatory process, as the anti-GM activists have learned to navigate it. In his 2005 Annual Report, the Chair of the Executive Council of the GMO Act reported that 'This is the first year that the anti-GMO movement have made active inputs into almost all trial, general and commodity clearance applications submitted to the department. Although this is a good indication that public participation has increased in the regulation of GMOs, it is concerning to note that applications take a considerably longer time to be processed, which impacts negatively on the regulatory process and industries' perception of Government's ability to manage public participation' (South Africa, 2005: 5). Ironically, the government tried to resolve the logjams in the process by passing legislation that would permit fast-tracking with even less oversight, but the legislation ran up against severe opposition, and indeed, the GMO Amendment Bill currently under discussion seeks to tighten regulation somewhat, partly in order to take greater account of the Cartagena Protocol. The anti-GM activists have, therefore, enjoyed some success.

Yet they have had little effective impact on public opinion: knowledge of GMOs remains scattered and spotty despite the polarised debate and the fact that all the participants in this boundary war are internationally networked. To make sense of the trajectory of this conflict, it seems clear that the context of apartheid-era struggles and postapartheid civil society mobilisation are highly significant in shaping the political terrain. On the one hand, it was the decline of apartheid and the period of regulatory flux in the early 1990s that opened the political space for a tightly networked, ambitious, and normatively cogent group of scientists to move to the centre of the regulatory system and to shape the discursive politics of GMOs[209]. On the other hand, the anti-GM activists, while gaining a boost from the democratic transition, occupy a rather odd political space on the South African landscape, which has inhibited their efforts to bring the National Environmental Management Act and the precautionary principle more centrally into the normative framework of biotechnology regulation.

In the first place, as several commentators have noted, the movement does not have a broad social base, for reasons of both race and class. Activists tend to be white, urban-based, middle-class people, often with a professional background. In a rather unsympathetic characterisation, Philipp Aerni has argued that 'The NGO coalition against genetic engineering in agriculture represents a mix of the

---

[208] Mail and Guardian Online: 'Parliament hears that Legislation hampers Biotech Research' (retrieved on February 24, 2007)

[209] It is in exercising this norming impact that networks can be seen to act as social *agents* rather than simply as *conduits* for ideas, interests, influence, etc.

earlier eurocentric conservation ideology and the fear of being subjected to a new form of colonialism by transnational corporations (TNCs). While reservations regarding the domestic expansion of TNCs are common among developing countries, the concern about the potential risks of GMOs to human health and biodiversity in combination with the outright discarding of any possible benefits for South African farmers seems to reflect a rather eurocentric attitude' (Aerni, 2001: 7). Aerni is concerned here, apparently, that white South African NGOs have given up social justice issues for environmental justice issues (and even environmental justice is conceived in a way that has very little concern for justice in it).

At the same time, this framing of environmental activism in South Africa has also been picked up by more sympathetic observers, such as Jacklyn Cock, who lauds the shift in notions of environmental justice 'away from this traditional authoritarian concept of environmentalism [by which she means "socially shallow with a mainly white, middle-class support base... predominantly concerned with preserving biodiversity" as represented by the EWT and WESSA], which was mainly concerned with the conservation of threatened plants, animals and wilderness areas, to include urban, health, labour and development issues' (Cock 2006: 205-07). (In Cock's conception, indeed, any inclusion of the role of 'science' seems anathema because it tends to inject notions of what is 'pragmatically possible' that undermine the purity of 'morally correct' positions). In South African environmentalism, then, 'biodiversity' tends to have a tainted name and environmental activism has a predominantly urban focus (Khan, 2002)[210]. This made it very difficult for anti-GM activists to galvanise the environmental movement, and without that action it was impossible to inject the National Environmental Management Act and its precautionary and participatory principles effectively into the norm-setting boundary wars of GMO regulation.

In this context, it is further ironic that the activists were unable to move farmers with their concerns about the impact of GMOs on biodiversity. There was, of course, little political ground to be made by taking up the cudgels on behalf of white commercial farmers (who would in any case have been generally unsympathetic). On the other hand, these activists had very limited entry to the world of African smallholders. Moreover, despite South Africa's brutal history of land dispossession and its very active civil society, there is no substantial or coherent movement for agrarian peoples' rights in South Africa. The political elite sees the future of agriculture in large-scale farming (deracialised) driven by the latest technology. The Landless People's Movement is weak and disunited (Greenberg, 2006). The politics of rural social justice is focused, understandably enough, on land reform. In this context, questions of rural *environmental* justice have been marginalised, both by the state and by the 'social movement community.' Threats brought by anti-GM activists against the Biotechnology Project seem to be effaced.

5. Return to Makhathini

What, then, does all this mean for the farmers of Makhathini and other resource-poor farmers? It was in the context of the emerging struggle over biotechnology in South Africa (and elsewhere) that these farmers were propelled into the global spotlight of the Biotechnology Project through the narrative of hope. But if that spotlight illuminated resource-poor smallholders, it also cast shadows over the Project itself. One of the most striking effects of the biotechnology boundary war is that all the actors

---

[210] As one Durban-based activist noted, the environmental justice movement itself was highly localised and poorly resourced, so that the Cape Town group had to drop anti-GM work in order to focus their resources on fighting the new nuclear power plant planned for Cape Town (Interview, August 2007).

involved have started to realise that in order to bring smallholders to their side they need to recognise smallholders more fully as agents and stakeholders in the deployment of agricultural technologies. Consequently, they have begun to shift their focus substantially towards the fate of smallholder farmers. Arguably, this has opened small but noteworthy windows of opportunity for *local* (and perhaps non-GM) innovation in agriculture.

Two effects are worth mentioning. One is what one might call the 'managed' effect. As an effort to create a viable alternative farming system for poor farmers, and to generate resistance to GM, BioWatch has set out to promote organic or 'low-cost' farming programmes with small-scale farmers, supplying local markets. Though the initiative is small, it has been successful in two provinces (KwaZulu-Natal and Limpopo), and has the effect of incorporating poor farmers in deliberations over appropriate agricultural technologies that may be alternative to GM. On the other hand, the biotechnology industry has set out a strategy of establishing of GM maize demonstration plots in order to demonstrate the power of GM technology to poor farmers. The guiding principle of this approach is that once farmers see the advantages of the technology, they will adopt it. In both these cases, farmer choice is a consideration in agricultural technology deployment that grows directly out of the boundary wars of biotechnology, and would most likely not have taken off absent those struggles.

The second effect is what one may call the 'unmanaged' effect in which the biotechnology industry loses control over the technology through the 'independent participatory action' of farmers. This effect seems most likely in the case of Makhathini farmers. In the early 2000s, shortly after Bt cotton was first planted on the Makhathini Flats, the narrative of hope was bolstered when a team of researchers from the University of Reading and the University of Pretoria produced a number of scholarly articles showing generally positive results from Bt uptake in Makhathini based on the results of surveys carried out in 2000 (Ismael *et al.*, 2001, 2002; Gouse *et al.*, 2002). This work expressed 'cautious optimism' about the role of agricultural biotechnology in improving the yields, incomes, and livelihoods of these farmers. Subsequent research, however, has been much more skeptical (Pschorn-Strauss, 2004; Hofs *et al.*, 2006; Witt *et al.*, 2006; Witt, 2007), and despite their prominent role in the global discursive politics of agricultural biotechnology, there is no strong indication that GM seed will resolve their farming challenges.

These farmers are very vulnerable. Because most cotton in South Africa is imported, the producer price is closely tied to the international price (because if the local price went up, local purchasers would simply import more of their cotton needs). And the international price, of course, is shaped significantly by subsidies in the global north. For farmers, however, input costs are a significant proportion of their production cost, and these costs are relatively fixed. In addition, farmers buying Monsanto's Bollgard GE seed must both pay a technology fee and sign a Technology Agreement before they can receive the seed. The technology fee more or less doubles the cost of seed (Ismael *et al.*, 2002: 3)[211]. The technology agreement imposes stringent restrictions on how it may be used: for instance, farmers must agree to use the seed for planting a commercial crop for only one season; not to supply any Bollgard seed to a third party; not to provide the seed to anyone for crop breeding, research or seed production; to plant a refuge as part of an insect management strategy; to allow Monsanto agents to inspect their fields (Pschorn-Strauss, 2004). Pschorn-Strauss has suggested that farmers sign these agreements without

---

[211] Ismael and colleagues (2002) argue that this cost increase is considerably offset by the reduced cost in pesticide application; Witt *et al.* (2006: 12), on the other hand, argue that reduced pesticide use for bollworm has been offset by increased pesticide applications against the jassid, which has increased since the advent of Bt cotton.

really taking in what they mean because they have no real options. The local company, the Makhathini Cotton Company does in fact offer farmers the option of buying either Bt or non-Bt cotton, but the seed company packages Bt cotton seed in a way that is more suited to farmers' cashflow constraints, so they tend to buy it (see Witt *et al.*, 2006; *Seeds of Hope(?)*). In recent years, the provision of credit has been taken over entirely by the Land Bank (previously it was supplied through the Cotton Company, Vusina, which was replaced by the Makhathini Cotton Company), and has become tighter.

In this context, the politics of contention around agricultural biotechnology are likely to have the paradoxical effect of *weakening* controls over the technology in the field. In the first place, as credit becomes tighter and indebtedness increases, Makhathini farmers are likely to start using, acquiring, and trading Bt seeds outside of the contract (such pirating of seeds is already a widespread practice in India and Brazil). In addition, farmers will not maintain refuges. Extension facilities are too underdeveloped to do adequate training and monitoring, and the state does not have the capacity to regulate at the local level. The seed companies, for their part, are unlikely to police the conditions of the contract rigorously because it will be very bad PR. They are being watched very closely by the activist organisations, for whom monitoring the company's actions is easier than monitoring the farmers' actions. The seeds are likely to spread uncontrollably. And the impacts on the environment and sustained livelihoods are anyone's guess.

6. Conclusion

The resource-poor farmers of South Africa's Makhathini Flats were drawn onto the global agricultural stage as an icon of the narrative of hope. As the first African adopters of genetically modified seeds, they would also provide a bridgehead for the Biotechnology Project into Africa. But they also became a lightning rod for a conflict over agricultural biotechnology, with both local and transnational dimensions, that invoked contending concepts of risk, rights, and authoritative knowledge. That conflict has become increasingly bitter (interview with activist, August 2007). But it has had paradoxical effects. Neither side has 'won' and perhaps both sides have 'lost'. For their part, activists have not been able to stop the deployment of transgenic technologies, to secure the stringent precautionary principle that they would like, or to tighten regulatory oversight of the technology significantly. Certainly, agricultural biotechnology will remain a key component of the agriculture and food sector. But the Biotechnology Project has been cracked open in significant ways. The industry has not been able to show that transgenic technology will systematically increase yields or solve problems of poverty and hunger. Agency of farmers in the agricultural system has been increased rather than diminished – and in some cases that might entail a weakening of effective control over the technology. And the question of alternative agricultural technologies has been put on the table. Thus, the politics of agricultural biotechnology has not derailed the Biotechnology Project. But it has shaken it, and perhaps thereby opened new avenues for innovative thinking about appropriate agricultural technologies for development.

References

Aerni, P. (2001). Public Attitudes towards Agricultural Biotechnology in South Africa. Final report, joint research project of CID (Harvard) and SALDRU (UCT).

Biowatch SA. (2004). Syngenta Appeal Summary. Available at: ; retrieved on June 16, 2004.

Bond, P. (2000). *Elite Transition: From Apartheid to Neoliberalism in South Africa*. Pietermaritzburg, University of Natal Press.

Boudreaux, K. (2006). Seeds of Hope: Agricultural Technologies and Poverty Alleviation in Rural South Africa. *Mercatus Policy Series* Policy Comment no.6.

Brenner, C. (2004). Telling Transgenic Technology Tales: Lessons from the Agricultural Biotechnology Support Project (ABSP) Experience. *ISAAA Briefs* no.31. Ithaca NY, ISAAA.

Bud, R. (1995). In the Engine of Industry: Regulators of Biotechnology, 1970-86. In: *Resistance to New Technology: Nuclear Power, Information Technology and Biotechnology* (Ed. M. Bauer), Cambridge, Cambridge University Press.

Burton, S. and Cowan, D. (2002). Development of Biotechnology in South Africa. *Electronic Journal of Biotechnology* 5 (1): http://www.ejbiotechnology.info/content/vol5/issue1/issues/03/index.html

Byrne, D. (2001). A European Approach to Food Safety and GMOs. Speech to National Press Club, Washington, 9 October 2001 (Speech/01/442).

Castells, M. (1996). *The Information Age, vol.I: The Rise of the Network Society*. Oxford: Blackwell.

Chan, T.M. (2004). The Richtersveld Challenge: South Africa finally adopts Aboriginal Title. In: *Indigenous Peoples' Rights in Southern Africa*, (Eds. R. Hitchcock and D. Vinding), Copenhagen, IWGA.

Cock, J. (2006). Connecting the Red, Brown, and Green: The Environmental Justice Movement in South Africa. In: *Voices of Protest: Social Movements in Post-Apartheid South Africa* (Eds. R. Ballard, A. Habib and I. Valodia), Durban: University of KwaZulu-Natal Press.

Dantas, E. (2005). The 'System of Innovation' Approach, and its Relevance to Developing Countries. Science and Development Network. Available at: http://; retrieved on January 14, 2007.

Freidberg, S. and Horowitz, L. (2004). Converging Networks and Clashing Stories: South Africa's Agricultural Biotechnology Debate. *Africa Today* 51: 3-25.

Gieryn, T. (1983). Boundary-Work and the Demarcation of Science from Non-Science. *American Sociological Review* 48(6): 781-795.

Gottweis, H. (1998). *Governing Molecules: The Discursive Politics of Genetic Engineering in Europe and the United States*. Cambridge MA, MIT Press.

Gouse, M, Kirsten, J. and Jenkins, L. (2002). Bt Cotton in South Africa: Adoption and the Impact on Farm Incomes amongst Small-scale and Large-scale Farmers. University of Pretoria, Department of Agricultural Economics: Working Paper 2002-15.

Greenberg, S. (2006). The Landless People's Movement and the Failure of Post-Apartheid Land Reform. In: *Voices of Protest: Social Movements in Post-apartheid South Africa* (Eds. Ballard, R. Habib, A. and Valodia, I.), Pietermaritzburg, University of KwaZulu-Natal Press.

Gupta, A and Falkner, R. (2006). The Cartagena Protocol on Biosafety and Domestic Implementation: Comparing Mexico, China and South Africa. EEDP Briefing Paper 06/01. Chatham House.

Habib, A. and Padayachee, A. (2000). Economic Policy and Power Relations in South Africa's Transition to Democracy. *World Development* 28 (2): 245-263.

Hisano, S. (2005). A Critical Observation on the Mainstream Discourse of Biotechnology for the Poor. *Tailoring Biotechnologies* 1 (2): 81-105.

Hofs, J-L, Fok, M. and Vaissaye, M. (2006). Impact of Bt Cotton Adoption on Pesticide Use by Smallholders: A 2-Year Survey in Makhathini Flats (South Africa). *Crop Protection* 25 (9): 984-988.

Ismael, Y, Bennett, R. and Morse, S. (2001). Farm Level Impact of Bt Cotton in South Africa. *Biotechnology and Development Monitor* 48: 15-19.

Ismael, Y, Bennett, R. and Morse, S. (2002). Benefits from Bt Cotton Use by Smallholder Farmers in South Africa. *AgBioForum* 5(1): 1-5.

Jasanoff, S. (2005). *Designs on Nature: Science and Democracy in Europe and the United States*. Princeton, NJ. Princeton University Press.

Joubert, G., Venter, M., Theron, G., Swanepoel, A., Eulitz, E., Schroder, H. and Macaskill, P. (2001). South African Experience with Bt Cotton. Available at: www.icac.org/cotton_info/tis/biotech/documents/techsem/SAexperience_tis01.pdf.

Khan, F. (2002). The Roots of Environmental Racism and the Rise of Environmental Justice in the 1990s. In: *Environmental Justice in South Africa*, (Ed. D. MacDonald), Athens OH, Ohio University Press.

Lodge, T. (2002). *Politics in South Africa*. Bloomington IN, Indiana University Press.

Marris, C. (2000). Swings and Roundabouts: French Public Policy on Agricultural GMOs 1996-1999. Centre D'Economie et D'Ethique pour l'Environnement et le Developpement. Cahier no.00-02.

Mayet, M. (nd). Analysis of South Africa's GMO Act of 1997. Biowatch SA. Available at: ; retrieved on November 11, 2006.

Mnyulwa, D. and Mugwaga, J. (2003). Agricultural Biotechnology in Southern Africa: a Regional Synthesis. In: *Biotechnology, Agriculture, and Food Security in Southern Africa* (Eds. S.W. Omamo and K. von Grebmer), IFPRI, pp. 14-36. Available at: http://www.ifpri.org/pubs/books/oc46.htm.

Monsanto. (2006). Biotechnology: A Decade of Biotechnology. Available at: http://www.monsanto.co.za/en/layout/biotech/10_years.asp; retrieved on November 11, 2006.

Munro, W. and Schurman, R. (2008). Chain (Re)actions: Comparing Activist Mobilization against Biotechnology in Britain and the United States. In: *Frontiers of Commodity Chain Research*, (Ed. J. Bair), Stanford University Press, in press.

Ofir, Z.M. (1994). Biotechnology in the New South Africa. *Biotechnology and Development Monitor* 20: 14-15.

Paarlberg, R. (2000). Governing the GM Crop Revolution: Policy Choices for Developing Countries. IFPRI. Available at: http://www.ifpri.org/2020/dp/2020dp33.pdf.

Padayachee, V. (Ed.) (2006). *The Development Decade? Economic and Social Change in South Africa, 1994-2004*. Cape Town, HSRC Press.

Pschorn-Strauss, E. (2004). Bt Cotton and Small-scale Farmers in Makhathini – A Story of Debt, Dependency, and Dicey Economics. GRAIN. Available at: http://www.grain.org/research/btcotton.cfm?id=100.

SAFeAGE. (2005). Monsanto Grabs southern African Seed Market but SA Farmers give GMOs Cold Shoulder. Available at: ; retrieved May 4, 2005.

Scoones, I. (2005). Contentious Politics, Contentious Knowledges: Mobilising against GM Crops in India, South Africa, and Brazil. IDS Working Paper 256.

Sell, S. (2003). Competing Knowledge Networks: the Quest for Global Governance in Intellectual Property. Paper prepared for SSRC Workshop on Intellectual Property, Markets, and Cultural Flows, October 23-24.

South Africa. (nd). *Innovation Made Easy*. Department of Science and Technology.

South Africa (1996). *White Paper on Science and Technology: Preparing for the Twentieth-First Century*. Department of Arts, Culture, Science and Technology.

South Africa. (1997). *GMO Act (no 15 of 1997) and GMO Application Process*. Directorate Genetic Resources, Department of Agriculture.

South Africa. (1998). *National Environmental Management Act* (no. 107 of 1998). Government Gazette 401, no.19519.

South Africa. (2001). *A National Biotechnology Strategy for South Africa*. Department of Arts, Culture, Science and Technology.

South Africa. (2005). *Genetically Modified Organisms Act, 1997: Annual Report 2004/2005*. Department of Agriculture.

Thackray, A. ed. 1998. *Private Science: Biotechnology and the Rise of the Molecular Sciences*. Philadelphia: University of Pennsylvania Press.

Touraine, A. (1971). *The Post-Industrial Society; Tomorrow's Social History: Classes, Conflicts and Culture in the programmed society*. (Eds. Tr. Leonard and F.X. Mayhew). New York: Random House.

Van Renssen, S. (2006). Innovation in South Africa: too much, too soon? *Science Development Network*, April 6, 2006. Available at: http://www.scidev.net/features/index.cfm; retrieved on January 14, 2007.

Van Roozendaal, G. (1995). Organizing Biotechnology Research in Africa. *Biotechnology and Development Monitor* 23: 12-15.

Witt, H. (2007). Cotton: Still the Mother of Poverty? University of KwaZulu-Natal, ms.

Witt, H, Patel, R. and Schnurr, M. (2006). Can the Poor help GM Crops? Technology, Representation and Cotton in the Makhathini Flats, South Africa. *Review of African Political Economy* 33 (109): 497-513.

Woodward, B, Brink, J. and Berger, D. (1999). Can Agricultural Biotechnology make a difference in Africa? *AgBioForum* 2(3&4): 175-181. Available at: http://www.agbioforum.org/v2n34/v2n34a05-woodward.htm

World Bank. (1997). *World Development Report: The State in a Changing* World. New York. Oxford University Press.

Wright, B. and Pardey, P. (2006). Changing Intellectual Property Regimes: Implications for Developing Country Agriculture. *International Journal of Technology and Globalization* 2(1/2): 93-114.

Biotechnology policy: the myth and reality in Sub-Saharan Africa

George O. Essegbey

1. Introduction

The compelling riddle of development is why the apparently most naturally endowed parts of the world such as Africa, should be the poorest and at the bottom rung of human development. Africa is known to have a rich complex of mineral, oil and gas deposits, an invaluable spectrum of flora and fauna and a wide range of natural habitat. The continent is known to harbour one of the ecological lungs of the world in its rainforests and minimal presence of emissions and effluents harmful to the environment (NEPAD, 2001). How to address the riddle of poverty comes in substantial part to the deployment of new technologies such as biotechnology. Nevertheless, the approach to that deployment and the conceptual underpinnings are bound to determine the outcome.

It is estimated that in Africa about half the population live on less than one US dollar a day. According to the Human Development Report of 2006, the mortality rate of children under five years of age is about 174 per 1,000 live births compared to 75 for the world average (UNDP, 2007). Table 1 gives the specifics for selected countries.

Table 1. Human Development (HD) of selected African Countries, 2006 (The World Bank, 2007b; UNDP, 2007).

Country	HD Rank	Population (millions)	GDP in current US$ billions	Life Expectancy at birth (2000–05) in years	Population under-nourished (2003) %	HIV Prevalence Age 15-49 (2005) %	Under 5 mortality rate per 1,000 live births (2004)
Nigeria	159	131.5	99.0	43.3	9	3.9	197
South Africa	121	46.9	239.5	49.0	-	18.8	67
Rwanda	158	9.0	2.2	43.6	36	3.1	203
Kenya	152	34.3	18.7	47.0	31	6.1	120
Namibia	125	2.0	6.1	48.6	23	19.6	63
Ghana	136	22.1	10.7	56.7	12	2.3	112
Developing countries				64.9	17	1.1	83
World				67.0	17	1.0	75
OECD				77.6	-	0.4	12
Sub-Saharan Africa		690		46.1	30		

The life expectancy for the African child at birth is only about 46.1 years compared to the average of 65.2 even for the developing countries. In a country of the OECD, the child expects to live up to about 77.8 years. The relatively high HIV prevalence, apart from other diseases such as malaria and respiratory diseases, has contributed to truncating life in Sub Saharan Africa. In the midst of the debilitating diseases, malnutrition is a major challenge. The percentage of the population under-nourished in Africa is 30% compared with 17% for the developing countries. The incidence of under-nourishment in, for example Rwanda, is as high as 43% of the population in 2004. However, mal-nourishment, diseases and the overall quality of life of the people can be addressed with technologies appropriate for the context.

This paper focuses on the potentials of biotechnology in addressing the basic challenge of enhancing the food sovereignty of the people and ameliorating the quality of life of the people in Africa. Usually the danger in such a focus is the tilt in discussion towards the benefits and not the risks of the technology. It then falls in the discordance of the pro-contra debate, which often does not lend itself to the full appreciation of the merits of the arguments on both sides of the debate. However there is need to focus on the technology as a tool for meeting one of the most critical challenges facing the African society. In doing so, the point needs to be emphasised that the mere application of biotechnology is not a sufficient guarantee for life improvement. There has to be radically innovative ways for biotechnology that goes beyond the orthodoxy in modern biotechnology application and development. This paper aims at discussing the options for some of these innovations.

2. The conceptual framework

The fundamental concept underpinning this paper is what essentially drives the TELFUN project, which pivots on the premise that biotechnology development and application is 'shaped by the social context in which it appears' and it is open to change (Ruivenkamp and Jongerden, 2005: 11). It is the opportunity for change, which validates a proactive position for a redesign and a reconstruct of biotechnology within the socio-cultural, economic and political coordinates of a given society. There is no exclusivity in any paradigm. Rather there are always continuums and overlaps. This paper adopts the proposition that the multi-dimensional facets of biotechnology justify a dismantling for re-assembling of biotechnology for a good fit in its context. The process for dismantling and re-assembling is neither simple nor singular. There are the pro and anti forces amplifying unnecessarily the superlatives of the best and worst of biotechnology. Taming the forces for positive neutrality is part of the process of arriving at an appropriate construct. There are many options in that process. For example in considering the options for maneuvering within the hegemonic political, economic and legal setting of biotechnology development, there is said to be the need to demonstrate the 'affluent possibilities of alternative biotechnologies to meet the needs of the resource poor' (Hisano, 2005: 102). This paper uses the case of the biotechnology policies coming out of Africa to restate the need for the multi-dimensional definition of biotechnology and the usefulness of re-designing, re-packaging and re-constructing biotechnology. The reason for coming to this conceptual underpinning is simply the fact that the context for agriculture is changing rapidly and the process of knowledge generation and its use has been transformed (The World Bank, 2007a). Just so that the resource-poor is not further marginalised and strangled in the unrelenting technological progression, there is the imperative of strategic technological choices that must touch base with the roots of society. Biotechnological innovation in context is one such choice.

The link between biotechnology and biosafety is more than conceptual. It fairly well illustrates the theoretical underpinnings of this paper that on one hand there are the technical facets of the technology and on the other hand there are the social facets of the technology. Biosafety represents the face of the social facets as framed by the voice of caution and the exigencies of risk management. In some of the national policies, the overall challenge of biotechnology development and biosafety are addressed together.

There is a fundamental question: is a biotechnology/ biosafety policy necessary? Given that it is really a generic question and not rhetoric, the answer may be inferred from the vision of the European Commission (EC) that 'the scientific and technological revolution is a *global* (emphasis added) reality which creates new opportunities and challenges for all countries in the world, rich or poor' (Commission of European Communities, 2001: 4). But that is as far as any one can go in inferring global applicability of the EC vision to the rest of the world. The policy was ostensibly to outline and communicate the EC's societal priorities and in particular the societal framework and the ethical basis for the development and applications of the new sciences and technologies (Commission of European Communities, 2001). A similar raison d'etre may be inferred for all national policies. In the adoption and application of biotechnology, there is the intention to encapsulate that which relates to the national interest and priorities economically, socially, politically and ethically. The question is whether it is always the case that priorities are captured in the policy enunciation.

Six African countries have been selected to illustrate the extent to which some efforts are made to address the need for capturing national priorities, analyse the degree of success and the precipitated failures. The key challenge for most African countries is the achievement of food sovereignty with the local communities exercising control of the food production-consumption nexus. Thus the challenge is not only in the production from the farming systems, but also the processing and distribution to establish a sustainable supply and demand regime, which is fundamentally independent of external props and ramifications.

2.1. Nigeria

Nigeria is the most populous country in Sub-Saharan Africa with about 131.5 million population. It professes democratic governance of its polity and society. But skepticism surrounds its democratic credentials as the last multi-party elections attracted concern from several observing entities[212]. However, Nigeria is a burgeoning free market economy with the immense potential of dominating, not only its West and Central African neighbours, but also much of the Sub-Saharan Africa. The incidence of a vast oil wealth on one hand, producing about 2.35 million bbl/day, and on the other hand stark poverty among the majority of its citizenry to the extent that the country ranks at 159 in human development, has seemed to suggest there is a resource curse in Nigeria and for that matter Sub-Saharan Africa.

Nigeria's mission statement in its biotechnology policy is that 'as a matter of priority, imitate appropriate steps to explore the use of Biotechnology for the benefit of Nigerians and thus ensure that Nigeria becomes one of the international leaders in Biotechnology' (Federal Ministry of Science and Technology, 2001: 3). There is a strong emphasis on the indigenous in the Nigerian policy document. It states for example that the policy is to address issues such as indigenous acquisition and development of easy and affordable requisite biotechnology and indigenous Research and Development to generate 'copious innovations' in biotechnology (Federal Ministry of Science and Technology, 2001: 4).

---

[212] For example, the 150-member European Union Election Observation Mission noted that the April 2007 election fell short of basic international and regional standards for democratic elections.

The policy document highlights clearly the particular objectives and outlines strategies for capacity building, R&D, the development of bioresources and the implementation strategies. It contains specific projections in relation to food demand and food self-sufficiency up to the year 2010 for food commodities such as yam, cassava, millet, sorghum, beans, oil palm fish and poultry. In the summary of the Biotechnology Entrepreneurship Projects highlighting production of planting materials, afforestation activities, production of a variety of biotechnology products such as mushrooms, biofertilisers, biopesticides biogas, diagnostic kits and vaccines, there is an indication of how the policy aims at facilitating impact on the ground though it is not very evident the linkages to the discrete units of the society.

Probably the best summary of the content of Nigeria's policy document is as stated by President Olusegun Obasanjo that 'as a matter of priority (to) initiate appropriate steps in the area of biotechnology and facilitate the effective utilisation of this new technology for the benefit of our people and use biotechnology as an example to show that we want to do things better than we have done in the past' (Federal Ministry of Science and Technology, 2001: 27).

2.2. Kenya

Kenya is relatively a more economically active than other African countries with an income per capita of about $545 (The World Bank, 2007b). About 80% of its labour force is engaged in agriculture with this sector contributing about 26% to GDP and generating about 50% of the total foreign exchange. Its major export commodities are tea, coffee, horticultural and dairy products. In recent years, the horticultural products of cut flowers, fruits and vegetables have been upscaled. Kenya is also the world's leading producer of pyrethrum, which is a plant (flower) used for biopesticide.

As stated in its National Policy on Biotechnology document, Kenya plans for biotechnology in targeting increased food production, poverty alleviation and environmental protection. The policy document focuses on their traditional food or export crops such as banana, millet, maize and cassava with emphasis on addressing challenges relating to diseases, pests, drought and nutritional traits. It takes on board the challenge of ensuring biosafety noting that there are about 35,000 known species of plants, animals and micro-organisms in Kenya, and that the active exploitation and utilisation of biodiversity will conform to the requirements of the Convention on Biological Diversity (Government of Kenya, 2005). With respect to the international ingredients determining the content of the policy, reference is made to among others, the standards of the Codex Alimentarius Commission, the Cartagena Protocol, NEPAD and the African Regional Intellectual Property Organisation (ARIPO). The policy document notes that the legislative response to biotechnology issues in the country 'has been characterised by fragmented and uncoordinated sectoral legal regimes, developed to facilitate resource allocation and to deal with the adverse consequences' (Government of Kenya, 2005: 24). It states that the Biosafety Law was a response to the deficiencies inherent in the sectoral approach to biotechnology management. The policy outlines approaches to public education, capacity building, resource mobilisation and management, monitoring and evaluation.

2.3. Namibia

Namibia's population is only 2.0 million despite the relatively large surface area of 824,300 square kilometers. It is a typically arid country with less than 1% being arable. It is highly dependent on South

Africa and other countries for food imports especially maize, sugar, fruit and vegetables. Yet about 70% of the population, are directly or indirectly, engaged in farming growing millet, sorghum and other grains. Its fishery and livestock sub-sectors are important for its agricultural activities with the former accounting for about 30% of its export earnings. This is a country, which clearly needs a novel strategy for achieving food sovereignty. The external dependence on food imports of about 50% of the national requirement undermines socio-political sovereignty.

The Biotechnology Policy of Namibia focuses on the safe use of biotechnology with an emphasis on among other things, striking an appropriate balance between biotechnology promotion and regulation in the sustainable development. The Namibian Biosafety Advisory Council (NBAC) is the established independent technical body operating under the Ministry of Higher Education, Vocational Training, Science and Technology. The membership of the Council is drawn from experts in human and veterinary medicine, agriculture, plant breeding, microbiology and environmental protection. The preponderance of expert representation on the Council derives from the seeming over-emphasis of the national policy on biosafety issues. The policy document lists the National Forensic Science Institute, Central Veterinary Laboratory of the Ministry of Agriculture, Water and Rural Development (MAWRD), the Division of Plant Production Research of MAWRD and Medical Laboratory Services as institutions to be strengthened in terms of research, development and biosafety capacities.

As in the case of most of the other countries, funding support for even the process of producing the policy document came from external sources such as the International Service for National Agricultural Research (ISNAR), USAID, GTZ and UNEP/ GEF. Namibia may be relatively better off than many Sub-Saharan African countries ranking 125 (2006) in human development and posting a total of some $6.1 billion in GDP as against a total population of only two million (The World Bank, 2007b). However, given the pressing development challenges, not the least being the struggle to feed the populace, investing in biotechnology development for Namibia, may not attract the priority required.

2.4. Rwanda

It is a relatively small country with a per capita GDP of less than $250 and the Rwanda Vision 2020 generally aims at enhancing agricultural productivity. Rwanda derives its main export earnings from tea and coffee. The leading food crops are bananas, beans, sorghum and cassava. Maize and rice have been singled out for priority attention in the agricultural programme. There is a robust livestock farming with a significant population of cattle, sheep, goats, poultry and other animals.

Rwanda's National Biotechnology and Biosafety Policy aims, at among other things, building and strengthening national capacity through R&D, ensure public and environmental safety and ethics in biotechnological research, development and application and establishing the framework for the conservation and sustainable use of biological diversity (Ministry of Lands, Environment, Forestry, Water and Mines, 2004). The policy emphasises training of the requisite expertise, the provision of adequate infrastructure and other facilities for biotechnology, a National Development Programme and the establishment of a National Biotechnology Advisory Committee with representation from Ministries such as those in charge of Environment, Science and Technology. Even though the Policy provides for Government commitment to a sustainable funding for biotechnology R&D, the inadequacy of the public resources is highly pronounced. Worse, the country having emerged from a traumatic experience of genocide in the 1994 civil war leading to the death of roughly one million Rwandans, the priority in resource allocation will remain the reconstruction of the nation and its socio-economic infrastructure. Nevertheless, the opportunity for rebuilding in general may also be an opportunity to reconceptualise Science and Technology generally and let biotechnology and other new technologies express their social dimensions.

2.5. Ghana

The Growth and Poverty Reduction Strategy (GPRS II 2006-2009) defines Ghana's development vision as middle-income status of at least $1,000 per capita income by the year 2015 (NDPC, 2007). Currently as derived from the statistics of The World Bank, Ghana's per capita GDP is less than $500 (The World Bank, 2007b). This is not surprising since typically the mainstay of the economy is the export of cocoa, timber and minerals particularly gold and diamond. Non-traditional exports such as pineapples and other horticultural commodities and vegetables are becoming important. The GPRS anchors on the priorities of continued macroeconomic stability, accelerated private sector-led growth, vigorous human resource development, good governance and civic responsibility. In this regard, there are details to improve sectoral performance in agriculture, industry and the other sectors of the economy.

Ghana's draft National Biotechnology Policy broadly aims at harnessing biotechnology to accelerate the envisaged socio-economic improvement. For the respective sectors, the following have been specified:
- *agriculture* – the challenge of enhancing productivity in crop and livestock;
- *health* – curtailing the incidence of human diseases especially malaria and HIV-AIDS;
- *industry* – improving production systems and the competitiveness of local industry;
- *energy* – use of cheaper and more reliable alternative sources of fuels;
- *environment* – ensuring sustainable environmental management.

Ghana in its Food and Agriculture Sector Development Policy (FASDEP) has outlined its objectives for agricultural development including to (Ministry of Food and Agriculture, 2002):
- ensure food security;
- facilitate the production of agricultural raw materials for industry;
- facilitate the production of agricultural commodities for export;
- facilitate effective and efficient output processing and marketing system.

It derives from the overall development framework of the Growth and Poverty Reduction Strategy 2006-2009 (NDPC, 2007) which broadly aims at wealth creation and poverty reduction to attain a middle-income status by 2015. It is explicitly stated that growth of Ghana's economy in the next few years will be agriculture-led with expected inputs for a 'vibrant agro-processing industrial sector' in the medium to long term (NDPC, 2005: 23). Much of this goes also for almost all Sub-Saharan African countries where the main challenge has been how to accelerate agricultural production to ensure food security and support industrialisation.

One of the fundamental reasons for a more revolutionary agricultural production is the strong link between agriculture and poverty reduction. In Ghana, the Ghana Living Standard Survey 5 conducted in 2005/2006 has shown that the proportion of Ghanaians described as poor in 2005/2006 declined from 39.5% in 1998/99 to 28.5% (NDPC, 2007: 16). It suggests there are possibilities that the Millennium Development Goal of halving the poverty rate could be met ahead of time. Such possibilities are attributable to growth rates in agriculture coming from high growth in cocoa production (NDPC,

2007). The agricultural sector therefore holds high promise for poverty reduction provided that more can be done to enhance its performance.

The policy thematic areas have spanned human resource development, infrastructural development, creating an enabling environment for biotechnology and development. The issues of biosafety take a whole chapter much in similarity with the other African biotechnology policies. Given that about 70% of Ghana's labour rely on agriculture, the objective strategies for biotechnology exploitation need to concentrate on this sector and in the rural areas where at least 65% of the population is ensconced.

2.6. South Africa

By all standards South Africa is an industrialised country and not exactly in the category of the selected countries in this paper. It is the largest geographically of the selected countries with a population of around 50 million and with a total GDP, which is almost twice the combined GDP of the five other countries. Its historical developments and the experience of apartheid have created a situation where the majority of South Africans live in poor communities. Life expectancy at birth in South Africa is only 49 years having one of the highest incidence of HIV/AIDS of 18.8% (of the population between ages 15 and 49). In the circumstances, there are vast portions of South Africa, which is much like the rural communities in the other African countries. It calls for strategies to enhance the totality of the quality of life and it is probably the reason to initiate steps to adapt a definitive policy that links the vast biotechnology capacity to the production-consumption needs in the rural communities.

The National Biotechnology Strategy for South Africa is a typical industrialised country document with the major thrust aimed at developing 'a viable and sustainable biotechnology industry' (Government of South Africa, 2001: 37). Chapter 4 of the Strategy details the objectives and intervention mechanisms for biotechnology application and development in South Africa. It encompassed the establishment of technology platforms where capital equipment and specialised expertise are shared by biotechnology programmes and industry and where the outputs of R&D are put into commercialisation. The linkages to the knowledge centres and the operationalisation of the biotechnology incubator system underscore the strong focus on the exploitation of South Africa's industrial capacities. In the process, the National Biotechnology Strategy addresses some of the pressing socio-cultural challenges of the people. For example there is the South African Aids Vaccine Initiative (SAAVI), which concentrates efforts to apply modern biotechnology know-how in the fight against HIV/AIDS. Apart from this there is the development of cheap, user-friendly and low-cost diagnostics for public health especially in rural clinics and hospitals and the development of vaccines against tuberculosis, cholera, malaria and others.

While the focus on high-technology application and development of the Strategy is only to be expected, it in no way underscores the urgency to orientate biotechnology to enhance the lives of the rural communities of South Africa. There are the relevant objectives such as improving food security and nutrition, improving plant or crop production in a changing environment and reducing agriculture on the environment and improving animal health and productivity. The applications of biotechnology in for instance, DNAmarker-based selection systems, diagnostic tools for early and accurate pathogen and pest detection, vaccine development for livestock diseases and pedigree determination based on DNA fingerprinting for livestock are applications with potential impact on the livelihoods of the marginalised. However the significance of impact is dependent on how the application is conceptualised and structured to extend to the grassroots. From the perspective of the marginalised rural communities, there appears to be a hiatus between the exigency of their circumstances and the stipulations of the Strategy.

3. The issues and the generalities of the policies

The point needs emphasis that biotechnology/ biosafety policies are useful as far as they meet the need of a bird's eyeview of the direction of technological flight. But they do not address the need for context-specific technology application and development which most Sub-Saharan African countries need. The analysis of the national biotechnology policies can be generalised in the model illustrated in Figure 1.

3.1. Orientation of policy

There seems to be a commonality of statement of purpose in the application and development of biotechnology in the direction of addressing relevant national interests. Invariably, [, improvement of agricultural performance and the exploitation of natural resources are the identifiable interests (except for South Africa, in the selected cases, where industrial capacity is the priority). However, there is ambivalence in the sense that some of these interests are outwardly oriented towards outside economies in the effort to earn foreign exchange. For example, the exploitation of natural resources may require the nation to invest in biotechnology capacities, which are mainly to support export industries. It is acceptable, if this is not at the expense of, for example, the required capacity to secure food self-sufficiency and the attainment of food sovereignty, or if there should be a link to the locally oriented objectives of food self-sufficiency. However, it is often the case that some of the public investments or even the articulation of the externally oriented objectives is made without any appreciation of the need for linkage with the locally situated objectives such as to ensure food self-sufficiency.

There are a number of factors, which also account for the enunciation of the goals and objectives in the policy document. The state of development of the country, the socio-economic aspirations, the sensitivity to cultural norms and mores, the available national capabilities and capacity for biotechnology application and development are some of the salient factors. These are captured in the National Environment as illustrated in Figure 1. The intangible factors however constitute some of the very influential forces determining the outcome of the policy content. The position of the country in the global political positioning, the strength and commitment to geo-political alliances and the influences coming from the International Environment (as in Figure 1), the dynamics of the country's socio-political hegemony are some of the intangible factors, which may colour the outcome of goal-objective definitions.

The specific goal of food sovereignty very well illustrates the point. For example, the incidences of drought expose Namibia's socio-political vulnerability as shown in the controversy surrounding the US-donated relief food aid. CARE, one of the world's biggest charities has taken a stand against distributing US food aid because it is said to undermine the capacity of the receiving countries to produce their own food. For CARE, distributing American food aid is worth about $45 million annually in federal funding. The charity organisation proposes that the US should rather subsidise American farmers to overproduce, should give the money to fund local purchases and support local food production (Dugger, 2007, Doyle, 2007). That really should be the guiding principle for donors' interventions in African food crisis; except in emergencies, the aim should be to facilitate the achievement of food sovereignty in terms of creating and sustaining the capacity for *localised* foods production, processing, utilisation

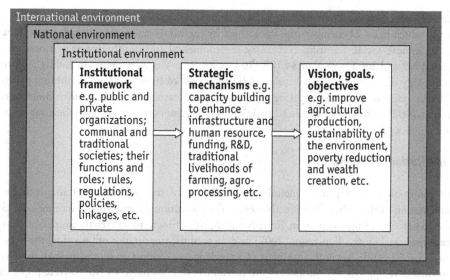

Figure 1. *Conceptual framework for biotechnology in context.*

and marketing. In this regard there is also the need to examine the existing innovations systems and strengthen their functionality (Puplampu and Essegbey, 2004). Achieving food sovereignty requires a holistic approach in going through the cycles of policy formulation, implementation, monitoring, evaluation and re-formulation.

3.2. Strategic mechanisms

For the achievement of the policies, each nation has to formulate the relevant mechanisms, which depend on the resources and technological capacity available. However the mechanisms often illustrate the extent to which the national context has been accounted for and the level of ownership of the public. For in the mechanisms there must be valves for public participation.

Before the drafting of the biotechnology policy, consultations are held with the identifiable stakeholders across the sectors of the economy e.g. farmers, doctors, nurses, public servants, mass media practitioners, consumer associations and environmental activists. The depth of the public participation not only in terms of the outlay of the consultation but also the veracity of the representation renders authenticity to the process. But quite often this is theoretical. In the institutional framework (in Figure 1), the traditional society and rural communities have specific roles to play in the policy formulation and implementation. But often this is only in theory.

Funding as a unilateral factor is at the base of every policy in Africa as it is in the rest of the world. The sources determine the extent of local ownership. In many of the processes to produce the policies in Sub Saharan Africa, funding was external. For example, in Kenya, Rwanda and Ghana, the Global Environment Facility of UNEP was a major source of funding. Obviously the question arises as to the level of national commitment to the policy. The implementation of the policy will invariably be externally dependent.

Other countries create their own funding sources. The Canadian Biotechnology Advisory Committee (CBAC), which has a 21-membership was formed by the Ministers to provide independent, expert advice on ethical, social, scientific, regulatory, environmental and health aspects of biotechnology. Between 1999 and 2002, a total of $4,183,470 was spent on CBAC activities including monitoring biotechnology developments, facilitating public awareness of biotechnology, maintaining a forum for citizen engagement and participating in a variety of outreach activities (Canadian Biotechnology Secretariat, 2002). African countries need a minimum of funding commitment from their own resources to their policies.

3.3. Institutional framework

Policy definition necessary must include the definition of the institutional framework for the implementation of the policy. The effectiveness of the institutional framework determines the extent of effectiveness of the implementation. In analysing the roles, functions and relationships of and among institutions, Essegbey and Puplampu (2007) underscored the importance of providing the necessary platform for the producers and consumers of knowledge in shoring-up biotechnology innovations for national development. But producers and consumers of knowledge are not discrete populations. From the perspective of the multi-dimensional concept of biotechnology and especially with reference to the social dimensions, stakeholders of the larger populace such as farmers, micro and small food processors, fisher folks, animal rearers and other agricultural functional groupings, though they may be set in marginalised rural communities can be part of the knowledge producers.

African governments should avail themselves of all the international goodwill and assistance complementary to their own strategies to build the technological capital needed to stimulate growth (Evenson, 2004). There are literally hundreds of multilateral, bilateral and non-governmental agencies offering a myriad of opportunities for countries to enhance local capacities in technological application and development. African governments must take advantage of these opportunities. However, these opportunities cannot be substitutes for home-grown and context specific strategies. More importantly, the people living in their respective environments may understand best how to make things work.

4. Tailoring biotechnology and the options

One innovative approach is what has been proposed by the High-level African Panel on Modern Biotechnology in the report *Freedom to Innovate* (Juma and Serageldin, 2007). The report emphasises the point that agricultural biotechnology holds the promise of improving food security and better nutrition. As a strategy for wide application of biotechnology, it recommends the creation of 'local innovation areas' involving universities, research institutes, private enterprises and other actors (Juma and Serageldin, 2007: 96). Local authorities are to be engaged in the application of the prioritised biotechnology programmes in the given localities. This fairly illustrates the general approach of tailoring biotechnologies to the local context.

Going back to the basics of the concept of tailoring biotechnology, the context of use of the technology should be the context of need definition and application. This cannot be done without articulating development visions and strategies that focus on the role of technological innovation in development (UN Millennium Project, 2005). In the broad development plans and the specific biotechnology policies, the vision and the goals primarily converge on an enhanced status of the socio-economic circumstances especially as encapsulated in economic statistics. But the attainment of the economic indicator is perhaps less important as enhancing the socio-economic conditions of the people deriving from their inclusion in technological applications and development. In the specific case of biotechnology, achieving inclusion is when biotechnology dovetails into the technological progression of the respective segments of the societies. The fundamental challenge is in how to realise this.

Firstly, no society is homogenous, much less the rural poor. The rural poor constitute a wide range of cultures with their own unique worldviews, indigenous knowledge systems and practices and social networks (Punkari, 2007). Although disaggregation into the discrete units is impracticable, the policy initiatives for technological inclusion should format processes on the principle of genuine democratisation of the technological choices and ownership of outcomes. One of the drawbacks (though not intended) is the seeming cosmetic public participation of the rural poor in the formulation of the national biotechnology/biosafety policies. There were farmers represented at the consultation workshops. But the validity and reliability of the representation is open to question given the millions of farmers and their remoteness (both culturally and intellectually) from their representatives at the consultative policy dialogues. Apart from the issue of representation, there are controversies surrounding the whole issue of popular participation in policy processes. For example there is the issue of trustworthiness of the participatory verdicts and the extent to which they can be relied upon in making policy decisions (Scones, 2002). A range of practical challenges also inhibits public participation in practice such as the cost of, time for and the complexity of the organisation of the participatory processes (Glover, 2003). Thus even when it is known that the policies emerged from a participatory process, there is need to go beyond the policy and institutionalise practical mechanisms to democratise the choices of biotechnology application and development in line with the needs of the respective units of the society.

Secondly, the summation of the goals and objectives of the national biotechnology policies fairly well fall within the express development vision, there is the need for elaboration to link to the respective contextual challenges. For example, it is not enough to aim at enhancing agricultural productivity. Though the extent of detail in a policy document cannot be exhaustive, there must be at least the addendum of the principle that each identifiable unit of the society takes up the challenge to decompose the national objectives to the respective socio-economic circumstances. As it were, each segment of the society writes its own biotechnology policy.

Thirdly, (and this relates directly to the writing phase), there are certain generalisable characteristics of biotechnology policies. The content of the policies usually highlight the orientation in terms of goals, objectives and guiding principles, the strategic mechanisms for addressing the stated objectives; and the outline of the institutional and legislative framework with reasonable details of the nature of responsible agencies and their functions; the expected linkages among the agencies and critical actors. Within the framework of decision-making at the respective levels of the segments of the society, all these details can be worked out in specific and *tailored* programmes or projects. The challenge however is in the linkage to the umbrella national policy and its implementation.

Fourthly, there are two other important characteristics, which may also be in the content of the policy or available elsewhere in related documents of importance to discussing the relevance and usefulness of policies in terms of context-specificity. These are the nature and extent of public participation in the process of formulation and operationalisation of the policy and the strategies for funding of the policy with particular reference to the proportionality of the internal funding *vis-à-vis* external funding. In the rural context, how much internal funding can be achieved? But this in itself shows the robustness of the in-context choices made for biotechnology application and development. The extent to which it can be independent of external sources of resources flows, to that extent, it will succeed.

5. Conclusion and recommendations

Technological choice is an option for those with commitment to the principles of sustainable technological development and application. Biotechnology as a generic and multi-dimensional technology must be rooted in the range of socio-cultural niches presenting diverse opportunities for economic livelihoods. In the actualisation of the concept of tailoring biotechnologies, the challenge is focusing and addressing particulate needs in the circumstances of the identifiable segments of the societies, especially the vast majority of the marginalised in Sub-Saharan Africa. Biotechnology policies are worth formulating, but they can be more useful if made context-specific programmes are drawn as the corollary.

There are three key areas in which context-specific programmes can be recommended namely the policy environment, human resource development and networking.

5.1. Policy environment

The policy environment encompasses the totality of the policy factors that stimulate the biotechnology activities including the national and institutional factors as illustrated in Figure 1. For Africa to move away from the myth of enacting biotechnology policies to the reality of institutionalising and implementing biotechnology policies, definitive programmes need to be developed on the basis of the national priorities. Most African countries prioritise agriculture in their development plans. In the national biotechnology policies reviewed in this paper, some attempts have been made to factor in the national priorities. However, even when significant public investments are made such as in agriculture, the necessary linkages e.g. to industry and markets, are not effectively done. Besides, with the emphasis on export strategies, the focus on the internal production-processing-consumption system is blurred. There is need to place significant emphasis on the production-processing-consumption system in the identifiable localities and ensure context-specificity. The National Innovation System concept definitively underscores how the production-processing-consumption patterns can be engineered to strengthen its roots in context. As Speilman (2006) has pointed out, in the operationalisation of the innovation system concept, linkages to the overriding development goals such as poverty reduction is important. For almost all the food commodities, this can be done.

For example in Ghana, it may be challenging to begin with cocoa, which still contributes at least a quarter of the national foreign exchange earnings and which still commands the highest government attention. A programme to harness all innovations in cocoa products – ranging from beverages to cosmetics – could aim at enhancing value addition and local consumption, without having to sacrifice on exports. For example, cocoa beverages used in the school feeding programme will contribute to greater food sovereignty in the localities. It is recommended that a project be done to produce action programmes in actualising context-specific production-processing-consumption systems for prioritised food commodities in Ghana (with potential extension to other African countries). The participatory methodology bringing in inputs from the critical actors should be used in this. Going beyond national biotechnology policies means having specific action programmes to translate the national objectives into reality.

5.2. Human resource development

One of the important steps in addressing the dysfunctions in human resource development for biotechnology application is the education of the critical knowledge actors in the conceptualisation of science and technology. Biotechnology training is mainly done at the tertiary educational level and in some of Ghana's universities, first and second degree programmes have been introduced in biotechnology. Yet the general view of science as external to society – especially the African society – has to give way to a more progressive concept with science viewed as embedded (or at worst, capable of being embedded) in the society. Especially at the tertiary level of education, which is attracting significant attention as the principal vehicles for constructing knowledge societies (The World Bank, 2002), there is need for programmes to re-orient the thinking and approach of graduates to science generally and biotechnology in particular. To start with, there can be short-term courses targeting the identifiable stakeholders and with specific expectations. For policy makers, up to a few days sensitisation courses to enable appreciation of the potentials of biotechnology and the strategic advantage for enhancing agricultural production, food security and food sovereignty will be desirable. For researchers and scientists, one or two weeks of orientation on various social science themes e.g. the concept of development, democratisation of technology choices, participatory technology development, science in society, will enable a re-think of their methodologies in technology application and development. These courses can be further developed into integral components of the existing tertiary courses.

5.3. Networking of critical actors

The systemic concept of innovation demands that policies and programmes are engendered to rope in all identifiable actors. Knowledge creation and dissemination need strong networks to ensure knowledge utilisation. But building networks needs conscious and deliberate efforts. For tailoring biotechnologies to be entrenched not only as a concept but as a practice, there is need for a well-drawn programme in the respective countries to build networks, which will entrench the concept in theory and practice. Currently the TELFUN aims at proving the concept with academic research and dialogue. But beyond this, there is the need for a programme that creates a fledgling network from identifiable organisations and communities. The key strength of the concept is the emphasis on grassroot participation in biotechnology application and development. That must be evidenced in the network or movement with sufficient visibility to articulate the tenets of the concept. The dynamism of the network will eventually produce spillovers in achieving the broad goals of developing a knowledge-based society with science as the fulcrum. The specific recommendation is for a programme to be designed to galvanise the network base of the tailoring concept. Such a programme will involve a variety of communications strategies including durbars, workshops, publications and mass media communication.

References

Canadian Biotechnology Secretariat (2002). Canadian Biotechnology Strategy Overall Performance Report 1999–2002. Available at: http://cbac-cccb.ca.

Commission of European Communities (2001). Towards a Strategic Vision of Life Sciences and Biotechnology Consultation Document. Communication from the Commission, Brussels.

Doyle, L. (2007). US Food Aid is 'Wrecking' Africa, Claims Charity. *The Independent/ UK*: 17 August 2007.

Dugger, C. (2007). Charity finds that U.S. food aid for Africa hurts instead of helps. *International Herald Tribune*: 14 August 2007.

Essegbey, G.O. and Puplampu, K. (2007). Reconceptualising the Role of Stakeholders in Agricultural Biotechnology. *Tailoring Biotechnologies* 3(1): 71-89.

Evenson, R.E. (2004). Making Science and Technology Work for the Poor: Green and Gene Revolutions in Africa. In: *Annual Lectures 2003* (Ed. J.J. Baidu-Forson), UNU-INRA & UN ECA, October 2003.

Federal Ministry of Science and Technology (2001). National Policy for Biotechnology and Programme Plan of Action. Abuja.

Glover, D. (2003). Public Participation in National Biotechnology Policy and Biosafety Regulation. IDS Working Paper 198, August, IDS, Sussex.

Government of Kenya (2005). National Policy on Biotechnology. Nairobi.

Government of South Africa (2001). A National Biotechnology Strategy for South Africa. Ministry of Arts, Culture, Science and Technology, Cape Town.

Hisano, S. (2005). A Critical Observation on the Mainstream Discourse of Biotechnology for the Poor. *Tailoring Biotechnologies* 1 (2): 81-106.

Juma, C. and Serageldin, I. (Lead authors) (2007). Freedom to Innovate: Biotechnology in Africa's Development. A report of the High-Level African Panel on Modern Biotechnology. African Union (AU) and New Partnership for Africa's Development (NEPAD), Addis Ababa and Pretoria.

Ministry of Food and Agriculture (2002). Food and Agriculture Sector Development Policy (FASDEP), MOFA, Accra.

Ministry of Lands, Environment, Forestry, Water and Mines (2004). The National Biotechnology and Biosafety Policy for Rwanda. Kigali.

NDPC (2005). Growth and Poverty Reduction Strategy. NDPC, Accra.

NDPC (2007). Implementation of the Growth and Poverty Reduction Strategy (2006–2009). 2006 Annual Progress Report, NDPC, Accra.

NEPAD (2001). The New Partnership for Africa's Development. Abuja.

Punkari, M., Fuentes, M., White, P., Rajalahti, R. and Pehu, E. (2007). Social and Environmental Sustainability of Agriculture and Rural Development Investments: A Monitoring and Evaluation Toolkit. Agriculture and Rural Development Discussion Paper 31, The World Bank, Washington D.C.

Puplampu, K.P. and Essegbey, G.O. (2004). Agricultural Biotechnology and Research in Ghana: Institutional Capacities and Policy Options. *Perspectives on Global Development and Technology* 3 (3): 271-290.

Ruivenkamp, G. and Jongerden, J. (2005). Tailor-made Biotechnologies: Between Bio-Power and Sub-Politics. *Tailoring Biotechnologies* 1 (2): 5-8.

Scones, I. (2002). Science, Policy and Regulations: Challenges for Agricultural Biotechnology in Developing Countries. IDS Working Papers 147, IDS, Sussex.

Speilman, D.J. (2006). A Critique of Innovation Systems Perspective on Agricultural Research in Developing Countries. *Innovation Strategy Today* 2 (1): 41-54.

UN Millennium Project (2005). Innovation: Applying knowledge in Development. Task Force on Science, Technology and Innovation, Earthscan, London.

UNDP (2007). Human Development 2006. UNDP, New York.

World Bank (2002). Constructing Knowledge Societies: New Challenges for Tertiary Education. The World Bank, Washington D.C.

World Bank (2007a). Enhancing Agricultural Innovation. The World Bank, Washington D.C.

World Bank (2007b). Data & Statistics. Available at: http://web.worldbank/WBSITE/EXTERNAL/DATASTATISTICS. Accessed 10th September 2007.

Reconstructing agro-biotechnologies in Tanzania: smallholder farmers perspective

Elibariki Emmanuel Msuya

1. Introduction

In April 2005, Tanzania became the seventh[213] African country to launch field trials of BT cotton. Researchers at Sokoine University of Agriculture (SUA) supervise these government-managed trials. The Confined Field Trials (CFT) started in the southern part of the country where cotton farming was stopped in 1968 in a government move then aimed at halting the spread of red-ball worm disease that affected cotton yield. Depending on the outcome of the GM cotton CFTs and the anticipated positive reception by the cotton farming community, GM food crops will be the next target (Nakora, 2005). With the establishment of the Agricultural Biotechnology Center (ABC) in 2006, and the bidding for the third International Centre for Genetic Engineering and Biotechnology (ICGEB) component in Africa to be hosted in the country (Sunday News, 2006), it is clear that Tanzanian government is not dismissing GMO as a threat. The government views agricultural biotechnology[214] as a tool to help Tanzania improve the nation's food security, raise agricultural productivity, increase farmer's income, foster sustainable development, and improve its competitive position in international agricultural markets. According to Dr. Jeremiah Haki, Tanzanian agricultural ministry's Director for Research, *'Tanzania, which largely depends on agriculture, cannot afford to be left behind in technologies that increase crop yields, reduce farm costs and increase farm profits'* (Nakora, 2005).

However, there is growing concern among development economists and policy makers regarding the impact of agro-biotechnology on Tanzania's agriculture and especially on smallholder farmer's livelihood. Since its inception, biotechnology has spurred a worldwide debate. The debate has been divided into bi-polar groups, those who see agro-biotechnology as the solution to every problem facing smallholders and those who see agro-biotechnology (the way they are currently packaged) as another way of worsening smallholders' livelihood. The debate has been going on for decades with little progress on how this technology can be re-appropriated for the benefit of smallholder farmers (see Hisano, 2005 and references therein for a review of the debate).

Tanzania smallholder farmers conduct their activities in complex, diverse and risky areas. They often have limited access to high potential land, and limited capital resources. As a result, the agricultural system practiced by smallholders is geared primarily toward subsistence and highly diverse to meet different coping strategies. Agro-biotechnologies, especially GM crops, bring a very different dimension, one that gives transnational corporations more control over farmers' seeds, whereby transforming their production system. This technology increases the dependency of farmers on outside actors (Hisano, 2005; Timothy, 2003; Kelemu, *et-al.*, 2003). Given that GM technology's major targets are the

---

[213] The other countries that have already started genetically modified crops trials are Tunisia, Zimbabwe, Egypt, Burkina Faso and Kenya. South Africa is the only African country that has commercialised GM crops.

[214] Agro-Biotechnology encompasses a broad range of research and development tools, including bioinformatics, micro propagation, molecular diagnostics, marker-assisted breeding and vaccine development. Such pursuits have applications in agriculture, environmental protection, industrial development and public health. The focus of this paper is GM crops and uses the term agro-biotechnologies interchangeably with GM crops.

commercial maize (main staple food of Tanzania) and cotton growing areas[215], the socio-economic impact to smallholder farmers in the country needs to be carefully studied and initiatives well planned and implemented. As abundance of food resources as a result of agro-biotechnologies would not necessarily ensure that smallholders have improved and sustainable livelihoods (Timothy, 2003), Tanzania, like many other developing countries, now faces a dilemma as to how to proceed on the further development of agro-biotechnologies.

Should Tanzania promote agro-biotechnologies and start to commercialise GM crops? What kind of technologies are currently needed by smallholder farmers and under what context will they work best for the smallholders; what will be the impact of alternative biotechnology policies on Tanzania's agricultural economy and trade? As there is very little (if any) research done on what the impact of GM crops will be in the Tanzanian farming community, answers to these questions are of critical importance for policy makers and agricultural industry, given that the agricultural sector plays an important role in the Tanzanian economy and possesses the potential to advance the country's objectives of growth and poverty reduction[216]. This paper is thus organised as follows: the next section uses Tanzania cotton sub-sector as a case to answer the question 'does Tanzania need agro-biotechnologies, specifically GM crops'. A general review of the main problems facing smallholder farmers in Tanzania is presented in section three and section four present a conceptualisation of the role to be played by agro-biotechnology in addressing smallholders' problem. Concluding remarks are provided in the final section.

2. Introduction of agro-biotechnologies in Tanzania

2.1. A brief profile of R&D and introduction of GM technologies

Tanzania was among the first early post-independence African states that recognised the importance of scientific and technological development in its quest for socio-economic transformation in the 1960s. The first national Science and Technology Policy for Tanzania was enacted in 1985. It was revised and published in 1996 by the Ministry of Science, Technology and Higher Education (MSTHE) with policy objectives including: promoting science and technology as tools for economic development, improving human, physical and social well being and protecting national sovereignty. The policy, which is currently under review, identifies some research priority areas such as materials, biotechnology, telecommunications and information technology. The Government has actively supported the development of formal plant genetic resource management institutions. However, as can be seen across most African countries, the private sector still does not perform much research. The links and interactions between local plant genetic resource management systems and formal plant genetic resource institutions have recently improved and are to some extent strong. Despite their lack of support, farmers' local plant genetic resource management systems are still responsible for providing seeds for the vast majority of food crops in Tanzania.

The 1990s structural adjustment programme (SAPs) had a great impact on the seed industry. They included the opening of the seed market to multinational private seed companies and scaling down of support for the parastatal seed company TANSEED. This is clearly shown in the Macro-economic Policy Framework for 2006-2007 to 2008-2009 budget, where the Tanzanian government admits that

---

[215] These crops already have well established commercial market structures.

[216] The sector contributes the most to GDP (over 45% of the GDP) and supports livelihoods of over 80% of Tanzanians living in the rural areas.

the total domestic R&D allocation in Tanzania is estimated at only 0.01% of the annual GDP and stated that this is insufficient funding to cater for the country's R&D needs. As a result, foreign support has become critical for research capacity building with some analysts showing that about 70% of R&D funding is through external donor assistance.

As in many other developing countries (with exception of China[217]), introduction of GM crops in Tanzania was an effort of multinational companies working behind USAID[218] external donor assistance. An advisor to USAID, C. S. Prakash who also serves as the principal investigator of a USAID funded project 'to promote biotechnology awareness in Africa' was in Tanzania in August 2002, where he 'met with Tanzanian scientists, describing the potential of bio-engineered foods for the benefit of the country' (The Express, Tanzania, Aug 21, 2002). Again in February 2005, Dr Sivramiah Shanthu Shantharam a member of the WHO/FAO Consultative Committee on Biotechnology-Food Safety and a former employee of Syngenta was also in Tanzania for a seminar on GM technology for East African plant inspectors. These and many other visits of representatives of the multi-national firms involved in promoting GM crops for 'poor resource farmers' lured Tanzania into accepting BT cotton CFTs. It is thus strongly argued that the introduction of GM cotton in Tanzania and later GM food crops is not an opportunity for smallholder farmers as the government heralds it, rather a threat to smallholder's livelihood as a result of not being participated in the whole process of constructing agro-biotechnologies in the country. And thus made receivers of a technology which main goal is to serve the interest of multinationals and has little (if any) interest to their well being.

2.2. Is There a role for GM cotton for smallholders in Tanzania?

In order to ensure household food security, increased levels of income, and improved livelihood for rural poor farmers, agricultural productivity needs to improve, to move agricultural diversity towards higher value products, and to ensure that farmers and enterprises are competitive and well integrated with rapidly growing urban and international markets. The main question now is what role agro-biotechnologies can play toward achieving this goal. This paper looks at the Tanzanian cotton sub-sector and asks whether Tanzania needs to adopt GM technologies in the manner they are presented today.

Being one of the largest cotton producing countries in SSA, the cotton industry sustains, directly or indirectly, the livelihoods of nearly 48% of the Tanzanian population, (currently estimated at 37 million). Most cotton is exported, contributing US$100 million to export earnings (2005/2006 season). Labour is the major input. Cotton is produced by smallholder farmers (estimated to be 500,000)[219] who cultivate between 0.2 and 2.0 hectares per year. Use of other inputs is limited (Box 1).

---

[217] To present, technology transfer in the case of GMOs is controlled by vested interests. China is the only country in which public sector investment has led to the release of locally produced BT cotton varieties (Baffes, 2002). Elsewhere in the world the technology is 'owned' by Monsanto with limited involvement by Delta and Pine Land Company. Dow Agrosciences and Syngenta are soon to release their own insect resistant transgenic cottons.

[218] The Agricultural Biotechnology Support Program (ABSP II) is a USAID funded project managed originally by the Michigan State University and more recently by Cornell (ABSP II). Its partners have included Asgrow, Monsanto, and Pioneer Hi-Bred. Promoting GM is an official part of USAID's remit - one of its roles being to 'integrate GM into local food systems'.

[219] The number of farmers varies depending on weather conditions and cotton market price trends. Droughts and downward shifts in cotton prices in the international market place compel some of the farmers to switch to alternative crops.

**Box 1. Cotton farm management practices in Tanzania
(Tanzania Cotton Board – TCB).**

Cotton is grown under the following farm management practices:
- Sown between November and February; harvested between May and July; and sold between September and November.
- 100% of cotton is rainfed.
- 95% of the farm's size ranges between 0.5 - 50ha; 5% between 50 - 100 ha.
- Around 60% of farm preparation is done using the hand-hoe; and 35% is done using animal traction. Only 5% of land preparation is by motor traction.
- 70% of all cotton is grown without applying any fertilisers. 30% is grown using organic or animal manure.
- Weeding is almost entirely done by hand hoe.
- Major diseases which are common include Fusarium wilt and bacterial blight. Insecticides include both water and oil based. Due to price differentials and environmental hazards, the use of oil-based insecticides is being discouraged in favour of water-based ones.
- 70% of the total area under cotton receives at least 2 sprays per growing season; while the remaining 30% is not sprayed at all. Early season insects include jassids and aphids. There are several late season pests like pink boll worms and cotton stainers.
- 100% of cotton is harvested by hand-picking.

Based on the these farm management practices, the cost of producing 1 hectare of rainfed cotton in Tanzania (using 2003/2004 data) is around US$ 178; compared to US$ 256 in South Africa; US$ 359 in India; US$ 484 in Benin; and US$ 932 in the USA. The cotton production costs are higher in the other countries because farmers there especially apply more inputs than their Tanzanian counterparts.

Cotton is cultivated in Tanzania in two areas, namely West Cotton Growing Area (WCGA) and East Cotton Growing Area (ECGA). These two areas vary greatly in terms of weather conditions, soil fertility and diseases and pests. The main insect pests are cotton bollworm (*Helicoverpa armigera*) (HBW) and spiney bollworm (*Earias* spp.) (SBW). The main pest control recommendation is for six insecticide sprays, applied at 2-week intervals from first flower. However, research work at Ukiriguru Research Institute has shown that in some seasons, when bollworm pressure is low, there is no significant difference in yield between sprayed and unsprayed cotton (Baffes, 2002). In practice though, few cotton farmers in Tanzania apply more than one, two or at most, three sprays in a season. The timing of the applications is based on visible evidence of damage to the bolls or just the presence of insects, some of which may be beneficial, rather than pest species. It is believed that among the cotton farmers in Tanzania who use insecticides, many apply their sprays at the wrong time, with poorly maintained equipment and sometimes with inappropriate chemicals (Baffes, 2002). Variation in weather conditions and other factors prompted the need to develop different varieties suitable for these very different areas. The current (from 2003/2004 season) commercial cotton varieties grown in WCGA and ECGA are UK91and ALAI 90 varieties respectively (visit http://www.tancotton.co.tz/production.html for more information).

Agro-biotechnology is considered the key to increasing productivity, food security and economic development. To quote just a few works, 'Agricultural biotechnology can assist crop-breeders to improve the yields and quality of crops and their production under strenuous conditions. Through genetic modification, desirable genes can be transferred to crops irrespective of species barriers. Plants can be made more tolerant of drought, heat and frost. They can be made more resistant to diseases and insect pests, reducing the input of agrochemicals. Genetic modification can also greatly facilitate the development of crops with improved storage properties and nutritional characteristics (e.g. proteins and vitamins). Agricultural biotechnology also enables development of thermo-stable and cost-effective vaccines to treat livestock diseases' (FAO, 2004). This biotechnological advancement has even been heralded as a third revolution in agriculture following chemicals and mechanisation (Marks *et al.*, 1995). Agro-biotechnology is often considered an extension of the Green Revolution, a paradigm that failed to address the needs of Africa's smallholder farmers. Despite the fact that little research has been done to address failures of the green revolution in Tanzania and SSA in general, and conditions under which agro-biotechnologies (GM Crops) can best help the smallholder farmer solve their problems, there is vigorous promotion of these technologies for smallholder farmers (FAO, 2004). What should be noted is that, like the green revolution before it, agro-biotechnologies have come to Tanzania as a result of developments in the North, bringing to question whether this new technology is appropriate, timely, in-demand and applicable in Tanzania agricultural systems.

GM cotton is being called the 'liberator' of Tanzania's cotton smallholder farmers by agro-biotechnology promoters. It is almost impossible to imagine, given the economic realities of cotton farming in Tanzania, how farmers could derive any benefit at all from the GM seed. It is claimed that the BT cotton that was supposed to be field tested in Southern Tanzania will re-open that part of the country to cotton cultivation. This is hailed as a good thing by members of parliament in Southern highland regions who have been calling on the government to find alternative means to re-start cotton production in the regions (Nakora, 2005). Tanzania is suffering from miserably low global cotton prices and its own increased domestic production (Figure 1 and 2). In 2004/2005 season after full introduction of UK91 and ALAI 90 varieties, the government introduced the region's first ever set of direct subsidies to cotton farmers. The price they were receiving was far below the cost of production[220]. According to the article 'Tanzania to subsidise cotton farmers', Tanzanian farmers were bracing themselves for hard times ahead, with an estimated 7.8 million tons of cotton, out of the 28 million tons produced worldwide, being surplus production. Tanzanian cotton production has also increased dramatically to 500,000 bales in 2004/2005 season, far higher than the 2003/2004 yield (The East African, 2004). And this increase is without the use of GM cotton.

Being labour intensive cotton is often an unpopular crop with smallholders. They grow it if it gives them a reliable source of income and when there is no alternative. In Tanzania like many African countries the government-backed Cotton Authorities used to provide incentives to grow cotton by providing access to subsidised fertiliser, seed, and transportation of harvested seed cotton. In 1992 structural adjustment policies (SAP) reached climax by complete removal of all subsidies. In 1999/2000 marketing season the Cotton Board established the Cotton Development Fund. A 3 percent levy on cotton exports is paid into a trust fund used to finance purchases of cotton seed (1.35 percent), chemicals (1.15 percent), and research and development (0.5 percent). The inputs are then distributed to registered cotton producers at below market prices, with the fund making up the difference (Baffes, 2002) as by selling cotton seed in fully market prices would introduce a further disincentive to cotton growing. However, under the above circumstances, smallholder farmers will be burdened with more costs for GM seed with the possible

---

[220] The subsidy allowed Tanzanian cotton farmers to earn $2.5 cents more per kilogram than they would ordinarily have received. The cost of production for cotton is above 30 US cents per kilogram.

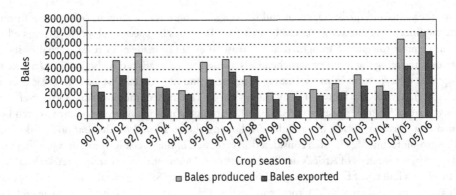

Figure 1. Tanzania cotton production and export trends from 1991/1992 – 2005/2006 (Tanzania Cotton Board – TCB).

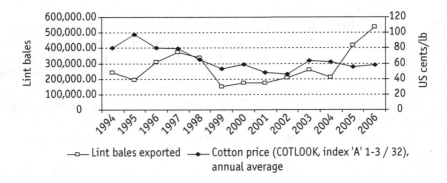

Figure 2. Tanzania cotton production trends and world cotton price trends 1994 - 2006 (FAO and Tanzania Cotton Board – TCB). Note: 1 bale is approximately 500 lb or 226.8 kg.

addition of a 'technology fee'[221] thus acting as a disincentive to smallholders. In addition to paying the 'technology fee', farmers will be expected to sign a Technology Use Agreement (TUA), containing tight regulations governing how they may use the technology. Included therein, is the condition that forces farmers to purchase seeds annually, and thus enhance corporate profits. This would again act as a disincentive to cotton production.

With two cotton growing areas (WCGA and ECGA), smallholders require seeds for the varied physical environments. It is difficult for formal sector seed breeders (often biotechnology corporations) to provide the kind of seeds needed by the majority of farmers. This is because the formal sector seed system is geared to generating a limited number of varieties (e.g. BT cotton), each of which is distinct, uniform and stable, displays a wide environmental adaptability, and has a potential in terms of high yield if grown with application of external inputs. Variation is dealt with by releasing a stream of new varieties

---

[221] Technology fees in South Africa have been roughly $50/ha. This is only slightly less than the cost of insecticide for the same area of cotton (Kuyek, 2004).

Reconstructing biotechnologies: critical social analyses

over time, each to replace the previous, rather than by generating a large range of varieties at any one time, among which farmers can choose (Timothy, 2003). Thus replacing UK 91 and ALAI 90 (varieties now in use), by BT cotton would not necessarily address the needs of varied physical environments and many attributes already included in these varieties would be lost.

On top of this, smallholder farmers, particularly those who do not use insecticides at present or who only use one or two sprays per season (Baffes, 2002; TCB, 2007); will find it more expensive to invest in GM cotton seed and especially so, if they still continue to suffer significant yield loss as a result of BT cotton vulnerability to secondary pests. It is apparent that seed technology is not what a cotton smallholder farmer in Tanzania needs or wants. With new non-GM varieties smallholder farmers have increased production dramatically but this is yet to translate to increased income and improved livelihoods. The problem in the cotton sub-sector is addition of value added in the cotton value chain (especially value added by famers themselves and rural local actors) rather than simple increase in production as argued for the case of BT cotton. Cotton consumption within the country is low thus forcing most of the crop to be exported. This is partly because of lack of enough textiles and spinning industry to process cotton. Due to this Tanzania has failed to export a value added product contributing to a lower producer price. Thus the introduction of GM cotton alone in the country would only if at all solve one problem i.e. opening up part of the country that was not suitable for cotton production (this, in itself as shown above, would be counter productive). The positive impact on income and poverty reduction is exaggerated.

3. Major problem facing resource poor farmers in Tanzania

As shown in the case of cotton above, the introduction of agro-biotechnologies in the country and in SSA in general has not taken into consideration the real situation facing the smallholder farmer. Aaron deGrassi (2003) using six criteria which are widely accepted in crop breeding (i.e. demand led, site specific, poverty focused, cost effective and institutional and environmentally sustainable) has shown that GM crops introduced in SSA are generally inappropriate for poverty alleviation in the region. However it might be irrational to completely abandon agro-biotechnologies in the country and in SSA. As proposed by Hisano (2005), 'what we need to do now is not to throw the baby out with the bath water, but to carefully evaluate the discourse of "biotechnology for the poor"'. In order to do so, the potential benefits of agro-biotechnologies have to be looked at in connection with the major problems facing smallholder farmers in the country. This section summarises the major problem facing smallholder farmers in Tanzania.

3.1. Performance of the agricultural sector

Apart from contributing 45 percent of total GDP (as of 2006), agriculture is the main source of income for most of the population in the country. Moreover, for most of the poor (majority of whom are smallholder farmers)[222] in Tanzania, agriculture is the main source of their livelihood. About 80% of the population directly or indirectly depends on agriculture for their livelihood. The performance of agriculture is thus a key factor in raising the income of the rural population and reducing poverty. As agriculture has strong linkages with the rest of the rural economy, a strong agricultural performance

---

[222] In Sub-Saharan Africa (SSA), where about 43 percent of its population is living below the international poverty line, the incidence of poverty is the highest among smallholder farmers residing in rural areas. Thus, if the war on poverty is to be won, developing countries need to place more emphasis on the agricultural sector (Mangisoni, 2006).

usually leads to investment and increasing economic activity in the rest of the rural economy, thus contributing to rural employment and further poverty reduction.

Over the past few years however, growth of agricultural GDP in Tanzania has been slow and highly variable (URT, 2006). In Tanzania the rural economy is almost completely dominated by agriculture. The low growth of agriculture explains the low development of non-farm activities in rural areas. It is argued that if clear results in poverty reduction are to be felt, agriculture must grow at a sustained rate of at least 6 per cent per annum. This growth needs to be broad-based and strategies that promote such broad-based growth must be well developed and implemented (URT, 2005). The past variable growth in Tanzania has been a reflection partly of its heavy dependence on weather (particularly rainfall patterns), and the small portion of agricultural trade in total trade. Even with the past decade's achievements in infrastructure development, institutional building, the establishment of a market economy, and sustained growth of the economy (annual average of 6.3%), agricultural growth continues to be highly unstable. Agricultural productivity has increased slightly but is still low[223] especially among smallholder farmers (URT, 2005: 82; Mbelle, 2005). As a result, the vast majority of smallholder farmers still face the challenge of overcoming poverty.

3.2. Major problem facing smallholder farmers

Figure 3 below, shows that the major problem facing smallholder farmers and thus the agriculture sector in Tanzania is to increase added value along the value chain. To achieve this, bottlenecks in the farming system, marketing system and those of linking smallholders to markets, need to be resolved. By tackling the core problem an impact will be felt on the income of smallholders, rural employment, and poverty reduction while improving food security.

---

[223] For the period 1986-2000, labour productivity growth declined marginally by 0.4% (Mbelle, 2005).

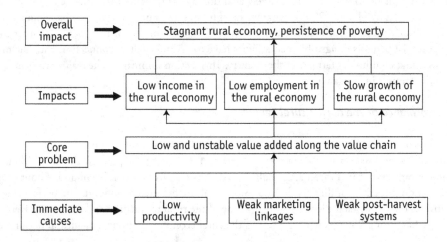

Figure 3. Major problem facing Tanzania smallholder farmers (agriculture sector).

As shown in the literature on Tanzanian agriculture (Djurfeldt *et al.*, 2006: 60; Nyange *et al.*, 2003), when agricultural productivity is low, then also added value is likely to be low. Low agricultural productivity in itself is a result of high costs of production (which will be increased when GM crops are adopted by smallholders), low yield, and high risk in agriculture that lowers investment (URT, 2005: 79-94; Mbelle, 2005). Weak marketing linkages among farmers and between farmers and traders and processors in the value chain translate to low prices, few market shares, and reduced volume of transactions, all of which impinge upon the low added value in the food crop productions value chain (Barrett and Mutambatsere, 2005: 7; RATES, 2003: 26-29; Cooksey, 2003). Thus, the introduction of GM crops alone would not be able to solve the smallholder's problems. On the other hand, weak post-harvest systems are responsible for losses in the value chain, lower quality of products, and reduced capacity for processing raw materials (Amani, 2004; Temu and Winter-Nelson, 2001; Kherallah *et al.*, 2000: 19-21). To some extent GM crops can play a role in increasing the quality of crops. This is because of the higher nutritional value of food crops, and improved taste, texture and appearance of food. This would however need to be balanced with the negative impact GM crops have on smallholders, such as increased cost of production as well as safety issues (potential foods and environmental risks).

The low and unstable added value along the value chain leads to slow growth, little employment generation, and low income, all of which contribute to a stagnant rural economy with persistence of poverty. As a result smallholder farmers' income remains low, employment opportunities for the poor are limited, and, in spite of food self-sufficiency for the country as a whole, many households remain food insecure and in poverty because of risks associated with rain-fed agriculture and limited income generation activities in production or post-production systems. The ultimate objectives of improving value chains for smallholders' are: First, to increase the total amount and value of products that the poor sell in the value chain. This is expected to results in higher absolute incomes for smallholders as well as for the other actors in the value chain. The second objective is to sustain the share of smallholder in the sector or increase the margins per product, so that they do not only gain more absolute income but also relative income compared to the other actors in the value chain. In this case also the piece of the cake hold by smallholders grows and they get less poor compared to the rest of the actors in the chain (M4P 2005). Therefore, if agro-biotechnologies are to be *'for the poor',* they must be reconstructed toward increasing agricultural growth and generating more income opportunities in rural areas. The reconstruction of agro-biotechnologies should then start from the smallholder point of view with the main stakeholder in the whole process being the smallholder farmer.

4. Re-appropriation of agro-biotechnologies in Tanzania

To make an impact, agro-biotechnologies must target crops that smallholder farmers traditionally know how to grow and address agronomic traits of significant importance to their needs. The currently available and widely commercialised GM crops (e.g. BT cotton currently on CFTs) are not good examples of technologies that will help smallholder farmers. Available GM crops are not designed for poor smallholder farmers and it is doubtful that large agricultural companies will ever design crops exclusively for the benefit of smallholders. Therefore, scientists, international agricultural research centers (IARCs) and other players need to join forces to re-appropriate agro-biotechnologies to tackle the specific problems that smallholder farmers face. This section presents a framework in which agro-biotechnologies can be used as a tool toward solving smallholders' problems.

4.1. Appropriate technologies needed by resource-poor farmers

It is now evident from the work of deGrassi (2003), Hisano (2005), Eicher *et al.* (2006), and the case of cotton described above that, most GM crops are presently not tailored for smallholders and that they will take at least 10 to 15 years or longer to reach smallholder farmers in Africa. The push for GM crops should be seen as part of a shift towards corporate-led agricultural research and development that has been happening in other areas of the world for some time now and is on Tanzania's doorstep. As already shown above, GM crops bring a different dimension, one that gives transnational corporations more control over farmer's seeds. With the patents they have on GM crops, corporations can prohibit farmers from saving seed from year to year. It should be remembered that, the vast majority of land in Tanzania is planted with seed of local species, developed and maintained by farmers themselves. For example in 2004/2005 season it is estimated less than 15% of the maize crop in Tanzania was planted with purchased certified seed of modern varieties. This is a very low level, even given that a considerably larger proportion of land is likely to be cultivated with retained seed of hybrid and composite maize varieties. Hence by introducing GM crops, potential benefits of improving food supply would be accompanied by overwhelming new risks to traditional farming systems and to ecological systems. Therefore, for the time being, special attention should be given to strengthening conventional plant breeding programs in National Agricultural Research Systems (NARS). Agro-biotechnology approaches must be nested and integrated into these plant breeding programs. On top of that smallholders need the support of rural development strategies that give farming communities control over their own resources and build on local knowledge and technology systems. As Timothy (2003) put it: 'Farmers must be able to choose to avoid a cycle of debt and dependency and that alternative strategies that rely to a greater extent on locally available inputs and provide farmers with tools to analyse what is happening in their fields, to make appropriate variations in their practices...' need to be pursued.

Thus initiatives from local researchers in finding the best solution to soil fertility and pest problems should be supported and encouraged. For example during the 2004/2005 crop season, in practical studies carried out by researchers, both at Ukiriguru's Lake Zone Agricultural Research Development Institute and on selective experimental farms (e.g. at Bukangilija Village in Maswa District), by use of integrated soil fertility management (ISFM), integrated pest management (IPM), application of appropriate pesticides and sprayers, and by use of other better farm and crop management practices, yields rose to between 530-750 kg/ ha. (TCB, 2007).

As pointed out earlier in 3.2, the problem in the cotton sub-sector (and agriculture in general) is added value along the value chain (especially at the farmers' level) rather than simple increase in production. Thus value addition should be at the center while addressing smallholders' problems. Initiative by BioRe® to introduce and support organic cotton in the country is a good example (Box 2). However, smallholders and the economy in general could benefit more if the spinning of organic cotton was done locally; as added value created in the processing industry (far away from the smallholder) does not automatically imply better prices for farmers. Thus, if alternatives are pursued well and fully, the pressure to adopt GM technology while the country is yet to set appropriate control measures might be reduced.

Box 2. Organic cotton in Tanzania (REMEI AG, 2007).

The project for organic cotton was initiated in 1994. Its aim is to support and improve the living condition of farmers by organic farming. By adopting organic farming practices, the farmer can produce his agricultural products environmentally friendly and without application of pesticides, in addition he receives a better price and a purchase guarantee for organic cotton, as well as a premium price above the local market price. These advantages allow the farmers to attain a more stable and long-term financial security. Farmers are integrated as partners into the bioRe® textile chain. BioRe® Tanzania is giving advice to farmers to change their production system to organic farming and to improve their farming techniques.

The number of contract farmers who collaborate with bioRe® Tanzania has increased from originally 45 farmers (1994) to a total of 1,700 farmers (season 2006-2007), who have decided to convert their farms to organic production. The production of organic cotton yield from the bioRe® contract farmers reached in the past season an amount of 4,581 t seed cotton yield (1,649 t lint cotton).

The organic cotton is ginned in the local ginnery. By land the cotton bales are transported to the port of Dar es Salaam at the coast of Tanzania. From there the cotton is shipped to spinning mills in Europe and India in order to make yarn. Thereafter the yarn is transported to various production suppliers.

4.2. Conditions under which agro-biotechnologies should be developed

Although smallholders have long been ignored by the formal plant genetic resource management system, smallholders' knowledge and management skills are alive and well at the local level. In many parts of the country, farmers continue to play a key role in the selection and maintenance of varieties for food resources. Their skills in selecting for different characteristics to meet environmental, household and consumer needs are evident and well-developed. While some benefits can be foreseen to farmers from the introduction of agro-biotechnologies, particularly in better-off, market-led economies, care should be taken to avoid a high price to be paid for the gene pool. New agro-biotechnologies should be developed in a manner that reduces dramatic impact on genetic erosion where the spread of a single variety can reach a point where community food security is under threat. In that context the links and interactions between local plant genetic resource management systems and formal plant genetic resource institutions need to be fostered. In order to provide a better basis for enhancing collaboration between these two parties, participatory projects have to be carried out to examine the social dynamics of farmers' plant genetic resource management. In other words farmers should be at the center of the new technological innovations; not only as receivers of the technological advancement but also as co-designers, implementers and monitors of these technologies. The demand driven technologies will give farmers more sense of ownership and thus act as an incentive towards adoption.

As observed in many previous works (e.g. Hisano, 2005; Eicher *et al.*, 2006) agro-biotechnologies are currently largely pursued by private multi-national companies which are profit oriented. For agro-biotechnologies to be 'for the poor' the public sector must play a greater role. Given that agro-biotechnology development in the country is in its infancy and capacity building is basically a step by step process that unfolds slowly and almost invisibly over time, donors and foundations can play a strategic role in supporting long term public sector investment in capacity building (human capital and infrastructure). This in no way denies the role to be played by the private sector in the development of agro-biotechnologies in the country. But rather it is a call for more collaboration between the public and the private sector in responding to smallholders technological and institutional needs. However, in this public-private partnership care should be taken not to avoid the question of what kind of technologies are really demanded by resource-poor farmers to solve their problems as pointed out clearly by Hisano (2005: 93). Given that we are talking about resource poor farmers and their biggest problem being low value addition, they require agro-biotechnologies that would address the issue of value addition. This is very specific to each locality and can not be generalised. In the case of Tanzania (which is also very diverse) agro-biotechnologies tailored to the value chain will play a bigger role.

Although it is evident that GM crops are still a long way from being a norm for smallholders in Tanzania, special attention should be given to raising public awareness of biotechnology, mobilising political support and commitment to strengthening the country's capacity in biotechnology, biosafety, food safety and IPR (Intellectual Property Rights) and mounting long-term training programs to train the next generation of plant breeders and GM crop specialists. On top of that, farmers should be made to understand that GM crops alone cannot deliver yield increases if basic crop management is poor. Thus pursuit of GM crops in the country should be promoted as a component of Integrated Crop Management (ICM) systems. With this in mind, agro-biotechnological development in the country will take the form of being only one part of the larger solution to problems facing smallholders.

4.3. Planning for the impact of agro-biotechnologies

Like all agricultural technologies, the socio-economic impact of agro-biotechnology outputs must be taken into account. Understanding their benefits, compared with those of alternatives, is crucial in deciding when and where it is most appropriate to invest in biotechnology. The potential benefits of biotechnology depend on the crop, the specific trait that is being improved and the gains that could be made from solutions other than biotechnological. The failure to anticipate and plan for differential response of farmers to varying opportunities and threats of agro-biotechnologies could lead to problems, as observed during the green revolution in Asia, where in-spite of a significant increase in yields as a result of high use of chemical inputs, the major problem of worsening income distribution is still felt today. This could have been avoided by better planning.

At this time, the government has not undertaken enough research to determine the impact of introducing agro-biotechnologies in the country. This raises the question of whether this whole rush to GM crops has been thoroughly analysed and planned or is it just another response for donor support. What smallholders do not need is another technological innovation that will see others (multinationals) benefit at their expense. According to Marks *et al.* (1995), by adapting Schumpeter's model of economic development, the effects of agro-biotechnologies can be anticipated. According to Schumpeter an innovation may involve the introduction of a new good or quality of good (product innovation), a new method of production (process innovation); a new market for existing goods (market innovation), a new source of supply for raw materials, and lastly, a new business organisation, all of which cover the main types of technology that could be introduced during the introduction of agro-biotechnologies (see Marks *et al.*, 1995 for further reading). Planning for the impact of agro-biotechnology is crucial and should definitely be embedded in the whole idea of introducing the technology in the country.

4.4. Challenges to overcome for agro-biotechnologies to benefit smallholders

The rise and decline of smallholder BT cotton in South Africa should be an example for the need of thorough planning before embarking on GM crops. BT cotton was rapidly adopted by smallholders in SA but after a few years the curtailment of credit to smallholders, drought and declining world cotton prices contributed to a decline in smallholder cotton production. 'Smallholder BT cotton in South Africa has recently been described as a technological triumph and an "institutional failure"', (Eicher *et al.*, 2006: 522). Even with proper planning, the reconstruction of agro-biotechnologies in the country faces two fundamental challenges. Firstly, most of the smallholder households and small and medium enterprises (SMEs) have few assets and are largely unorganised. As a result they do not reach the economy of scale to access the proposed agro-biotechnologies, markets, and capital required for rapid growth. The second challenge is the presence of policy and institutional weaknesses related to poorly defined property rights, inadequate research and extension, and high cost of credit, to mention a few. These policy and institutional weaknesses result in a difficult business environment for the growth of agriculture and added value even with the introduction of agro-biotechnologies. Therefore for successful reconstruction of agro-biotechnologies these two challenges need to be addressed. The challenge of economies of scale can be addressed by promoting the development of value chains and the strengthening of value chain linkages among smallholders, SMEs/agribusinesses, markets, and service providers. This would probably require institutional mechanisms and interventions that are in the public interest and therefore justify public investment. The second challenge of policy and institutional weakness can be addressed by promoting reforms through policy dialogue. Addressing these two fundamental challenges is expected to have a positive impact on smallholders' ability and thus, access to the whole process of reconstructing agro-biotechnologies in the country.

5. Conclusions

Tanzania should not give up the dream to end poverty and improve smallholder farmer's livelihoods by means of acquiring, developing and commercialising GM crops. The future of the country's agricultural development is to a great extent dependent on how well the agro-biotechnology industry would grow. However, this paper argues that there is a need to reconstruct agro-biotechnologies in Tanzania so that they can have an impact in solving the smallholders' problems. First, appropriate technologies which address smallholders' problem must be identified. We have seen that the present package of agro-biotechnologies is not tailored for this goal. Hence, for the time being, conventional plant breeding programs are advised. Resource poor farmers are expected to be the major stakeholder in the development of these appropriate technologies. This would mean that technologies are developed by smallholders for smallholders. Given that most agro-biotechnologies are pursued by the private sector, this study suggests that, it will be for the interest of 'the poor' if the public sector is assisted in developing capacity in agro-biotechnologies. Lessons from other countries which have introduced GM crops should be carefully assessed. This will help the government plan agro-biotechnologies which have a positive impact on the socio-economical situation of smallholders. More research is needed to give a comprehensive understanding of this subject in Tanzania. To have a positive impact, agro-biotechnologies must be 'pro-resource-poor farmers and pro-women and children, target crops that farmers traditionally know how to grow and address agronomic traits of significant importance to their needs' (Kelemu *et al.*, 2003). In their present form, agro-biotechnologies are likely to have very small impact (if any) on improving the livelihood of resource poor farmers in Tanzania, unless reconstructed.

Elibariki Emmanuel Msuya

Acknowledgements

I wish to acknowledge with gratitude the invaluable guidance from my supervisor, Prof. Shuji Hisano. The views presented here are my own and I take full responsibility for omissions or mistakes.

References

Amani, H.K.R. (2005). Making Agriculture Impact on Poverty in Tanzania: The Case on Non-Traditional Export Crops. Presented at a Policy Dialogue for Accelerating Growth and Poverty Reduction in Tanzania, ESRF, Dar es Salaam. Available at: http://www.tzonline.org/pdf/makingagricultureimpactonpoverty.pdf.

Baffes, J. (2002). Tanzania's Cotton Sector: Constraints and Challenges in a Global Environment. Africa Region Working Paper Series No. 42 Available at: http://www.worldbank.org/afr/wps/index.htm.

Barrett, B.C. and Mutambatsere, E. (2005). Agricultural markets in developing countries. Available at: http://aem. cornell.edu/faculty_sites/cbb2/papers/cm_agriculturalmarkets.pdf; visited January, 2007.

Cooksey, B. (2003). Marketing Reforms? The Rise and Fall of Agricultural Liberalization in Tanzania. *Development Policy Review* 21 (1): 67-91.

DeGrassi, A. (2003). Genetically Modified Crops and Sustainable Poverty Alleviation in Sub-Saharan Africa: An Assessment of Current Evidence. Third World Network- Africa.

Djurfeldt, G., Holmen, H and Jirstrom, M. (2006). Addressing Food Crisis in Africa; What can sub-Saharan Africa learn from Asian experience in addressing its food crisis? SIDA. Available at: www.sida.se/publications.

Eicher, C.K., Karim, M. and Idah, S-N. (2006). Crop biotechnology and the African farmer. *Food Policy* 31: 504-527.

FAO (2004). The State of Food and Agriculture 2003/2004 – Agricultural Biotechnology: Meeting the needs of the poor? FAO, Rome.

Hisano, S. (2005). A Critical Observation on the Mainstream Discourse of 'Biotechnology for the Poor'. *Tailoring Biotechnologies* 1 (2): 81-105.

Kelemu, S., Mahuku, G., Fregene, M., Pachico, D., Johnson, N., Calvert, L., Rao, I., Buruchara, R., Amede, T., Kimani, P., Kirkby, R., Kaaria, S. and Ampofo, K. (2003). Harmonizing the agricultural biotechnology debate for the benefit of African farmers. *African Journal of Biotechnology* 2 (11): 394-416.

Kherallah, M., Delgado, C., Gabre-Madhin, E., Minot, N., and Johnson, M. (2000). The Road Half Traveled: Agricultural Market Reforms in Sub-Saharan Africa. Food Policy Report (IFPRI). Washington, D.C.

Kuyek, D. (2004). Contaminating cotton. *BBC Focus on Africa* (July-September): 48.

Marks, L.A., Kerr, W.A. and Klein, K.K. (1995). Planning for the impacts of agrobiotechnologies; A guide for public administrators. *International Journal of Public Sector Management* 8 (1): 35-47.

Mangisoni, J. (2006). Markets, Institutions and Agricultural Performance in Africa. African Technology Policy Studies Network (ATPS); Special Paper Series No. 27. Nairobi, Kenya.

Mbelle, A.V.Y. (2005). Productivity performance in developing countries, country case studies; Tanzania. UNIDO Publications. Available at: http://www.unido.org/doc/4899.

M4P. (2005). Making value chains work better for the poor: A toolbook for practitioners of value chain analysis. Available at: http://www.markets4poor.org.

Nakora, H. (2005). Tanzania Jumps on GM Bandwagon. *Arusha Times*: 14 March 2005. Available at: www. africabiotech.com.

Nyange, D., Senkondo, E., Mundo, N. and Ngohero, H.O. (2003). Impact of Macroeconomic Policy Reforms on Agricultural Productivity, Food Security and Poverty in Tanzania: A Case of Njombe and Rungwe Districts. TARP II; Sokoine University of Agriculture, Morogoro.

Regional Agricultural Trade Expansion Support Program (RATES). (2003). Maize Market Assessment and Baseline Study for Tanzania. Nairobi, Available at: http://www.tradeafrica.biz/downloads/tanzania_report. pdf. visited February, 2007.

REMEI AG (2007). Information about bioRe® Tanzania Ltd. Available at: www.remei.ch.

Sunday News (2006). Tanzania is to establish biotech centre. TSN Daily News, 9 December 2006. Available at: http://www.dailynews-tsn.com/.

Tanzania Cotton Board (2007). TCB corporate strategic plan 2007/08-2009/10. Available at: http://www.tancotton.co.tz.

Temu, A.A. and Winter-Nelson, A. (2001). Institutional adjustment and agricultural markets: following the transaction costs in the Tanzanian coffee system. UIUC Working papers. Available at: http://web.aces.uiuc. edu/wf/workingpapers/coffeeinstitutions3.pdf.

The East African (2004). Tanzania to subsidize cotton farmers. 11 August 2004. Available at: http://www.tralac. org/scripts/content.php?id=2818.

The Express (2002). 21 August 2002.

Timothy, B. (2003). The Fallacy of Genetic Engineering and Smallholder Farmers in Africa. *Global Pesticide Campaigner* 13 (1).

URT (2005). Poverty and Human Development Report 2005. Mkuki and Nyota Publishers, Dar es Salaam. Available at: www.povertymonitoring.go.tz.

URT (2006). Agricultural Sector Achievements of the Third Phase Government 1995–2005. Ministry of Agriculture Food Security and Cooperatives. Available at: http://www.kilimo.go.tz

Part VI. Regulating technologies

Recoding life in common: a critical approach of post-nature

Eric Deibel

1. Introduction

In the novel 'Elementary Particles' Michel Houellebecq has one of his characters explain why it is that Huxley's vision of a Brave New World is not a totalitarian nightmare:

> 'I know, Bruno continued... Huxley's world is usually described as a totalitarian nightmare, an attempt to pass the book as a vicious accusation. That's most hypocritical. On all points, genetic control, sexual freedom, the struggle against old age, its culture of leisure – Brave New World is a paradise to us, in essence it is exactly the world we are trying to achieve, until now without success' (Houellebecq, 1999: 123).

Houellebecq carries this charge of hypocrisy to its full conclusion. Unlike those trans- or post-humanist enthusiasts whose aim it is to welcome the genetic age, his focus on the embrace of the scientific paradise which Bruno speaks of, is based on a ruthless exposition of the emptiness of the contemporary popular culture that is to be left behind. The point of entry for this exercise is that Houellebecq's novel thereby precedes the choice Huxley's main character could not make, how John Savage ultimately gave up on his choice between the Savage Reservation where he came from – a last remnant of life in nature – and the man-biology rationality of the Brave New World; to be approached as an echo of how Rousseau's famous savage that 'is born free, and everywhere he is in chains', as it says in the first line of the Social Contract.

Rousseau's savage: 'knows no law but his own will; he is therefore forced to reason at every step he takes' and when entering society he looses his natural liberty and equality, which are destroyed by how society's institutions perpetually set man in 'contradiction to himself' (Rousseau, 1993a [1762]: 89, 303). What the subtitle calls a critical approach of post-nature is then about this turn of the tables, this modern inversion of the nature-society relation of Rousseau's predecessors – those political theorists who would make civilisation into an exclamation of reason based on life in nature. Specifically how ownership in the life sciences can be seen against the background of their descriptions of the state of nature in terms of its property rights - as a nature-property relation.

By analogy to John Savage's tragic choice, also the man-biology relation should include incorporate a critical perspective on ownership in the life sciences with as its counterpoint Rousseau's remarkable observation on property in chapter nine of *'The Social Contract'* – called 'Real Property' when he writes:

> 'The right of the first occupier, though more real than the right of the strongest, becomes a real right only when the right of property has already been established... [T]he right of the first occupier which in the state of nature is so weak, claims the respect of every man in civil society. In this right we are respecting not so much what belongs to another as what does not belong to ourselves' (Rousseau, 1762: Book I chapter nine – Real Property).

Firstly this passage shows Rousseau's disagreement with a natural law of life that derives from the right of the strongest in nature – a reference to Thomas Hobbes' prescription of the natural condition of mankind as state of war of all against all (Hobbes, 1985 [1657]: chapter xiii). Secondly the passage also moderates John Locke's famous labour law of property, elaborated as a part of his state of nature theory in his famous fifth chapter on property in the 'Second Treatise of Government'. Rousseau, by contrast, emphasises the distinction between property in nature and society as key to its establishment in civilised society – the respect for what does not belong to us.

It is in this sense that 'recoding life in common' - as the title reads – is intended, following Rousseau's counterpoint to how Locke famously claimed that: 'tis allowed to be his goods who hath bestowed his labour upon it, though before, it was the common right of every one' (Locke, 1970 [1689]: § 30). This is not about the usual abbreviation of Locke's labour law to the claim that: if you make it you own it – it refers to how Locke's state of nature theory corresponded to a world that was given to 'Mankind in common' (*ibid.*: §26). In turn, Rousseau's critique of Locke focus on how intellectual property is established and recognised as a cornerstone of trade harmonisation in close relationship to the changing interpretation in the life sciences of how to own something that used to belong to man in common. Intellectual property has to be justified in terms of its compatibility to social imperatives – like the duty to conserve, cultivate and make bio-safe biodiversity as well as to the scientific requirements for a shared development of the incredible quantities of DNA, molecules, genes, proteins, cells and the like. It is as a shift in common property relations that it can be understood how genetic materials are being rendered as accessible, shareable and tradable genetic information and genetic information technologies in a variety of common projects.

On the one hand, the echo of Locke's world in common can be specified along the lines of how: 'nature is undergoing an involution much as space did – from the expansive to the interior aspects of the natural world and from an extensive to an intensive project of exploration is nowhere more evident than in the emergence of contemporary biotechnology (Perry 2006: 49). On the other hand, John Savage's choice refers to the critical theory of the Frankfurt school's thesis what Max Horkheimer – in *'Dialectic of Enlightenment'* – called a 'retreat to the state of nature' – a 'calling of itself, as in its pre-modern-times'; whereby 'enlightenment is more than enlightenment – nature, becoming visible alienation' (Horkheimer, 1983 [1944]: 55). To 'recode life in coming' is then about how the biotechnological take on the interior aspect of the natural world correspond to an interiorisation of power to the inner space of individual life – what Ruivenkamp has called the 'technologisation of life' (Ruivenkamp, this volume).

The theoretical background consists of three components. Firstly the state of nature theories of Hobbes and Rousseau will be approached from the perspective of Michel Foucault's writings on, in *'the order of things,* the painting 'Las Meninas' by Velázquez. This famous painting has been described by Foucault as the 'manifest essence' of the classical space of representation (Foucault, 2002 [1966]: 16-18). Secondly, this allows for a perspective on a 'science-nature' relationship by drawing on a key text from the sociology of science – *Leviathan and the Air-pump: Hobbes, Boyle and the experimental life.* Hobbes' controversy with the scientist Robert Boyle over the nature of scientific fact, is often regarded as evidence for the so-called separation of science and society (Shapin and Schaffer, 1989, see also Latour, 1993, Haraway, 1997). Thirdly, the critical perspective is elaborated further by approaching Locke's labour law in relation to Marx' concept of the species-being – a Locke-Marx relation whereby to approach ownership in the life sciences.

The second part of the exercise proposes a distinction between a 'biodiversity market' and a 'free market for genetic information' along the lines of these three relations (nature-society/society-science/nature-property). As already indicated, it looks at the convergence of biology and informatics and the capacity to translate genetic material into genetic information and *vice versa*. 'Recoding life in common' is considered specifically by also emphasising an analogy of open source informatics and open source in the life sciences – a post-modern form of the linkage of life in common to the respect in civilized societies, as Rousseau observed. In the last section this is considered from the standpoint of a 'Marcusean design critique' and in the conclusion it is linked to the twentieth century re-interpretations of Las Meninas by Dali and a Picasso. From this point of view John Savage's tragic choice can not only be approached in terms of the respect for what belongs to someone else, as Rousseau would have it, but also appears as a fact of living together and its shifts in common property relations.

2. Las Meninas and the state of nature

Velázquez' 'Las Meninas' (Figure 1), as any observer notices, has at the centre of the painting the reflection in a mirror of two frail images. These are Philip IV and his wife. This centre space is surrounded by a self-portrait of the painter at work on a canvass, an astonished royal household observing the monarchs; in the background an incidental observer is looking through a door in the background.

The exceptionality of the painting is, as Foucault explains, that 'three points of view come together': the gaze of the observer of the painting, of the painter as he was painting this scene, and of the reflection of the Spanish sovereigns. Taken together these points of view form:

> 'a point exterior to the picture, an ideal point in relation to what is represented, but a perfectly real one too, since it is also the starting point that makes the representation possible. Within that reality itself, it cannot be invisible'.

What matters here is how the picture leaves invisible, – 'despite all mirrors, reflections, imitations, and portraits', 'the gaze which has organised it and the gaze for which it is displayed'. In other words: both the observer of the painting and the sovereigns are drawn into the picture while remaining foreign to it. What it demonstrates, Foucault says, is 'an essential void: the necessary disappearance of that which is its foundation' (Foucault, 2002 [1966]: 16-18). Furthermore the mirror image of las Meninas, as described by Foucault, also draws in the state of nature theories of Hobbes and Rousseau.

Especially in Hobbes' theory of society, the mirror image returns. The sovereign and the civilised society mirror each other in the sense that society is ruled by an absolute sovereign. Its essential void is then

Figure 1. 'Las Meninas' also known as the maids of honor. Painting by Velázquez (1656).

about Hobbes' state of war, about life in society is an escape from the natural condition of man in nature – from the war of all against all. Only when there is a transfer of man's natural right to life to a sovereign can the multitude by united in a social body, as the cover of *Leviathan* demonstrates (Figure 2). It is by sword and scepter that sovereigns escape from nature – 'where there is no common power there is no law' (Hobbes, 1985).

One implication of such a 'natural condition' is that sovereigns get to be less social then natural men in a state of nature; sovereigns become much more likely to behave as if in a state of war. Furthermore how its scope was directly applicable to life beyond the boundaries of civilised societies follows from how he wrote:

> 'It may per adventure be thought, there was never such a time, nor condition of warre as this; and I believe it was never generally so, over all the world: but there are many places, where they live so now. For the savage people in many places of America... have no government at all; and live at this day in that brutish manner.' (Hobbes, 1985: 187)

In other words, it is not only Hobbes' experience of civil war that characterises nature – to Hobbes most of the world outside of civilised societies – Hobbes' world – fits this description.

How the early modern descriptions of the natural condition of mankind comprise their preferred end states can also be interpreted in the sense wherein Foucault writes about the mirror image – between the sovereigns and their court. He argues that it is: 'essential to observe that the functions of nature and human nature are in opposition to one another:

> 'in the classical episteme: nature, through the action of a real and disordered juxtaposition, causes difference to appear in the ordered continuity of beings; human nature causes the identical to appear in the disordered chain of representations' (Foucault, 2002: 336/7).

This applies to state of nature theories. In Hobbes' theory of absolute rule all 'disorder' and 'difference' is located in nature, while Rousseau's 'savage', when leaving nature, does the same for the 'state of society'.

Inversely put, Hobbes 'fearful savage' enabled a naturalisation of the 'ordered continuity of beings' while Rousseau's savage takes fault with this order. The transfer from nature to society, in both accounts, demonstrates how 'human nature causes the identical'. In other words, the classical space of representation comprises the characteristics of the both state of nature theories whose relationship to society appeared as exact opposites – either Hobbes' state of war and his absolute sovereign or Rousseau's savage and his social contract. If there is a difference with Rousseau, then it would have to be that he replaces the vague reflection of the sovereigns in the painting by the accidental passenger who thereby gets to represent

Figure 2. Book cover of Leviathan by Hobbes (1985).

the centrality of his isolated savage – as one of the two points of view that constitute the essential void of the image. However, what should be kept in mind is that Foucault's essential void thereby comprises both Hobbes' and Rousseau's savage, which, after all, only retains his nobility when by himself – when encountering others he demonstrates that 'by nature men are not sociable – the claim which all contemporary readers associated with Hobbes' (Tuck, 2003: 132, 20).

Lastly, the state of nature theories can also be defined as relationship to science – for example when characterising as its contemporary equivalents the insecurities of climate change, biodiversity, hunger and disease. Such global themes might be seen to require some sort of Hobbesian solution, as functions of the disorder of nature to be solved by science and technology or, in turn, reflect a state of war that is a consequence of society, which, as Rousseau would have it, comprises science and technology.

On the one hand, this latter relationship to science would juxtapose to the given character of the 'state of nature of states' as a consequence of the need to escape from perpetual war. For example the important neo-realist Kenneth Waltz argues that: 'the state of nature, among men is a monstrous impossibility. Anarchy breeds war among them, government establishes the condition for peace' (Waltz, 2001 [1954]: 227). On the other Rousseau's savage echoes his view of science. In *a discourse on the origin of inequality* (1755) – argues that assumptions about nature should be seen as hypothetical:

> 'the investigations we may enter into, in treating this subject, must not be considered as historical truths, but only as mere conditional and hypothetic reasoning, rather calculated to explain the nature of things, than to ascertain their actual origin, just like the hypothesis which our physicists daily form respecting the formation of the world' (Rousseau, 1993a [1755]: 50, 51).

Are the hypotheses of physicists that hypothetical? Probably they are when comparing what Rousseau knew about life in nature with Hobbes' description. Looking at the latter's colonial connotations or as a foundation for a modern state of war, Rousseau was right to argue that views of nature *should* have been hypothetical. Also to his savage applies that it became as hypothetical as Hobbes' state of nature of states – it should have been hypothetical because the literal interpretations of his predecessors were not; inversely put, Hobbes' society, as the mirror image of Hobbes' world, is closely related to how he lost his argument over the establishment of scientific fact from Robert Boyle – of Boyle's laws of gases and paradigmatic in the origins of the experiment and scientific fact.

3. Boyle's Victory

As Shapin and Schaffer explain, in *Leviathan and the Air-pump: Hobbes, Boyle and the experimental life*, it would be a mistake to reduce the controversy between Hobbes and Boyle to the question of who was right about the scientific method - neither had ideas that would pass the today's criteria.

More importantly, what Hobbes' objected to was Boyle's organisation of a special community of practice organised around technical devices, in particular how his air-pump was argued to produce a vacuum as a scientific fact. He could not accept how Boyle reserved a privileged philosophical relationship to nature for scientific fact. Not only did Hobbes, also after repeated improvements by Boyle, continue to reject that there was no invisible ether present in the air-pump. He did so on political and philosophical grounds because he found that also scientists should be subjected to the rule of society, no different then communities of lawyers, clerics and especially priests. Hobbes lost – as is most immediately illustrated by the cover of Shapin and Schaffer's book where the Leviathan no longer carries a scepter and a sword but a sword and an air pump (Figure 3).

For Shapin and Schaffer, Boyle's victory illustrates 'the origins of a relationship between our knowledge and our polity that has, in its fundamentals, lasted for three centuries' (Shapin and Schaffer, 1989: 3). Accordingly they conclude in the very last line of their book that:

> 'the form of life in which we make our scientific knowledge will stand or fall with the way we order our affairs of state'... knowledge as much as the state, is the product of human actions. Hobbes was right' (*ibid.*: 344).

It is at this point in their argument that the anthropologist and philosopher of science Bruno Latour intervenes in '*we have never been modern*' (1993). Like Shapin and Schaffer he thinks that the dichotomy of 'the separation of science and society' is: the 'entire modern paradox'. However, it is not only Boyle's paradigm that is at the origins of the problem, also Hobbes' society plays its part. He answers Shapin and Schaffer's question with:

> 'No Hobbes was wrong. How could he have been right, when he was the one who invented the monist society in which knowledge and power are one and the same thing? How can such a crude theory be issued to explain Boyle's invention of an absolute dichotomy between the production of knowledge of facts and politics?

Any sociology of science that sees only politics as a source of explanation is too crude, Latour argues - it would imply that 'knowledge has a place only in support of social order'. Hobbes' rejection of the vacuum is motivated by a belief that the phenomenon of the empirical experiment cannot 'be produced on a scale other then that of the Republic as a whole' (Latour, 1993: 26). As a consequence, Latour argues, Shapin and Schafer end up privileging society over science, while Latour replaces such an answer to the separation of science and society with a focus on a 'division of power' – how 'Boyle is creating a political discourse from which politics is to be excluded, while Hobbes is imagining a scientific politics from which experimental science has to be excluded' (*ibid.*).

Lastly, Latour only briefly mentions that Hobbes continues to represent the intellectual tradition that is 'at the root of the entire modern Realpolitik' (*ibid.*). This raises the question about how the state of war has its part in his division of power between science and society – how Hobbes' criticism of Boyle also bears on questions of perpetual peace and war.

Figure 3. Book cover of Leviathan and the Air-pump by Shapin and Shaffer (1989).

State of nature theories were, at least until after Rousseau, near universal components of political theories. For example Louis Althusser mentions Vico and Montesquieu as the only political theorists who did not subscribe to this method (Althusser, 2007 [1959]: 25). Similarly also its literal interpretation considering the entire world as a state of war has enduring consequences. This is not primarily about how there is no more outside to the state of nature of states, it follows from how Richard Tuck argues, in his brilliant *'the rights of war and peace'*, that there is:

> 'no powerful theorist of a rights-based liberalism' [like Kant and Rousseau] 'who has not subscribed to the basic account of the liberal agent which appeared in the early seventeenth century' (Tuck, 2003: 14).

The expansive characteristics of liberal agency are derived from the early modern interrelation of the early modern formation of the individual and the state – as a mirror image like that of 'Las Meninas'. This is the case in Hobbes' literal and Rousseau's hypothetical versions while also Kant, whose categorical imperative is independent from such theories, made a proposal for a 'perpetual peace' that 'remained hemmed in by the Hobbesian dilemma about international relations' (*ibid.*, 2003).

For Tuck, this implies, however, that the origins of liberal agency characterise the contemporary predicament of the modern individual and his society. By contrast to the independence of the societies of Hobbes' world, when the behavior of sovereign states gets more interdependent – either bound within international agreements or moving towards perpetual peace – its consequence is that there are going to be implications for how the rights and private sphere of sovereign individuals can be imagined. To illustrate the predicament he quotes Max Weber's to say that: 'in our whole economic life even today this breeze from across the ocean is felt...but there is no new continent at our disposal (Weber, 1906; cited by Tuck, 2003: 15).Of course, it might be that the matter is not whether there are new continents – but about a science-society relationship that has got Boyle's victory as its one leg and Hobbes' world as its other.

To imagine society and the individual is then about their early modern interdependence, as a mirror image of 'Las Meninas'. The matter is not only that science is also based on social institutions, as was already Latour's observation; when not distinguishing Hobbes' natural and political philosophies anymore a perspective becomes possible on the natural world of scientific privilege – Boyle's victory – that increasingly coincides with that of sovereign right – Hobbes world and Rousseau's inversion. To say that its a-social definition got entangled with science is to consider science-property relations.

4. Locke's labour and Marx' species being

Like Hobbes natural law of life, also Locke's property involves plenty of assumptions about life in nature and similar to Hobbes' assurance that his state of war exists in the Americas, also Locke's state of nature theory draws in the entire world.

His view that property exists in nature follows from the right to use what before was assumed to belong to someone else, as he explains when he says:

> 'I shall endeavor to shew, how Men might come to have a property in several parts of that which God gave to Mankind in common, and that without any express Compact of all the Commoners'.

There must by necessity be some way to appropriate from nature for it to be of any use to man The state of nature involves a duty to cultivate the gift to men in common – a world in common that can only be useful if there is a natural right to property. Furthermore he also anticipated objections against an unlimited account of ownership by arguing that the same gift to men common binds the property too. This appears most immediately by how Locke distinguishes the 'world in common' from the English commons.

Locke carefully describe land that when is common in England it is not a state of nature – it is 'common in respect of some man, it is not so to all mankind', he writes (Locke, 1970: §35). He continues by saying that what is, on the other hand, common to all mankind is the case in the occupation of the 'in-land, vacant places of America'. This occupation Locke maintains gives the inhabitants 'no reason to complain, or think themselves injured by this Man's Encroachment' (*ibid.*: §36). Consequently the state of nature and the state of society can be distinguished also by their common properties.

On the one hand, Locke does argue that the Indian knows property, for instance:
> 'the fruit or venison, which nourishes the wild Indian who knows no inclosure, and is
> still a Tenent in common, must be his' (Locke, 1970: §26 see also §30).

However such ownership is bounded by nature because what is useful to 'Americans now, are generally things of short duration; such as, if they are not consumed by use will decay and perish of themselves' (*ibid.*: § 46). It is the labour of settlers that cultivates waste lands and earns gold, silver, money etc. – he assumes that rational cultivation and money are characteristics of civilised societies that are unknown to anyone in the state of nature[224].

When Locke argues that property should exist in nature, because the Indians know it too, this is about binding property in nature while still continuing to allow settlement and the defense of what is occupied without any permission required –if permission is not granted, ownership can be defended. As long as enough is left in nature, his sufficiency provision, it is by natural right of property men from civilised societies are allowed to claim ownership over what used to be common to everyone. What matters about this is how also Locke's theory of the state of nature is something that is to be left behind –Locke asks about nature:
> 'why would anyone give up this Empire? For all being Kings as much as he, every Man
> his equal, and the greater part no strict Observers of Equity and Justice, the enjoyment
> of the property he has in this state is very unsafe, very insecure (Locke, 1970: §123).

What is important to observe about Locke's escape from an Empire is that in his theory of nature a return is something desirable when it is about more civilised notions of property. Hence, with Locke it is property that is the cause to escape nature and to return later, while for Hobbes' this would merely imply a return to a state of war. As such it is now possible to consider Rousseau's critique – that there

---

[224] The full quote says 'let him plant in some in-land, vacant places of America, we shall find that the Possessions he could make himself upon the measures we have given, would not be very large, nor, even to this day, prejudice the rest of Mankind, or give them reason to complain, or think themselves injured by this Man's Encroachment (Locke, 1970: § 36). Further illustration: 'The greatest part of things really useful to the Life of Man, and such as the necessity of subsisting made the first commoners of the world look after, as it doth the Americans now, are generally things of short duration; such as, if they are not consumed by use will decay and perish of themselves, that Fancy or Agreement hath put the Value on, more then real Use, and the necessary Support of Life' (Locke, 1970: § 46). See also Locke (1970: §6, 26, 31). For detailed discussion, I follow Tuck (2003), on Locke and the commons see Macpherson (1962), Tully (1980, 1991), and Arneil (1996).

was no return from the insecurity of civilised societies – from the perspective of the species-being, which brings this to its full conclusion.

Marx's concept of the species-being, as Eugene Thacker explains, can be read as a lens whereby to see biotechnology with the aim of: 'frustrating any attempt to isolate a state of nature separate from society' (see Thacker, 2003: 29-36).

The concept is a part of Marx' criticism of classical economics in his early work, the unpublished *economical and philosophical manuscripts* of 1844 wherein he wrote:
> 'do not let us go back to a fictitious primordial condition as the political economist does, when he tries to explain. Such a primordial condition explains nothing. He merely pushes the question away into a grey nebulous distance' (Marx, 1961 [1844]: 94).

In other words, although the state of nature theories became hypothetical after Rousseau, still its assumptions about life in nature found their way into economics. The species-being is then, as Erich Fromm puts it, about how the alienation of labour is about how capital takes away: 'the object of production from man, it also takes away his species life, his real objectivity as a species-being, and... his inorganic body, nature is taken from him (Fromm, 1961: 49).

As Fromm explains, Marx' species-being has as its aim to demonstrate a commitment to 'emancipation of the individual and the overcoming of alienation and the restoration of his capacity to relate himself fully to man and to nature' (Fromm, 1961: 13, 24). Furthermore, to Marcuse the text thereby provided a philosophical basis for the economics of Marx' Capital, demonstrating how essence and existence have moved apart too far in modern capitalism. Existence realises man's essence because man is never immediately one with his life activity, but rather distinguishes himself from it and relates to it (see Wiggershaus, 1994: 102-104).

How the species-being links nature to capitalism by way of man's labour can also be seen against the background of Rousseau's critique of Locke. When Rousseau makes the nature-civilisation distinction hypothetical, this also comprises the observation that assumptions about property in nature are the basis whereby respect for property is established in society. With Marx' species-being this reappears as a nature-capitalism relation – whereby it is the establishment of property is entangled with how man's relation to nature gets defined in capitalism. As a lens whereby to observe biotechnology, Marx emphasises the naturalisation of capitalism in the sense that biological organisms are 'put to work' in various biotech industries.

In the *'the global genome'* he focuses on the 'species-being and man's 'inorganic body' as well as Marx' later references to the metaphor of social metabolism to ask: 'if there is no human subject or wage, is there still exploitation and alienation?' To naturalise capitalism is then immediately about how value in the biotech industry is 'not human labour, but the specific labour power or life activity of cells, molecules and genes' – which he calls 'living labour without subjectivity' and 'living dead labour'. On the one hand, the sale of labour power on the market place is increasingly about capturing the cellular, enzymatic, genetic character of life activity to generate capitalist value – on the other hand Marx' focus on the potential of living labour, his emphasis on the sovereign individual, increasingly becomes biological potentiality – reconsidering the:

'complex' properties of biological 'life itself' – metabolic networks, bio pathways, single-point mutations, immuno knowledge, protein folding – offer a resistance to the genecentric and reductionist approaches taken by the biotech and pharmaceutical industries (Thacker, 2005: 40, 205, 206; 2003: 29-36).

Lastly, Thacker's emphasis on the biological character of labour power coincides with Sunder Rajan's suggestion, in *Biocapital*, that Marx can be seen as a 'methodologist' of the 'rapidly emergent political economic *and* epistemic structures' of the life sciences. He argues that 'what Foucault does explicitly is what... Marx does implicitly, which is consider political economy as consequential because it is a foundational epistemology that allows us the very possibility of thinking about such as system as a system of valuation' (Sunder Rajan, 2006: 7, 13).

In *the order of things* Foucault explains Marx' radical economics as the other side to the coin of a bourgeois economics of a system of labour and scarcity because: he reserves a specific domain that is proper to man, while 'any possibility of a Classical science of man' had been inconceivable. Foucault argues that Marx appears similar to the economists that he criticised in the sense that 'it would not be possible to discover the link of necessity that connects the analyses of money, prices, value and trade if one did not first classify this domain of wealth'. Marx then is not to be understood as distinct from the classical economists of the nineteenth century, like Smith, Malthus and Ricardo, but by how he is part of the 'general epistemic re-arrangement' of the nineteenth century. It is the belief in the 'primary, irrefutable, and enigmatic existence of man whereby the place of the king in Las Meninas was taken by the: 'difficult object and sovereign subject of all possible knowledge' (Foucault, 2002: 336-338).

What is important about the Locke-Marx relation from this stand point is that it is able to comprise 'the necessary disappearance' of the human body as the foundation of labour power against the background of Locke's world as a common property. As a foundational epistemology whereby systems of valuation are formed these are then the background to how in the next sections a 'biodiversity market' is distinguished from a 'free market for genetic information' – closely related with the complex and layered view of gene behavior in the open and distributed networks of modern biology.

5. A biodiversity market

One of the contentious aspects of the patentability of genetic resources is how it suggests that there is also a 'biodiversity market' wherein nature is the basic resource for biological techniques and its conservation as a scarce resource is something commercially interesting.

The first example to illustrate the business of a biodiversity market was supposed to be the 'Diversa' corporation, which, however, in the course of writing this article changed into a 'bio-fuels' company. Before, its mission statement (see Box 1) was about biodiversity in the sense that it perceives of ecosystems as 'mimicking' industrial conditions and as something to mine with the aim of using and claiming proprietary and patented technologies. The biological entities that serve this purpose, particularly enzymes, can be optimised for targeted application using Diversa's DirectEvolution® technology.

Box 1. Diversa Mission statement *(accessed May 2007 – off-line july 2007)*.

'Diversa accesses a wide variety of extreme ecosystems, such as volcanoes, rain forests, and deep sea hydrothermal vents, to collect small samples from the environment to uncover novel enzymes produced by the microbes that dwell there. Because the harsh temperature and pH conditions in which these "extremophiles" live often mimic those found in today's industrial processes, they are a rich source of potential products. Through the use of proprietary and patented technologies, Diversa extracts microbial DNA directly from collected samples to avoid the slow and often impossible task of trying to grow the microbes in a laboratory. We then mine this huge collection of microbial genes, numbering in the billions, using ultra high-throughput screening technologies with the goal of discovering unique enzymes'

In July, however, the website immediately linked any user to that of Verenium – the nature of energy"[225]. This is about developing, the website explains, 'bio-fuels' 'from low-cost, abundant biomass' and its specialty enzyme products. With the name change there was, therefore, a shift in what is considered to be nature – when nature is about abundance it is no longer the nature characterised by a diversity that is particularly difficult to patent. As the website claims it is in contrast to the scarcity of oil that bio-fuels are derived from widely available biomass resources on a recurring, sustainable basis. The new website speaks of: 'new, greener methods to produce the fuels and other industrial products that are the lifeblood', 'the building blocks' of 'our 20th century way of life'.

The disappearance of the 'Diversa corporation', however, could just as easily be seen against another background – not in relation to the black gold but in relation to the 'green gold' where to be rich in biodiversity is about the mega-diversity of seventeen 'like minded' counties[226]. While the new site mentions that bio-fuel would replace how a lot of oil comes from U.S unfriendly countries, it might more appropriately be seen in relation to the geo-political mega-diversity of countries like China, oil-rich Venezuela and bio-fuels advocate Brazil – three of the seventeen members of the like-minded mega-diverse countries with an interest in capitalising on the access to genetic resources. In other words if it is a valuable resource that is the aim of the Diversa corporation there are others that are trying to claim it – not only sovereign states but also the closely related hassle of having to deal with all kind of groups that might want compensation.

Not entirely coincidental these seventeen countries are also the places where access to genetic resources also involves the claims for recognition of those individuals, groups and peoples who have only a very limited international legal personality. A range of overlapping international mandates is being arranged that represent the disproportionate contribution to the conservation and cultivation of biodiversity of groups like farmer communities and indigenous peoples. Their entanglement with biodiversity markets then immediately coincides with how international organisations perpetually emphasise that further action is necessary to enable 'access and benefit sharing' – when this involves the claims of groups that are often not recognised by their own governments there is immediately an exemplary reason for

---

[225] http://www.verenium.com/index.html. The announcement reads: Verenium was formed in June 2007 through the merger of Diversa Corporation, a global leader in enzyme technology, and Celunol Corporation, a leading developer of cellulosic ethanol process technologies and projects. It works on subsidised ethanol plants in the US and Japan.
[226] http://lmmc.nic.in/

multilateral 'gene regimes' to take charge of the equitable distribution of the potential funds from the biodiversity market.

Consequently: mining for patentable genetic material in some exotic park is likely to involve the entire spectrum of categories and discourses on offer[227]. On the other hand when changing the basic resource from biodiversity to bio-mass there no longer appear to be complicated compensation requirements for a patent to be justified. In other words, such a redefinition of genetic resources is immediately about decoupling patents from benefit sharing claims. Especially when considering that intellectual properties are a cornerstone of global trade harmonisation, it appears that how 'gene regimes' try to regulate the patents of a biodiversity market is being eclipsed by a shifts the basic resource of post-nature. Its scope appears even more from a second company called DNA 2.0 Inc.

The website welcomes you with the message DNA 2.0: 'the building blocks of life made for you' and it advertises itself as the fastest provider of synthetic genes and offers 'DNA-2-Go'™ and Planet Gene. In this example there is no more genetic resource that is being considered, neither the mining of natural parks, nor is an amorphous bio-mass involved in the 'path from virtual sequence to physical gene'. On offer in the business of synthetic biology is a collection of synthetic genes that are designed to improve, optimise or alter one or more properties of the gene sequences found in biological organisms (Box 2).

Box 2. DNA 2 Inc Website information *(http://www.dna20.com/; last checked September 2007).*

DNA 2.0 Inc. is a leading provider for synthetic biology. With our gene synthesis process you can get synthetic DNA that conforms exactly to your needs, quickly and cost effectively. [...] The massive amounts of publically available genomic information can be condensed using our ProteinGPS™ algorithm into mathematical rules on how proteins function. These rules are subsequently converted into physical entities using our efficient DNA-2-Go™ process.

Rush Order Gene Synthesis – DNA-2-Day™ is the fastest gene synthesis service available anywhere. Thanks to DNA2.0's advanced, innovative technology, and a highly optimised gene synthesis process we can produce genes up to 1 kb in five days. This is twice as fast as our already industry-leading DNA-2-Go gene synthesis speed of eight to ten days.

To order DNA-2-Day synthetic genes, please click on the 'Order a Gene' button on the left side of the page. Please check the 'Rush' option on our online order page.

Furthermore what its public domain might be like, on the other hand, might be understood from how the M.I.T. Bio-bricks registry. It similarly aims to make available strands of synthesised DNA in a standardised catalogues but in addition it also aims at the shared development of biological modules in an open source project. These modules can then be designed and combined as the components of more sophisticated biological circuitry, with more complicated and adaptable patterns of behavior then previous genetic engineering. It is open source in the sense that similar to informatics it is about claim the right to develop and control the basic resources in common projects that are freely accessible and

---

[227] See for instance the description of Shaman pharmaceuticals as well as the Merck-In-bio agreement in Perry (2004).

adaptable for anyone on the internet. This is considered more efficient then closed source – where small groups of pubic or private scientists are motivated by patent incentives (see De Vriend, 2006; ETC, 2006; Rai *et al.*, 2007)[228].

'Such an analogy between open source code and open genetic code is then about the pragmatic feasibility of various and distinguishable development models to those of the business of the biodiversity market; more or less closely tied to patentability. In either case, however, the instantaneity whereby the information provided by the DNA sequence can used for a much broader range of purposes – access, downloads, circulations, copies, modifications etc. – is represented as an alternative to the patentability of genetic resources. It appears as much more valuable then the material DNA sequence or the possession of the material molecule and coincides with a progressive 'identification of source code [that] opens the biomolecular body to the principles of design' (Thacker, 2005: 74).

In the next section the convergence of biology and informatics will be elaborated further as a *free market for genetic information,* which juxtaposes to the perpetual strengthening of intellectual property rights the sound free market logic of circulating and operationalisation the immense quantities of genetic information.

6. Free market for genetic information

It might not be in terms of patents that one makes sense of the incredible amounts of sequence data that is being produced.

Quite a lot of publicity has gone to biotech and genomics companies which decided to release their genetic sequence data – to release it into the public domain rather then putting it into subscription databases or trying to get patents. This might this appear as a victory of the public scientists' who complaint about the corporations that where interested to be the publisher of the most interesting pages of the book of life – the human genome – and translators of the most spoken languages on Rosetta stone of crops – rice. What this public-private distinction obscures, however, is how remarkable is – about systems biology – how much more it is invested in the business of informatics then with the business of strengthening biotechnology patents.

The immediate example is then the DNA or gene chips, also called micro-arrays, that involves a series of ordered small glass slides with thousands of DNA sequences and markers that allows for the identification of gene or protein expressions. It provides the input that is needed for the construction of databases and quantitative models that run on algorithms and allow statistical interpretation that 'normalise' the data – 'just as an algorithm is called a gene, the collection of algorithms is referred to as the genome or genotype' (Keller, 2002: 280). It is hoped that these algorithms and numeral coordinates can describe a higher dimensional space that constitute a model for the expression of genes, proteins and enzymes.

Most importantly, however, DNA chips are closely related to a shift in the understanding of life, as explained by Dennis Noble when he writes:

---

[228] See http://openwetware.org/wiki/The_BioBricks_Foundation.

'the genome is like an immense organ with 30,000 pipes. The larger the organ, the more pipes it has, and the wider the range of pitch, tonality, and other musical effects it can be made to produce. The music is an integrated activity of the organ. It is not just a series of notes, but the music is not itself created by the organ. The organ is not a program that writes, for example, the Bach fugues. Bach did that. And it requires an accomplished organist to make the organ perform'.

Contrary to more gene centric views, his human physiome project intends to model the approximately 200 cells in virtual representations that can help to understand the main organs of the human body. However, what Noble is explaining is also about how systems biology involves dealing with how even the smallest organs have millions of cells, as well as folded proteins and molecules of different sizes and shapes - beyond even the tetra-bytes of IBM's Blue Gene computers (see Noble, 2006: 31).

Looking at it this way, it is possible to qualify how it follows that system biology is about the: 'ultimate integration of biological databases [which] will be a computer representation of living cells and organisms, whereby any aspect of biology can be examined computationally (Kanisha and Bork, 2003). Furthermore when also considering the previous speed wherein the production capacity of bio-informatics and genome sequencing projects increased in its recent history, it can be expected that it is soon possible to model, quantify and also make useful more complex organisms then unicellular bacteria and yeast. This is further backed up by the popular expectations of a market for bio-informatics tools and services runs on estimates in tens of billions and the hardware market for the life sciences estimated in single digit billions.

Even though such numbers are not very reliable what matters about system biology here is its illustration of how speaking of a biodiversity market is intricately linked to making sense of the quantities of data on sequences of genes and genomes. This is increasingly about a business of calculating, quantifying and modeling countless cells, folded proteins and molecules of different sizes and shapes as a changing business practice that is not centered around the patented biological techniques of the Monsanto's, Syngenta's and Du Pont's of this world, but involves supercomputers and DNA-chips that has much more affinity to the business of Microsoft, Google and even Linux.

Of course there are still patents on biological techniques, gene patents, software patents, database patents and so forth – while IBM's Blue Gene supercomputer runs on Linux simultaneously there are countless patents involved in its life science business. What matters is that in the open and distributed network of modern biology there is a changing approach to what is life - in systems and synthetic biology this is immediately about ownership in the sense that its notion of what is its basic resource is progressively less that of a biodiversity market. Genetic material is either not scarce or it is not there at all as a basic resource for the trade around the exchange of biological entities – like DNA, molecules, genes, proteins and the like.

This also applies to open source in the sense that it is not sufficient when considering the life science to emphasise the sharing and shared development of genetic information and genetic information technologies. Its qualification as a quintessential alternative to patenting is the logical end result of departing from analytical or critical histories of intellectual property – by comparison to the scope of patent systems any alternative appears as exceptional.

Shifts in property-nature relation in the life sciences cannot easily be arrived at by quantitative estimates of when the amount of patenting might become a problem for innovation – either to argue there is no problem or to reveal a fragmentation of the circulation of knowledge in line with Heller and Eisenberg's tragedy of the anti-commons (1998). To show how open source might become an efficient alternative to patenting in the life sciences then begins from how also intellectual property is based on the integration of knowledge. A body of studies in information law and economics then speaks of the intellectual commons as being in between the formalised private property system and a free market for information (Benkler, 2002; Hess and Ostrom, 2003; Rai and Eisenberg, 2003; Reichman *et al.*, 2003; Rose, 2003).

In '*a philosophy of intellectual property*' Richard Drahos (1996) defines the intellectual commons as an independently existing set of discoverable abstract objects which are open to use. While an intellectual commons might not be accessible to everyone, for instance because it requires a certain type of knowledge or culture, it is inexhaustible, its usage doesn't deplete the resource like the occupation of land (see Drahos, 1996: 55, 56). It is against such a background that open source is also frequently seen as the quintessential example of collaborative innovation in modern biology (see Hughes, 2005; Maurer, 2004; Burk, 2002; Boettiger *et al.*, 2004; Opderbeck, 2004; Hope, 2006).

If considered in terms of an intellectual commons that belongs to everyone, like Locke earlier defined the entire world, it implies that open source is about the right to be included. It is in this sense that open source and not patents would be the answer to Locke's question of how are individuals able to use the (intellectual) commons: 'without having to obtain the consent of all the commoners' (see, Drahos 1996: 55, 56). From the perspective of a nature-property relation, however, the focus on access, sharing and an intellectual commons is closely tied to a changing understanding of life and nature. When recalling Rousseau's critique of his predecessors the respect for patents 'as something that does not belong to ourselves' gets established in how DNA, molecules, genes, proteins, cells and the like are being rendered as shareable, translatable and tradable genetic information technologies.

In fact to speak of genetic information technologies follows immediately from the economic necessity to streamline the patent-based market logic with an 'excess' of genetic information, like junk DNA, feedback loops and so forth. Strictly analytically speaking a 'free market for genetic information' then refers to how supply and demand in the perfectly free market implies a free circulation of information, introducing intellectual property, for instance to encourage the production of information, is then a distortion of the free market. Its immediacy, however, follows from Perry's observation that genomics performs two intertwined functions: the production of an excess of genetic information and the development of new technologies for managing that excess, should be understood. If this excess and its management is a growing industry that is being streamlined across many fields of applications, then: 'it is not difficult to understand genomics as the medical, technological and economic management of information – that is, as a political economy' (Perry 2004: 95 see also Thacker, 2005).

Lastly it is then as a characteristic of this shift from a biodiversity market to a free market for genetic information that there are also examples of open source in biology that have explicitly humanitarian aims.

Probably the best known of these is the 'Biological Innovation for Open Society (BIOS) initiative, an organisation that offers a plant biotechnology package free of patents. In medicine there is the 'Tropical

Disease Initiative (TDI), which integrates the collaborative character of genomics with releasing under-developed drug candidates for tropical diseases. Already a recognisable linkage to international mandates is also under construction in the Generation Challenge Programs of the Consultative Group for International Agricultural Research (CGIAR). It similarly involves the aim of improving the germ plasm of crops relevant to the third world by releasing biotechnologies and sharing genetic information (Thomson, 2001; CAMBIA, 2003; Herrera, 2005; Maurer *et al.*, 2004; Louwaarts, 2006; see also Deibel, 2006).

Such projects can be compared to open source methods because of their emphasis on the effectiveness of collaborative research and with free software by how its aims apply biology to a humanitarian objective. These examples begin to provide some of the necessary contrast to open source as a business model for the life sciences – to be able to distinguish and support such ends from other ways of sharing genetic information that perpetuate the same patent ridden policies also for the genetic information society. Nevertheless its idea of a 'freedom to operate', either by involving researchers in the development of tropical medicine or crop scientists in developing countries, leaves obscured how its release of genetic information and genetic information technologies still coincides with the convergence of biology and informatics.

This is to ague that the relevance of these humanitarian examples is closely related their ability to raise the question how open source in biology might be conceived as something more then an echo of the common projects in informatics. Its implication would then be that also common projects are only the humanitarian exception to the boundary between patents and a free market for genetic information would only be an exception – a Lockian sufficiency provision – with their approach to biotechnology as a part of the further justification of the patent system. To reconsider this property-nature relation from a potentially more principled standpoint would then have as its obvious counterpoint the free software foundation (FSF).

The FSF in the early eighties, at the time when patentability was introduced to both software and biology, began to emphasise the shared development and release of source code. A decade later, at the time when patents became a cornerstone of international trade harmonisation, the release of source code entered a new stage when the kernel was released that enabled the Linux operating system – a functionally equivalent to patented systems. The FSF has always linked the efficiency of its common projects to how source code should be shared, free of intellectual property requirements, as a matter of principle – as a part of an ethic of sharing (see Berry, 2004).

What is different then the efficiency approach is then how Eben Moglen, a legal adviser of the FSF, explains free software as a new kind of common property. By analogy to the management of 'fisheries, surface water resources, and large numbers of other forms of resource beyond human production', he explains free software by comparing it to Faraday's law from physics. He writes:

> 'Wrap a coil around a magnet; spin the magnet. Electrical current flows in the wire. One does not ask, "what is the incentive for the electrons to leave home?" It's an inherent, emergent property of the system, we have a name for it: we call it induction. The question we ask is, "what is the resistance of the wire?"..., "What is the resistance of the network?" Moglen's Corollary to Ohlm's Law states that the resistance of the network is directly proportional to the field strength of the intellectual property system. The conclusion is: Resist the resistance. Which is what we do' (Moglen, 2003).

Moglen's call for resisting the resistance can also be applied to biotechnology patents – open source biology not only a different development model to patenting but is an emergent property that is proportional to the strength and scope of intellectual property protection. The question, of course, is then still what a life science equivalent for the FSF's active release of source code and its shared development into a free operating system might look like?

Its answer as it stands is that this emerging property appears mainly in systems and synthetic biology – as a scientific necessity to share genetic information and genetic information technologies that is as much big business as it is open source. A 'recoding' of life in common – as the title described it – is then about resisting that the emerging property is only about the justification of patentability – about uncoupling the regulation of a biodiversity market from a free market for genetic information that manages its own boundaries with intellectual common resources.

7. Towards a recoding of life in common

The re-interpretation of 'las Meninas' by Salvador Dali (Figure 4) would suggest that the displacement of the natural order of the original is decisive – that Boyle's victory over Hobbes and Rousseau is complete.

In the other re-interpretation by Dali, the focus had been on the glory for the painter and the main figure of the original painting (la infanta). The mirror image of Velázquez's original was exaggerated, obscuring everything else. In this version, however, the original is only recognisable by the incidental observer at the door, still casting a casual glance through a door that might just be opened or is about to close. The painter, the sovereigns and the maids of honor are all gone. All that is left for him and us to see are the strings, nails, many other materials and a last window of the original painting, about to disappear in a vortex. Indeed, the space of classical representation is disappearing from view. Earlier Dali saw Velázquez in the light of his own glory, this time the sovereign, his society and the painter himself are being displaced, decisively.

Figure 4. Las Meninas, painting by Salvador Dali.

What appears is that there is no restoration of man's relationship to nature, only a definite rupture with early modernity. What Dali's vortex represents is that the observers, both us and the accidental passenger, are just in time to see the natural order disappear. When considering such a view of the classical space of representation described by Foucault, it might also be emphasised that any alternatives to the international recognition of intellectual property are closely entangled with was called a free market for genetic information; in the future alternatives to patents have little to do with the relation of patents to genetic material as its basic resources.

It was already mentioned in the introduction how critical theorist Max Horkheimer wrote about modernity as a 'retreat to the state of nature' (Horkheimer and Adorno, 1983 [1944]: 55). As such how Horkheimer wrote of modernity as characterised by a declining ability for critical thought appears against the background of the space of classical representation 'las Meninas'. He explains that this is immediately about the mutation of nature in terms of a scientific rationality – a transformation of the world from one of ends into one of means, as he wrote in the *'Eclipse of Reason'*. This transformation is about abilities of science and technology to conceive of man's actual behavior in such a way that is made into something instrumental, a subject of the coordination of means with ends as given – wherein 'end-states' are given (Horkheimer, 2003: 7).

Like Dali's vortex this would suggest a technological dystopia – what would follow Boyle's victory is then something like Zygmunt Bauman's 'Liquid Modernity'. Bauman concludes about Marx and Engels' 'melting of the solids' that the 'wish to discover or invent solids of – for a change – lasting solidity', like the 'communist manifesto', is no longer possible when the 'overall order of things is not open to option' (Bauman, 2000: 3, 5). Instead of a new solid modernity, the given end-state that appears is then a 'liquid modernity' and what appears would seem to recall how Marcuse wrote about science and technology in *'one dimensional man'* that:

> 'the prevailing modes of control are now technological in a new sense... [as] the very embodiment of reason for the benefit of all social groups and interest – to such an extend that all contradiction seems irrational and all counteraction impossible' (Marcuse, 1991: 11).

Also Marcuse was well aware that the technological transformation of ends into means corresponded to a declining ability for critical thought – in terms possible end-states this implied that technological modes of control were juxtaposed to the Marxist hope of social revolution, as a solid natural order.

However, also Marcuse echo of the early modern state of nature theories with a threat of dystopian technological control has its own mirror image – his utopian hope for a new science of nature. From that perspective, his idea about societies that are ruled by technological modes of control can also be approached by how the state of nature theories became hypothetical – replaced by the radiance of the scientific laws of nature. In that sense Marcus's hope for a science of nature is similar to Rousseau's savage – it is a purposefully idealised outside, that is aimed at showing the oppression of life in society – his aim for a 'new science-nature relation' turns the tables on the relation of nature to science.

Such utopianism is then not immediately disqualified as an approach to science and technology. The critical philosopher Andrew Feenberg has stressed that while it is certainly the case that Marcuse never really considers technological culture 'on its own terms', his idea of a science of nature is still a useful entry point for a 'Marcusean design critique'. What it has to offer, as an end to the means, is the aim of trying to introduce 'a third term in between anti- and pro-technology positions: the idea of a future change in the structure of technological rationality itself' (Feenberg, 1999: 166, 178; Marcuse, 1964; see also Kellner, 1991).

Here it is in this sense also that a Marcusean design critique can be considered in relation the convergence of biology and informatics by means of an open source analogy. Two interrelated aspects are important – that at its basis is a more principled approach to humanitarianism and that its approach to modern biology includes the explicit aim to reveal and make negotiable the values that get to reside in nature, not only as a question of its patentability but also including its representation of DNA sequences, cells and genetically modified organisms in some digital or electronic form that is amendable to the action of other information technologies. As Perry observes that what gets 'expressed in a digital or electronic form, and only when in that form, information derived from biological material can be acted upon and interacted with in ways that would not otherwise be possible' – 'standing in for particular materials resources henceforth to be absent' (Perry, 2004: 65. 95).

Consequently the future potential to change technological rationality is then to reverse or invert the translation of genetic materials into genetic information – for example by making available open source genetic information technologies with a humanitarian purpose. The matter is not to create a dichotomy between nature and technology, or the body and its representation but to suggest, as Thacker does, that biological potentiality can also be understand as a chance: 'to optimise, enhance and renormalise what counts as biological' (Thacker, 2003: 76). As such a Marcusean design critique that is based on Marx' species-being would bring together again the philosophical basis of Marx' economics with the biological dimension to how capitalism gets naturalised. It is against this background also that biological potentiality would comprise how Ruivenkamp considers a 'reconstruction of food genomics research' – to free the science-nature relation of food genomics from its involvement in the crisis of agriculture (Ruivenkamp, 2003, 2005, this volume).

This crisis can be understood by how Jack Kloppenburg, in his important book 'First the Seed' from the late eighties, wrote that:
> 'we have not historically had the power to alter 'species-being'. That is, we have not had this capacity until very recently' (Kloppenburg, 1988: 2).

This is to say that the 'alteration' of the 'species-being' is about how biological techniques alter man's relation to nature by their capacity for a further commodification of agriculture, food production and a control by external institutions of farming practice. For example the formation of markets for modified seed further alienated the relation of farmers to the natural basis of agriculture in the sense wherein already external research had become responsible for plant breeding. Farmers who used to breed crops became the customers of a market that earns its profits by selling seeds, pesticides and insecticides and with the introduction of biotechnologies not only did farm practices become 'externalised' even more, also bio-chemical substitutes made it possible, henceforth, to make farm products mutually interchangeable (see Ruivenkamp, 2003).

The task is then to disconnect the science-property relation from the exchangeability of farming products and the interchangeability of the basic materials of food production. This is about a freedom to operate for scientists but not primarily, Ruivenkamp emphasises that the focus is on engaging scientists in region-specific agriculture – for instance by focusing on nutritional traits, health qualities, and so forth in order to refine the local characteristics of crop systems and produce specialised for specific geographical conditions. More generally speaking, the involution from Locke's 'world in common' to the interior aspects of the natural world, – as a fact of life in common – requires that the excessive availability and shared development of an increasingly complex and layered molecular body is renormalised. How Ruivenkamp speaks of a 'technologisation of life' can be interpreted as the tension between what gets to count as biological, – as is being encoded and recoded already – with the potential to even consider its biological potential in terms of regionally specific and local biological characteristics. It is at the intermediate level, in between open source-like scientific change and its humanitarianism – that biological potential can be defined in terms of shifting common property relations.

8. Conclusion

At one point in 'Elementary Particles' one of Houellebecq characters observes about Huxley that the main mistake of the Brave New World is its emphasis on genetic disposition in relation to work. Why would there be any need for your assigned biological constitution to determine the status and type of work you would be fit for, as is the case in the Brave new World, when technologies – some of them biological techniques – can do the work? After all there is 'soma' – the rationed happiness drugs – to mind control the population. A genetic class society, as Huxley described it, would not be necessary. Indeed it would not be necessary – yet to embrace the Brave New World, as in Houellebecq's novel, does not thereby overcome John Savages' impossible choice between nature and biological rationality.

This can be illustrated by way of the re-interpretation of 'Las Meninas' by Salvador Dali (Figure 5).

In the image only the infanta looms large in the light, still admired at its location in the Prado Gallery. The sovereigns and maids of honor of the original have disappeared and only the silhouette of the painter is left, seen in the light of the glory of his creation. Dali leaves only the painter and la infanta

Figure 5. Salvador Dalí: Velázquez Painting the Infanta Margarita with the Lights and Shadows of His Own Glory.

– with light radiating from the princes. When seen in relation to the mirror image as it was, involving the sovereigns and their court, again Boyle's victory appears in relation to Hobbes' Leviathan – rather then the image of the sovereign hovering over the city the shining princess recalls Dona Haraway's description of scientific fact as involving 'clear mirrors, fully magical mirrors, without once appealing to the transcendental or the magical' (Haraway, 1997: 24).

Haraway, like Latour, also discusses the Boyle-Hobbes controversy noticing that it reveals the formation of the laboratory as a 'private, or even secret and not a civil public space' – a 'culture of no culture', she calls it that was 'a founding gesture of modernity' (Haraway, 1997: 23-27). Her magical mirror almost echoes how Foucault wrote about the mirror image and 'Las meninas'; it reveals the origins to a science-nature relation that reappears in how he concludes, in *'the order of things'*, that man no longer appears in the space of classical representation but in an:

> 'ambiguous position as an object of knowledge and as one that knows: enslaved sovereign, observed spectator, he appears in the place belong to the king, which was assigned to him in advance by Las Meninas... and as though by stealth, all the figures... (the model, the painter, the king, the spectator) suddenly stopped their imperceptible dance, immobilised into one substantial figure, and demanded that the entire space of the representation should at last be related to one corporeal gaze' (Foucault, 2002: 340)

Dali's shining princess is there, in the position of the sovereign that man was assigned too – as the substantial figure that replaced Hobbes' sovereignty by becoming a subject of knowledge. This mirror image needs no early modern state of nature theories. They necessarily disappear as the foundation to the relationship of man and society which is permeated by the persuasive belief in a 'the strange nature of a being whose nature (that which determines it, contains it, and has traversed it from the beginning of time) is to know nature, and itself, in consequence as a natural being' (Foucault, 2002: 336-338).

Indeed modern liberal and humanist theories usually dismiss such a-historical methods as 'weak' or 'irrelevant' additional moral claim about the absence of natural subordination (see Kymlicka, 2002: 60). For example Rawls' 'original position', is defined as 'a purely hypothetical situation characterised so as to lead to a certain conception of Justice, not as an actual historical state of affairs, much less as a primitive condition of culture' (Rawls, 1999 [1971]: 12). What this obscures are again the interdependent origins of the just society and its individual with Hobbes' and Locke's worlds as well as with Boyle's victory. Both Huxley's savage reservation as well as the theorists of societies isolated from the state of nature of states echo how Rousseau wrote in 1762 that 'the philosophers, who have inquired into the foundations of society have all felt necessity of going back to a state of nature; but not one of them has got there' (Rousseau, 1993: 50). Of course, this also applies to himself – his savage cannot be hypothetical like the hypothesis of scientists – also his own division of power between nature and society got to be closely entangled with modern modes of ownership.

His critique of Hobbes, in the introduction, was that property in nature was 'more real' then Hobbes' right of the strongest. As a critique of property-nature relations this can be traced to the typical pro-contra positions on biotechnology. For example Hobbes' emphasis on an escape from nature's insecurities can be characterised by, on the one hand, the pursuit of the improvement of nature beyond the insecurities of age, sickness, famine and so forth (as Houellebecq does), and, on the other hand, the need to escape from the 'bio-un-safety' or 'bio-insecurity' to man's relationship to nature. Regardless of which of these is to be embraced - this life in nature is threatened in the sense wherein Hobbes thought it was characterised by a 'continual fear, and danger of violent death; and the life of man, solitary, poor, nasty, brutish, and short' (Hobbes, 1985 [1657]: chapter xiii).

Consequently also the 'post-human condition' can be understood in terms of Hobbes' mirror image – both irreducible positions emphasise the insecurity of the man-biology relation – either to be attached to post-nature or to proto-nature. Also Locke's view of nature should have been hypothetical in this sense – because it is not; Rousseau's critique, for example, clearly echoes in the persuasive accusations of piracy. On the one hand, patent piracy turns the recognition of patents as a cornerstone of trade harmonisation into a matter of global security. It pretends to be a near universal claim in order to make all other possible property-nature relations into theft. On the other hand the verdict that 'everything under the sun made by man' is to be patentable – if you patent it you own it – can also be based on Locke's Empire. Bio-piracy is about how the patentability of genetic materials requires compensation to a proliferation of sovereignty claims and common property tied to the interests of indigenous peoples and farmers in a variety of international mandates.

To put it differently – avoiding the binary distinctions between a nature and biotechnology – when in the early eighties chemical compounds, microbes, genes etc., no matter how interdependent and complex, were declared patentable, its suggestion of something that '*belongs to no one*' was immediately resisted by claims that emphasised how before there patentability its subject matter was already part of how nature was owned and shared by others. When genes get money signs, they do not remain a matter of humanity in general – biodiversity was declared a state responsibility and closely followed by the claims for recognition of all kinds of groups. It was not that patents made life a matter of ownership – as artificial inventions – its extension to genetic resources made these into something that had to be regulated *as if they belong to everyone* – to humanity, all countries, all peoples, groups and individuals at the same time.

What Rousseau's critique can reveal is then that there are 'shift in common property relations', what it obscures – by analogy to Huxley's savage reservation – is the closing of the gap between computers and biological organisms. In a not too distant future this will imply that drawing an analogy with open source informatics will comprise the abilities of DNA-chips, computer simulations of cells, DNA computers that use base-pair binding for more powerful calculations and pre-designed biological modules that might be activated in bacteria, microbes and so forth. It is in this sense that to speak of common property relations is about how the various and incommensurable ideas about property and life in common are already coming together and about what they are becoming it should be realised, as James Tully observes, that 'the solutions advanced by Hobbes and the other modern theorists are now part of the problem' (Tully, 1997: 15). To put it stronger, it is as a part of the problem that how a 'free market for genetic information' becomes a system of valuation disappears behind its redefinition of nature as something that belongs to everyone – an idealised form that takes the appearance of a duty to cultivate an intellectual commons and the responsible shared development of genetic information and genetic information technologies.

How to go beyond this establishment of the respect for the free market for genetic information can be illustrated best with a primary example of Picasso's fifty-four re-interpretations of 'Velázquez's painting (Figure 6).

Figure 6. Las Meninas, painting by Pablo Picasso (1957).

The image shows clearly that Picasso's versions very much resemble the original even though they bear no traces of the classical space of representation that Foucault described. The painting has a horizontal form filled with what are almost cubist characters, which include the sovereigns and the maids of honor. These latter are seen in light from the windows and include the painter himself, shown as a huge figure that mirrors only himself; not the observer or the sovereign. Unlike the previous interpretations of Las Meninas by Dali it does not suggest that the mirror image or the natural order that it represents are about to disappear. Picasso's space of representation by its very modernity demonstrates the possibilities wherein to conceive of manifold ways whereby to recognisably re-arrange the painting around the reflection in the mirror.

As such the Picasso can be said to have retained the Hobbes-Boyle controversy – the division of power between science and society as shown with Dali's princess and his image of the disappearance in the vortex. What occurs, however, in Picasso's image takes place in an entirely changed space of representation with limitless possibilities for variation. Its science-society relation is still vaguely visible in the mirror image but there is no longer any strict analogy with the original painting. Similarly, Hobbes, Locke and also Rousseau's views of property-nature relations are still there, as critical heuristics whereby to re-arrange the circumstances that would imply a mandatory entry into the Brave New World and the kind of tragic choice that John Savage could not make.

What this suggests is that also contemporary science-property relations might be seen against the background of the natural order of the original painting – the necessary disappearance of its foundation reappears in terms of an entirely new space of representation. The point of entry is then to look for the greatest possible of variety for property-nature relations – to 'recode life in common' and a Marcusean design critique that might suggests another kind of open source analogy. It might begin to be possible to consider a future change of biotechnological rationality as a part of conceiving an open source operating system for the life sciences that would be about some sort of reconstruction, as Ruivenkamp put it, of what is to belong to everyone. What comes to mind is how there are a variety of ways wherein to define the ongoing shift – how the translation of genetic material into genetic information might be reversed or inverted without disconnecting from the variety of possible linkages to nature as its basic resource or as a subject for conservation, cultivation and bio-safety.

In other words: embracing post-nature, along the lines of Houellebecq's prescription, is something to be accommodated already to the widest possible scope for recognising other property-nature relations as a matter of mutual consent within common projects. Already the boundaries between patented biotechnologies and shareable genetic information technologies are being negotiated within the open and distributed networks – no different from open source software. When departing from the existing humanitarian open source examples the problem is how to avoid their role as a 'sufficiency provision', meaning that even when there are not enough genetic materials left, those that who would like to address the insecurities of post-nature are enabled to use humanitarian biotechnologies. Furthermore the more immediate question would be: what might it take to add an additional characteristic to its role as an alternative to patented biological techniques and gene expressions – how to make explicit that what is at stake, also in shifts in common property relations, is that by the time man's relation to post-nature has to be considered it might turn out to be already owned and ruled by someone else.

It is not socially or scientifically determined that property-nature relations have to look like patents or state sovereignty requirements – for instance why could the release of biological techniques, as an example of a surplus in biological potentiality, not already make existing markets for hybrid seeds – modified or not – something of the past. What is worthwhile then, for those that would support a world of open source societies, is to consider what it might take to make man's future relation to the science of post-nature's into something that is hypothetical – like how Rousseau already argued – it should be something that might happen, something that might just make John Savage's choice unnecessary.

References

Althusser, L. (2007). *Politics and History: Montesquieu, Rousseau and Marx*. London, Verso.

Arneil, B. (1996). *John Locke and America: the defence of English Colonialism*. Oxford: Clarendon Press.

Bauman, Z. (2000). *Liquid Modernity*. Cambridge, Polity Press.

Benkler, Y. (2002). Coase's Penguin or, Linux and the nature of the firm. *Yale Law journal* 112 (3): 369-446.

Berry, D.M. (2004). *The Contestation of Code: A Preliminary Investigation into the Discourse of the Free Software and Open Software Movement*. Critical Discourse Studies, Volume 1(1). Available at: http://opensource.mit.edu/papers/berry1.pdf

Boettiger S. and Burk, D.L. (2004). Open Source Patenting. *Journal of International Biotechnology Law* 1: 221-231. Available at: http://papers.ssrn.com/sol3/papers.cfm?abstract_id=645182.

Burk, D. (2002). Open source genomics. *Journal of Science and Technology law* 8: 1. Available at: http://www.bu.edu/law/scitech/OLJ8-1.htm.

CAMBIA (2003). Australian genetics pioneer recognised in global top 50. Available at: http://www.cambia.org/daisy/bios/164/version/live/part/4/data (last checked april 2006).

De Vriend, H. (2006). *Constructing Life: Early social reflections on the emerging field of synthetic biology*. Rathernau Institute, The Hague.

Deibel, E. (2006). Common Genomes: Open source in biotechnology and the return of common property. *Tailoring Biotechnologies* 2 (2): 49-84.

Drahos, P. (1996). *A philosophy of intellectual property*. Aldershot, Ashgate Publishing Limited.

ETC group (2007). Extreme Engineering: An introduction to synthetic biology. Available at: http://www. etcgroup.org/upload/publication/602/01/synbioreportweb.pdf (last checked October 2007).

Feenberg. A. (1999). *Questioning Technology*. London, Routledge.

Fromm, E. (1972). *Marx's concept of man.* New York, Ungar.

Foucault, M. (2002). *The order of things: an archaeology of the human sciences.* London, Routledge.

Haraway D.J. (1997). *Modest_Witness[at]Second_Millennium.FemaleMan(c)_Meets_OncoMouse(TM): feminism and technoscience.* New York, Routledge.

Heller, M. and Eisenberg, R. (1998). Can patents deter innovation? The anticommons in biomedical research. *Science* 280: 698-701.

Herrera, S. (2005). Richard Jefferson. *Nature Biotechnology* 23(6): 643.

Hess, C. and Ostrom, E. (2003). Ideas, Artifacts, and facilities: information as a common pool resource. Available at: http://www.law.duke.edu/journals/lcp/articles/lcp66dWinterSpring2003p111.htm (last checked april 2006).

Hobbes, T. (1985). *Leviathan: edited with an introduction by C. B. Macpherson.* London, Penguin Books: London.

Horkheimer, M. (2003). *Eclipse of reason.* New York, Continuum.

Horkheimer, M. and Adorno, T.W. (1983). *Dialectic of enlightenment.* London, Allen Lane.

Houellebecq, M. (1998). Les Particules élémentaires. Paris, Flammarion.

Hope, J. (2004). *Open Source Biotechnology.* Thesis submitted to the Australian National University. Available at: http://cgkd.anu.edu.au/menus/PDFs/OpenSourceBiotechnology27July2005.pdf (last checked april 2006).

Hughes, S. (2005). Navigating genomes: the space in which genes happen. Available at: http://www.centres.ex.ac. uk/egenis/staff/hughes/documents/Navigatinggenome1final.pdf (last checked april 2006).

Kanisha, M. and Bork, P. (2003). Bioinformatics in the post-sequence era. *Nature Genetics* 33: 305-310.

Kant, E. (1983). *Perpetual Peace and other essays.* Hackett Publishing Company, Indianapolis.

Keller, F. (2003). *Making sense of life: explaining biological development with models, metaphors, and machines.* Cambridge Massachusetts, Harvard University press.

Kellner, D. (1991). Introduction. In: Marcuse H. (1991) *One-dimensional man: studies in the ideology of advanced industrial society.* London, Routledge.

Kloppenburg, J. (2005). *First the Seed: The Political Economy of Plant Biotechnology.* University of Wisconsin Press.

Kymlicka, W. (2002). *Contemporary Political Philosophy: an introduction.* Oxford, University Press.

Latour, B. (1993). *We have never been modern.* New York, Harvester Wheatsheaf

Locke J. (1970). *Two Treatises of Government: a critical edition with an introduction and apparatus criticus by Peter Laslett.* Cambridge University Press.

Louwaarts, N. (2006). Controls over plant genetic resources. *Nature Reviews Genetics* 7: 241-241.

MacPherson, C.B. (1962). *The political theory of possessive individualism: Hobbes to Locke.* Oxford University Press.

Marcuse, H. (1991). *One-dimensional man: studies in the ideology of advanced industrial society.* London, Routledge.

Marx, K. (1972). Economic and philosophic manuscripts of 1844. In: Fromm, E. (1972). *Marx's concept of man.* New York, Ungar.

Maurer S., Rai M.A. and Sali, A. (2004). Finding Cures for Tropical Diseases: Is Open Source an Answer? *Plos Medicine* 1 (3): 180-182. Also available at: http://www.tropicaldisease.org/documents/MauRaiSal_PLOS2004.pdf (last checked april 2006).

Moglen, E. (2003). *Anachism Triumphant: free Software and the Death of copyright, 4.* Available at: http://www. firstmonday.dk/issues.html (last checked april 2006).

Noble, D. (2006). *The music of life: biology beyond the genome.* Cambridge University Press.

Opderbeck, D.W. (2004). The Penguin's genome, or Coase and open source biotechnology. *Harvard Journal of Law & Technology* 8: 167-227.

Perry, B. (2004). *Trading the genome: investigating the commodification of bio-information*. Colombia University Press.

Rai, A.K. and Boyle, J. (2007). Synthetic Biology: Caught between Property Rights, the Public Domain, and the Commons. *PLoS Biology* 5(3): e58 doi: 10.1371/journal.pbio.0050058.

Rai, A.K and Eisenberg, S. (2003). Bay-Dole reform and the progress of biomedicine. *Law & Contemporary Problems* 66: 289-315.

Rawls, J. (1999). *A theory of justice.* Oxford University Press.

Reichman J.H. and Uhlir, P.F. (2003). A contractually reconstructed Research commons for scientific data in a highly protectionist intellectual property environment. *Law & Contemporary Problems* 66: 315-462.

Rose, C. (2003). Romans, Roads, and Romantic Creators: traditions of public property in the information age. *Law & Contemporary Problems* 66: 89-110.

Rousseau J.J. (1993a). *The social contract and discourses.* Everyman, London.

Rousseau J.J. (1993b). *Emile.* Everyman, London.

Ruivenkamp, G.T.P. (2003). Genomics and food production - the social choices. In: *Genes for your food-food for your genes* (Eds. R. Van Est, L. Haassen and B. Gremmen), The Hague, Rathenau Institute, pp. 19-44.

Ruivenkamp, G. (2005). *Wetenschap in de samenleving: de ontwikkeling van biotechnologie en genomics op-maat.* Inaugurele rede Vrije Universiteit Amsterdam.

Shapin, S. and Schaffer, S. (1989). *Leviathan and the air-pump: Hobbes, Boyle, and the experimental life.* Princeton University Press.

Sunder Rajan. K. (2006). *Biocapital: the constitution of Postgenomic life.* Duke University Press.

Thompson, N. (2001). May the source be with you: can a band of biologists who share data freely out-innovate corporate researchers? *Washington Monthly.* Available at: www.washingtonmonthly.com/features/2001/0207. thompson.html.

Thacker, E. (2003). What is bio-media? *Configurations* 11: 47-79.

Thacker E. (2005). *The Global Genome: biotechnology, politics and culture.* Cambridge MA, MIT Press Books.

Tuck, R. (2003). *The Rights of War and Peace.* Oxford University Press.

Tully, J. (1980). *A discourse on property: John Locke and his adversaries.* Cambridge University Press.

Tully, J. (1991). Introduction. In: On *the duty of man and citizen according to natural law* by S. Pufendorf. Cambridge University Press.

Tully, J. (1997). *Strange multiplicity: constitutionalism in an age of diversity.* Cambridge: University Press.

Waltz, K. (2001). *Man, the state and war: a theoretical analysis.* Columbia University Press.

Wiggershaus, R. (1994). *The Frankfurt School: its history, theories and political significance.* Cambridge, Polity press.

The wiki way: prefiguring change, practicing democracy

Kate Milberry

1. Introduction

What democratic potential does the Internet hold? This is a much-asked question, both within and outside academia. And yet the question remains unanswered, in part because the Internet remains an unfinished and evolving technology. The duality of science and technology – on the one hand its promise for a more humane and just society, on the other, its potential to dominate nature, and therefore humanity - reflects a similar tension between status quo power relations and alternative visions of the future. This tension plays out in the way recent progressive social movements have engaged with new information and communication technologies, in particular the Internet, within a framework of global capitalism. As such, it is not clear whether cyberspace will be fully colonised by corporate forces or whether it will be preserved as a virtual public sphere that can enhance 'real world' democracy. Neither has it been determined if the Internet will be controlled by the state, by its corporate partners or by citizens, although a decidedly less open Internet protocol, IPv6, is currently being tested.

Today, various actors compete for dominance on the web, as the commercialisation of cyberspace continues apace. Among them, activists in the global justice movement[229] (GJM) have appropriated Internet technology in their struggle against the negative impact of corporate capitalism on a planetary scale. Since the eruption of the GJM at 1999's Battle of Seattle, much has been made about the impact of the Internet on progressive activism. Of particular interest have been the ways in which activists have used the Internet as a communication medium, as a forum for information dissemination and as a tool for organising (Deibert, 2000; Kahn and Kellner, 2004; Meikle, 1999; Smith, 2001). Applications like Websites, email and Internet Relay Chat (IRC) have largely facilitated the new movement as a global phenomenon (Bennett, 2004; Van Aelst and Walgrave, 2004). Cyberactivism – political activism on the Internet – is a new mode of contentious action, and new practices such as virtual sit-ins, online petitions and email campaigns have enhanced the repertoire of contention (McCaughey and Ayers, 2003). But what impact have activists have had on the Internet? 'Tech activists' – programmers, coders, and hackers who subscribe to the philosophy of the free software movement yet are committed to the pursuit of a just society – are largely responsible for facilitating the novel combination of interactive digital technology and activism. They are responsible for the design of the virtual infrastructure used by activist groups. But in addition to building and maintaining websites, wikis, web logs, email accounts and mailing lists, these self-described geeks customise free software to meet the needs of activists engaged in the new global activism. In using and developing technology that augments the notion of cyberspace as a virtual public sphere, tech activists enhance the democratic potential of the Internet. Their work, therefore, alters not only the way people 'do' activism; it is changing the face of the Internet itself.

How do we evaluate such a claim? I approach the problem by acknowledging first and foremost that technology is political – both in design and use. I further contextualise the problem historically, considering the origins of critical thought on the interrelation between modern technology and society, noting the inherent tension underlying the human-machine bond. Through the lens of critical constructivism, I then trace the rise of tech activism, which has roots in the free software movement

---

[229] Variously called the anti-globalisation movement, anti-corporate globalisation movement, pro-democracy movement and sometimes simply 'the movement'.

but has cultivated its own ethically grounded and socially informed focus. Finally, I examine how and why tech activists have appropriated wiki technology, using it as a space and tool for democratic communication in cyberspace. In turn, this has enabled the realisation of new communicative practices offline, implying a dialectical relation between the technological and the social. In other words, democratic practice online is prefigured by the fundamental desire for a more just society; actualised as democratic interventions into the development and use of technology, it then manifests in alternative modes of social organisation in the physical world. Feenberg (2005: 49) affirms the dual nature of technology. Where technical action is an exercise of power, it is manifest in designs that reflect and help reproduce capitalist hegemony. However, the wielding of technological power provokes a reaction from those 'who suffer the undesirable consequences of technologies...'. Will 'opening up' technology to a broader range of interests and concerns inspire a radical reform of the technical sphere, as Feenberg suggests? It is with this question in mind that I consider the implications of tech activism for the generative process of Internet technology.

2. (Hu)Man against machine or the duality of science and technology

From the dawn of modernity, the promise of technological advancement has inspired awe and dread in seemingly equal parts. It signaled either humanity's triumph over nature or, conversely, humankind's impending doom. Francis Bacon was the first Western thinker to cut through the mystique of science with his formulation of the scientific method and his identification of the obstacles or 'idols' that confounded the 'true' understanding of nature in the 17th century. More than 200 years later, Bacon's approach finally prevailed. But popular opinion was divided: extravagant claims about the beneficent impact of technological 'progress' competed with fearful predictions of societal breakdown and the corrosion of traditional institutions and values. Eager to shed its religious, social and ethical skins, the new 'scientism' dispensed with the ethos of science identified by Bacon in favour of the promise scientific innovation held for the domination of nature. Leiss (2005: 4) calls this the 'two-sided significance' that science and technology hold for society; it is this duality that produces the 'essential, internal tension in the epoch of modernity' between *inventive science* and *transformative science*.

Leiss (2005) defines *transformative science* as the progenitor of cultural change, encompassing the diffusion of the ethos of the modern scientific method throughout society. It is vigorously challenged and its outcome is highly uncertain. *Inventive science* aspires to conquer nature and is the originator of scientific change. Uncontested in any meaningful way, it promotes a vision of the continual flow of new products and technologies that improve the material conditions of life. Inventive science also gave rise to the 'idols of technology'. Evocative of Bacon's idols, which were rooted in devotion to magic, religion and irrational social convention, these are 'the false notions that have grown up around modern society's fervent commitment to technological progress' (Leiss, 1990: 5). Transformative science – innovation's better half – endured through the end of the 1800s in European culture, maintaining harmony within the project of science. Up until then, the new scientific methods were considered important not only as a toolkit for better understanding nature, but for their potential to positively influence social policy and social institutions (Leiss, 2003).

3. The society-technology disconnect

Today, however, the two sides of the internal tension within science and technology have become unhinged; thus separated, they no longer support and enhance one another. What Leiss (2005) calls the 'cultural mission' of science has faltered. Marcuse (1964) recognises this disconnection between modern technology and social values in his concept of the one-dimensional society. Here dialectical contradiction (the crux of true reason) is flattened and the Platonic *logos* of a technology – its rationale or reference to the good served – is lost. 'The totalitarian universe of technological rationality is the latest transmutation of the idea of Reason' in which logic has become the 'logic of domination' (*ibid.*: 123). Thus technological rationality triumphs as reason - the basis for scientific thought and technical action - becomes unreason in the 'closed operational universe of advanced industrial civilisation' (*ibid.*: 124). For Marcuse, the only way to transcend this situation, this closed universe, is through a 'catastrophic transformation' of society that is at once technological and political. 'The political change would turn into qualitative social change only to the degree to which it would alter the direction of technical progress – that is, develop a new technology' (*ibid.*: 227). Such a qualitative change would facilitate the transition to a more advanced level of civilisation if technologies were designed and used for the 'pacification of the struggle for existence' (*ibid.*). What would emerge, Marcuse posits, is a new idea of reason, one opposed to modern scientific and technological rationality.

Feenberg (2005) similarly acknowledges the imbalance in modern times between the transformative and inventive sides of science – or technology and values – and the resulting tendency of technical action toward domination. 'Technical action is an exercise of power', he argues.

'Where, further, society is organised around technology, technological power is the principle form of power in the society. It is realised through designs [that] narrow the range of interests and concerns that can be represented... This narrowing distorts the structure of experience and causes human suffering and damage to the natural environment' (*ibid.*: 49).

What is necessary, therefore, is a reorientation of the basis for technical production, as indeed Marcuse (1964) notes, in order to create technologies that meet the full range of human needs. In response to the limited interests that feed technical design, a broader inclusion of human concerns into technology development would aid in the 'pacification of the struggle for existence', as well as challenge unequal power relations in society.

4. A critical theory of technology

While Marcuse (1964) suggests a reordering of the technical principles undergirding technological rationality as a way to enhance peace, freedom and human fulfillment, Feenberg (1991) offers another response. He draws from a number of Western critiques of technology, including Heidegger and the Frankfurt School, deftly interweaving technology studies and philosophy of technology. The result is a hybrid theory – critical theory of technology, also known as critical constructivism – that rests on the central theme of democratising technology to enhance human ideals of liberation, equality and justice. Technology, according to Feenberg's thoroughly historicised approach, 'embodies the values of a particular industrial civilisation and especially of its elites, which rest their claims to hegemony on technical mastery' (p. v). He dismisses the technologically determinist insistence on the neutrality of technology, stating that the real issue is not technology itself, but the variety of choices involved at the level of technical design and the numerous potential outcomes of the design process. At the same time, Feenberg points out the asymmetry of power relations between human and machine, or actor and object, suggesting modern technology embodies political values that promote hierarchy and domination. Here, he invokes Marcuse's claim that 'technological rationality has become political rationality' (*ibid.*: 16). But, following the constructivist position, he asserts that technology is subject to conscious social control. Openings for democratic intervention appear during the various stages of the design process, making possible a radically different technology that serves more broadly the needs of humankind.

Democratic control of technology suggests the possibility of an alternative industrial civilisation based on values different than those that currently underwrite global corporate capitalism. The critical theory of technology 'charts a difficult course between resignation and utopia', seeking to explain how modern technology can be redesigned to support a freer society (Feenberg, 1991: 13). Feenberg retains the Frankfurt School insight that the domination of nature - or technological progress – is achieved through social domination. Indeed, as Langman (2005: 48) points out, critical theory is useful as an emancipatory discourse that roots social injustice and human immiseration firmly within the 'rationalised, reified, commodified culture of modern capitalism'. The only remedy postulated by the Frankfurt School is democratic advance, leading to the conclusion that 'the liberation of humanity and the liberation nature are connected in the idea of a radical reconstruction of the technological base of modern societies' (Feenberg, 1991). But critical theory lacks a concrete conception of a 'new technology'; Feenberg's approach seeks to rectify this.

5. Technology as a scene of struggle

According to Feenberg (1991) the technical order is not merely a sum of tools but instead acts to structure the social world in a rather autonomous way. 'In choosing our technology we become what we are, which in turn shapes our future choices. The act of choice is technologically embedded and cannot be understood as a free "use"' (*ibid.*: 14). But critical theory is not fatalist and Feenberg retains this thrust; the future of civilisation is not determined by the 'immanent drift of technology' therefore, but can be, and is, influenced by human agency. Political struggle continues to play an important role, however tenuous and uncertain of success.

In societies organised around technology, such as modern Western nations, technological power is key to the exercise of political power. Feenberg (1991) explains how the ruling elite preserve their power through his concept of the *technical code*. Whereas earlier constructivist notions, like momentum (Hughes, 1987) and path dependency account for certain technological trajectories, the technical code is the embodiment of dominant social principles at the level of technical design. In other words, the technical code translates what are typically ruling class objectives into technical terms; it 'invisibly sediment[s] values and interests in rules and procedures, devices and artifacts that routinise the pursuit of power and advantage by a dominant hegemony' (Feenberg, 1991: 14). A technology reaches closure when disputes over its definition are settled by privileging one over any number of possible configurations; these disputes often involve conflicting ideological visions. Their outcome often aligns the technology with dominant social forces, rendering the technical code a direct reflection of status quo power relations (Feenberg, 1999).

However, the exercise of technical power engenders political resistance as disenfranchised or disempowered users react through resistance or protest. Here the technical code reveals an opening in the closed system of total domination envisaged by the Frankfurt School. Rather than a reified 'thing', technology is cast as an 'ambivalent process of development' (Feenberg, 1991: 14), one pregnant with both liberating and oppressive possibilities. If technology is a process and not a series of finished products, the chance for intervention, and hence change, exists. The ambivalence of technology differs from neutrality in that it finds social values embedded in the design of a technical artifact, as well as in its possible uses. In this way, 'technology is not a destiny, but a scene of struggle. It is a social battlefield', wherein outcomes with weighty implications for civilisation are decided (*ibid.*). Technical devices and systems are indifferent to power; that is, there is no necessary, pregiven correlation between technology and social dominance. This highlights the ambivalence of technology: it can be used just as easily in alleviating the 'struggle for existence' as in dominating humanity.

6. Toward liberation? The internet considered

Feenberg's program for reforming technology to create a freer, more humane society calls for a more inclusive design process. This requires broad democratic participation, which suggests a deeper societal transformation rooted in technology itself. Feenberg (1991) posits an alternative conception of modern industrialism, one that does not rely solely on the current capitalist version of technology: 'A fundamentally different form of civilisation will emphasise other attributes of technology compatible with a wider distribution of cultural qualifications and powers' (*ibid.*: 19). When applied to the Internet as a communication technology, these ideas form an exciting project: the construction of a virtual public sphere, developed and maintained by users, with important implications for democratic practice offline. Indeed, the creation of community through democratic practice in cyberspace prefigures alternative conceptions of social organisation offline. As with many new communication technologies, the Internet was heralded by cyberoptimists as potentially revolutionary, holding new promise for civic participation – even a democratic utopia – online (Rheingold, 1993). This initial euphoria has been tempered by critical analysis, as well as the ever-encroaching corporate presence in cyberspace; nonetheless a community model of the Internet that envisions a virtual space for the development of democracy suggests commercial domination is not inevitable (Feenberg and Bakardjieva, 2004).

These discordant models of the Internet indicate that it is an unfinished project (Feenberg and Bakardjieva, 2004); that is to say, conflicts over its design and meaning have not been resolved. Herein lies the 'two-sided significance' or the dual nature of technology. With its potential to be both inventive and transformative, the future direction of Internet remains dynamic and very much contested. It is unclear whether cyberspace will be sold off to the highest bidder or whether it will be preserved as a place for public communication and interaction. The turf war in cyberspace is still being waged, and actors with competing goals, values and interests continue to battle for supremacy. Thus Internet has not reached closure, nor have the dominant norms of modern western capitalism sedimented into a technical code; both the social and technical definition of the Internet remain at stake. Many possible outcomes are visible on the horizon of the future, making this is an opportune moment to investigate the Internet's emancipatory and democratic potential.

7. Interpreting the internet

Viewed through the lens of critical constructivism, the Internet's contingent nature is apparent. Its development is characterised largely by interpretive flexibility, and the concomitant notion of user agency in the arena of technological design. The Internet was originally conceived as a means for connecting government researchers at various military and academic institutions, enabling them to share expensive computing resources (Abbate, 1999; Ceruzzi, 2003). But it quickly developed into a medium for human communication, demonstrating interpretive flexibility. The designers of ARPANET, the progenitor of the Internet, were also first generation users, and as such, they intervened in the design process in ways that strayed from the official vision of military computer networking. What makes the Internet unique in the history of communication and information technologies is the openness of its design principles – in its standards, its software and its engineering – and the prospects this offers for user agency. This was a deliberate choice of its originators with profound impact on the Internet's social meaning. 'From the very beginning these principles have been understood to have a social as well as a technological significance. They have, that is, been meant to implement values as well as enable communication' (Lemley and Lessig, 2004: 44). The value of openness that characterised the Internet's birth has endured, despite increasing contestation, and is apparent in its ongoing 'invention' (Abbate, 1999).

Alternative conceptions of society such as those sought by the global justice movement will value other attributes of technology, beyond those currently upheld. These include 'the vocational investment of technical subjects in their work, collegial forms of self-organisation, and the technical integration of a wide range of life enhancing values, beyond the mere pursuit of profit or power' (Feenberg, 1991: 19). We see this in the continual development of Internet at its 'content' layer (e.g. applications), undertaken by programmers in the corporate arena and hackers in the free software and open source movements. Tim Berners-Lee, who wrote the original prototype for the World Wide Web in 1990, designed into his application a value contrary to the norms endorsed by capitalist hegemony. 'This space was to be inclusive, rather than exclusive' (in Ceruzzi, 2003: 302). Ironically, it was with the privatisation of the Internet in 1995, and its subsequent release from the exclusive domain of universities and research facilities, along with the popularisation of the personal computer, that this vision of inclusivity seemed ever more likely.

8. Tech activism's radical roots

The Internet is arguably well suited to the task of facilitating alternative, progressive conceptions of society and tech activists in the global justice movement are at the fore of the push to mold it into a medium for democratic intervention. They take seriously the idea that 'another world is possible'[230], and that their activism in the realm of digital communication technology will aid in the quest to democratically reorder modern industrial society (Feenberg, 1991). The current strain of tech activism is the second wave of a movement that emerged in the 1960s as a digital counterculture. Hackers working in the Artificial Intelligence laboratory at Massachusetts Institute for Technology developed the habit of sharing source code based upon a cooperative spirit and a belief that information should be free (Stallman, 1999). They were part of a student culture that took up computer networking as a tool of free communication (and later, a tool for liberation), which included graduate students who largely designed the protocols for ARPANET. As Castells (2001) observes, most of these students were not part of the countercultural movement in the same way as many radical activists of the day. 'And yet they were permeated with the values of individual freedom, of independent thinking, and of sharing and cooperation with their peers, all values that characterised the campus culture of the 1960s' (*ibid.*: 24).

---

[230] Taken as the official slogan for the 2001 World Social Forum, this phrase has become something of a rallying cry for the global justice movement. It is not a vision of a specific other world, as Naomi Klein (2001) astutely observes, simply the idea that, in theory, another one could exist. This contradicts the truism of capitalist hegemony, which states that the current socio-economic configuration of modern Western society is the only possible one, whatever its flaws.

By the 1980s, these values were increasingly marginalised as the computer industry became more and more proprietary. One of the MIT hackers, Richard Stallman, quit the AI lab in response to this change and founded the free software movement in 1984. This was, arguably, the formalisation of a long tradition of openness in the computing community. Ceruzzi (2003) traces the custom of sharing source code as far back as 1955, to the forming of SHARE, a disparate group of programmers who banded together to tackle upgrading their IBM systems. Stallman (1999) took the ethical stance that proprietary software was antisocial and unethical, rejecting the assumption that 'we computer users should not care what kind of society we are allowed to have'. He began developing an operating system, GNU (Gnu's Not Unix) that became complete with the addition of the Linux kernel in 1992 (gnu.org). The movement was based upon four essential freedoms: the freedom to run a program; the freedom to modify a program; the freedom to redistribute copies (gratis or for a fee); and the freedom to distribute modified versions of the program. Because freedom is considered in the context of liberty rather than price, the ability to share source code, and sell a finished program are not necessarily incompatible. The crucial point is that the source code always remains freely available – in proprietary and free software.

9. Free software vs. open source

Freedom, and not simply program development and use, is the central concern of the free software movement, making it an explicitly political project[231]. In this way, it suggests 'a digital revolution that is social before it is technical' (Obscura, 2005). But some in the tech community have purposely avoided the subversive potential of free software. In 1998, Eric S. Raymond launched the Open Source Initiative (OSI) in response to the value-laden approach of the free software movement. Although it assumes an apolitical stance, this movement reveals its bias in its support of the status quo.

The Open Source Initiative does not have a position on whether ideas can be owned, whether patents are good or bad, or any of the related controversies. We think the economic self-interest arguments for open source are strong enough that nobody needs to go on any moral crusades about it... (OSI: *FAQ*).

While the two projects share a similar definition of what constitutes free software, their objectives are different. Activists in the free software movement focus on the user-technology relationship, founded on an implicit critique of corporate capitalism. Proponents of the open source project strive to facilitate the development of superior software through access to the source code, in alliance with capitalist hegemony.

In an effort to appear business-friendly, the Open Source Definition 'logically abandoned all reference to the social and ethical means and motives of free software, not to mention the fight for freedom as a primary aim' (Obscura, 2005). The Open Source Initiative does not disguise its efforts to make free software more compatible with capitalist discourse, describing itself as 'a marketing program for free software. It's a pitch for 'free software' on solid pragmatic grounds rather than ideological tub-thumping. The winning substance has not changed, the losing attitude and symbolism have...' (OSI, *FAQ*). For free software advocates, however, it remains about the ethics surrounding software use and development – what Stallman (1999) calls community practice and values. This vision extends beyond

---

[231] Another political project founded in defense of freedom on the Internet is the Electronic Frontier Foundation. Begun in 1990, the EFF works to protect the public interest in legal battles over digital rights in cyberspace. A discussion of this group, however, is beyond the scope of this chapter. See www.eff.org.

the computer industry and embraces the ideal of a better world. According to Stallman, some perceived a threat in this challenge to the status quo:

'Talking about freedom, about ethical issues, about responsibilities as well as convenience, is asking people to think about things they might rather ignore. This can trigger discomfort, and some people may reject the idea for that. It does not follow that society would be better off if we stop talking about these things.'

Despite its broad political program, the free software movement represents a minority of the tech community, which drifted away from its more radical origins and is today largely apolitical.

This divide within the tech community recalls the 'essential, internal tension' of modernity, the disconnect between *inventive* science and technology, with its focus on innovation, and its *transformative* counterpart, which inspires cultural change. The Open Source Initiative, operating on the linear model of progress, supports the development of software technology based on a proprietary system that underwrites modern capitalist hegemony. The free software movement, however, offers a working example of an alternative social model, one based on decentralisation, volunteerism, cooperation and self-empowerment, with the ultimate goal of creating a freer society. It is an example of what Feenberg (1999) calls *democratic rationalisation*, the use of new technology (software) to undermine the existing social hierarchy. Put another way, democratic rationalisation highlights the political implications of user agency for technical design, suggesting the possibility of organising society in ways that enhance democracy, rather than capitalist efficiency and control. In this case, democratic control of software suggests a different Internet and, broadly considered, a different world.

10. Second wave tech activism: repoliticising technology

The resurgence in tech activism in the early 2000s rested firmly on the foundation laid by the free software movement. It is unsurprising, then, that a similar rift exists between tech activists in the global justice movement and the generally apolitical advocates of open source. While both projects share an affinity for collaboration and coordination, with geeks often moving easily between the two, their political, philosophical and technical motivations differ. Programmers working on open source projects are rewarded by the creative expression, intellectual stimulation and improvement of technical skills acquired through programming (Lakhani and Wolf, 2005). Similar rewards may also inspire tech activists in their work but there is no question as to their overarching motivation: 'technical means are directed toward political ends' (Coleman, 2004). These political ends include the pursuit of social, economic and environmental justice under the auspices of the GJM. This shift in focus signals a return to the radical tradition of the free software movement and the repoliticisation of computer technology.

The reclamation of computer technology as a political frontier for contentious action is a hallmark of the global justice movement. The GJM comprises the latest wave of social justice activism, and seized the world's attention at the 'Battle of Seattle', 1999's massive street protest against the World Trade Organisation. Here, upwards of 50, 000 activists from a variety of cultural, ethnic and political backgrounds formed an unprecedented alliance, united by their common opposition to the debilitating effects of neoliberal globalisation, a world economic policy that has generated massive profits for a minority of the world's population at the expense of labour and human rights, environmental sustainability, democratic practice and national autonomy (Langman, 2005). In the face of increasing corporate dominance, there was increasing resistance, and a movement of movements swelled, embracing the vision of a people's global justice. This movement also produced an analysis that historicised their struggle, and therefore denaturalised neoliberal globalisation: the global march of capital was not inevitable (nor inevitably 'good') thus human intervention was possible. Activists quickly realised the potential and power of the Internet for their burgeoning movement, beginning in Seattle, and continuing with other major citizen protests and people's summits at subsequent meetings of institutions of global economic power brokers such as the G8, International Monetary Fund and World Bank. The Internet facilitated the organisation of campaigns and movements into 'super movement spheres' (Morris and Langman, 2002), and enabled activists to communicate and mobilise without previous time, space and cost barriers. The importance of the Internet for the new global activism was further underscored by the creation of the Independent Media Centre (IMC), a web-based network of radical media making collectives that went live for the Seattle protest.

Tech activists have been central to the global justice movement since its inception, facilitating the novel combination of interactive digital technology and social justice activism, and bridging the divide between geek and activist communities. While their programming skills distinguish their contribution, tech activists share in the movement's overarching goals of social justice. One IMC geek summed it up this way: 'I belong to a movement which strives for equal rights (not the written but the real ones) and conditions for all humans (and partially other beings, too) on this planet'[232]. IMC – also called Indymedia – was initially founded to give voice to activists' concerns during the anti-WTO demonstrations; indeed, Indymedia's mission statement reflects its origins in the GJM:

> '[IMC is] committed to using media production and distribution as tools for promoting social and economic justice. Through this work, we seek to...illuminate and analyse local and global issues that impact ecosystems, communities and individuals. We are dedicated to generating alternatives to the corporate media and to identifying and creating positive models for a sustainable and equitable society (Seattle IMC, nd).'

While IMC was the dream of media activists, it was the geeks in the movement who developed and implemented the code to realise that dream. In particular, the innovation of open publishing software enabled anyone with an Internet connection to upload stories and images to the website, bypassing the gatekeeping function of editing and subverting journalistic norms. Indymedia thus emerges as more than an experiment in radical media making: it is clearly an example of the democratic rationalisation of the Internet – activists appropriating Internet technology to not only challenge the dominant ideology (neoliberal globalisation), but to foster alternative visions of social organisation.

11. The birth of Indymedia

There are numerous examples of tech activism, such as the construction and maintenance of activist websites (including mailing lists, email accounts and other functionalities), the refurbishing of old computers for distribution in technology poor areas/nations, and the hosting of hacklabs[233] and other tech training events. Tech activists are also responsible for setting up media centres for major street

---

[232] Personal communication with Alster, 2 December 2005.

[233] Hacklabs are political spaces (often temporary) that provide community computer and Internet access. They are used for independent media, the promotion of free software and other emancipatory technologies. Here tech activists share skills with one another and the broader public. For example, see www.hacklab.org.

demonstrations and during natural disasters, such as Hurricane Katrina[234]. But Indymedia is arguably the most prominent, and perhaps best, example of tech activist work done under the banner of the global justice movement. The building of the first IMC in Seattle now approaches legendary status. The inaugural post, by founding geeks Manse Jacobi and Matthew Arnison, acknowledges the novelty of the new movement; on 24 November 1999, they wrote: 'The resistance is global... a trans-pacific collaboration has brought this web site into existence'[235]. But it was activists' prior use of the Internet as a communication tool that enabled the global resistance to unite in one locale.

Another geek, Evan Henshaw-Plath, took part in the birth of Seattle IMC, which he had heard about from a friend of a friend at a pre-protest party. He describes the scene as 'packed and hectic', with techs scrambling to shore up the server and code before the protests began:

'Almost the instant I walked in to the Indymedia Center I had caught the IMC bug. Without knowing the organising structure, extent of the projects, political background, I could experience the energy. I worked all night on the server and throughout the day of the protests. My experience of the protests was just a half hour when I managed to escape in to the streets ...'[236].

Since helping found seattle.indymedia.org, Henshaw-Plath, has been involved with dozens of IMC locals and wrote some of the code that would be incorporated into the open publishing platform that made Indymedia (in)famous. The first open publishing tool, Active, was originally coded by Australian tech activists. 'Open publishing is the same as free software', notes Arnison (2002: 329), one of Active's developers. They're both (re)evolutionary responses to the privatisation of information by multinational monopolies'. As with free software, open publishing enables the free distribution and exchange of information – in its case, news stories. The process of creating news, like that of developing code, is made transparent by open publishing software; readers can contribute and redistribute stories, see and get involved in editorial decision-making, or copy and develop the software to address a shortcoming.

The choice of free software for the implementation of the global site, indymedia.org, was deliberate, and suggests a philosophical inheritance from the free software movement, if not direct lineage. It also shows with clarity the project's political objectives. At present, all the software on the global network, which includes more than 130 'nodes', is by charter free software. Throughout Indymedia's nine-year history, free software has enabled the IMC tech collective to develop applications 'that encourage cooperation, solidarity, an equal field of participation' in their brand of radical media making (Henshaw-Plath, 2002). In late 2001, the IMC Tech Collective discussed the rationale for committing to free software: 'It's clear that the technology we use and process by which it's constructed and articulated is deeply political. We are creating the technical systems that prefigure the change we want to see in society' (Henshaw-Plath, 2001).

Tech activists thus understand coding as technical process with social implications. While they make an explicit attempt to imbue software with ideals that mirror their social justice goals, tech activists

---

[234] In Houston, Indymedia and low power FM radio activists set up a disaster information radio station. New Orleans IMC offered breaking coverage and activists set up a media centre in Algers, a portion of the city that did not flood from the levee breaches. IMC USA created a topical site, Katrina.indymedia.us.org, which carried news from across the Indymedia network (http://www.anarchogeek.com/articles/category/indymedia).

[235] For the full transcript, visit http://seattle.indymedia.org/en/1999/11/2.shtml.

[236] Interview with Evan Henshaw-Plath, 28 July 2003.

never lose sight of the social purpose of the software, nor of the user-technology relation. In the case of the continual hacking of Active[237], 'the geeks of IMC-Tech were keenly aware that each technological design or set of features creates a particular publishing structure and, as a result, empowers users ... in an equally particular way' (Hill, 2003: 2). Here we see how users can intervene in technical design to transform a technology, making it more inclusive of human values and needs, which is central to democratising technology. Thus in their software development, tech activists demonstrate insight into the power asymmetries inherent in capitalist socio-technical systems, as well as the knowledge that such asymmetries are both socially constructed and reflective of inequality in the broader social context. With Indymedia, it is apparent that the social and technical are tightly coupled; IMC geeks consciously attempt to create a technical environment that promotes equality and democracy and that, in turn, supports the social change goals of Indymedia, as well as the broader global justice movement.

12. Wild wild wikis: the latest frontier

Tech activists combat power imbalances in the technical sphere through their development and use of free software. Thus they carve out their own virtual terrain oriented toward the community model of the Internet, which is based on democratic practice (Feenberg and Bakardjieva, 2004). Recognising communication as key to achieving the goals of the global justice movement, activists created their own media system. Indymedia's philosophy is summed up in the now-famous slogan: 'Become the media'. However, it soon became apparent that the importance of communicating movement ideals of social, economic and environmental justice through a global digital newswire depended upon internal communication within Indymedia. The IMC tech collective initially communicated by email lists and Internet Relay Chat (IRC). By 2002, however, a number of wikis were set up in an effort to create a sustainable system for documenting IMC's history and ongoing activities. As one member of the Docs Tech Working Group observed: 'Getting a functioning and used wiki is really vital for the network ... Email lists just aren't cutting it for the level of organising and information exchange and growth we need to help facilitate'[238]. Techs maintaining the global site needed a virtual workspace with a constant online presence, where they could jointly yet asynchronously work on common projects and tasks. In addition to facilitating workflow, the wiki had the benefit of constructing and cohering an online community of programmers interested in contributing their skills to the global justice movement.

Wiki software originated in the mid-90s in the design pattern community as a means of writing and discussing pattern languages. Ward Cunningham invented the name and concept and implemented the first wiki engine in 1995. Because of its speed, he named the system wiki-wiki, a Hawaiian term meaning 'quick'. According to Cunningham and Leuf (2001: 14), 'a wiki is a freely expandable collection of interlinked Web 'pages', a hypertext system for storing and modifying information – a database where each page is easily editable by any user with a forms-capable Web browser client'. Plainly put, it is a series of linked, dynamic web pages that can be created, edited and deleted by any logged-on user. All changes are recorded; thus the wiki documents its own history, and stores it for future viewing. By the end of the 1990s, the business community had embraced wikis as a 'conversational knowledge management solution' to foster an efficient and collaborative work process (Gonzalez-Reinhart, 2005: 5). In the business environment, wiki use can eliminate the need for conference calls, emails, discussion forums and instant messaging. As with physical communities, the virtual community facilitated by a

---

[237] See Hill (2003) for a history of open publishing software development within IMC.

[238] John Windmueller posting a comment to the Indymedia Documentation Project Wiki, http://docs.indymedia.org/view/Sysadmin/ImcDocsReplaceWikiEngine.

wiki fosters socialisation and information exchange, which in turn encourage collaborative knowledge creation (*ibid.*).

For tech activists, building a community that jointly created and maintained knowledge via wiki technology was a breakthrough. But the implications of this new social software went beyond quick communication, increased productivity or cost/time savings. In essence, what IMC geeks discovered in the wiki was a new mode of communication. The concept of the wiki rests on the notion of collaboration, which in turn is based on trust. According to Cunningham (nd), trustworthiness is a principle that inspired his initial wiki design, and is built into the software's technical code. 'This is at the core of wiki. Trust the people, trust the process, enable trust-building'[239]. Wikis encourage trust because their ability to function is based on the assumption that participants have good intentions; the open-ended power to add, delete or alter content makes a wiki vulnerable, and dependent upon ethical conduct. Thus, as with any well-functioning community, a wiki is heavily reliant on norms of social behaviour.

Wikis can be used to communicate and exchange information with others in much the same way as online discussion forums and email lists. Uniquely, however, wikis create a virtual arena for project organisation and documentation. Open editing allows for the collective authorship of material as well as co-production of the website in a way that other conversational Internet applications do not. The intent is to foster communal development in a virtual space that is jointly owned by all users, and for which all users are responsible. This accounts for the organic nature of a wiki page, where content changes as users add missing information, correct mistakes and delete erroneous or unnecessary material. In this way, the knowledge jointly produced in a wiki improves and grows over time. The 'link as you think' feature, whereby a contributor creates links to existing and potential pages in a wiki, is one example of this organic collaborative knowledge production. It is a critical and deliberate design element that fosters the creation of a shared language. This shared language emerges instinctively and is fundamental to effective communication within a wiki (Kim, 2005). According to one tech activist, the 'link as you think' feature is 'a way of building a community-specific vocabulary that allows you to easily formulate complex thoughts by using the terms your community thinks are important' (Schroeder, 2005).

Importantly, the wiki enables them to enact the social change they seek in the broader society. Here, democracy, equality and justice switch from being abstract ideals to concrete social practices. At the same time, wiki software is part of the digital infrastructure tech activists build and maintain in order to achieve more immediate movement goals, and as such is represents only one tool in the activists' repertoire of contestation. Considered thus, wikis emerge as an ideal mode of communication for distributed networks like Indymedia and the global justice movement, where participants from disparate geographical locales, with varying skill and commitment levels, as well as ethnic, class and technical backgrounds, work together toward a shared vision of a better world.

13. IMC meets TWiki

Indymedia made early use of wiki technology for the Global Indymedia Documentation Project, which gathers collective knowledge about IMC's history, its current role(s) and its short and longterm goals. Documenting their project is vital to the success of Indymedia; not only does it provide a public record, it creates a fluidity that facilitates participation at varying levels. 'The Indymedia Documentation Project

---

[239] For more on Ward Cunningham's wiki design principles, see http://c2.com/cgi/wiki?WikiDesignPrinciples.

looks like a normal Web site... except that it encourages contribution and *editing* of pages, questions, answers, comments and updates' (IMC: *Welcome*). Importantly, participants are not required to know how to code in order to add, change or delete content. Because Indymedia is predominantly a web-based project, implementing a wiki addressed the persistent problem of how to organise communication within the disorganised environs of cyberspace. While mailing lists facilitated information exchange, and IRC enabled real time discussion, neither application provided a collaborative space where Indymedia volunteers could work asynchronously on common projects. Wiki technology appealed to IMC geeks because of its ability to facilitate information flow, which allowed distributed teams to work together seamlessly and productively, and eliminated the one-webmaster syndrome of outdated content.

In 2002, IMC techs adopted TWiki, a free software wiki clone aimed at the corporate intranet world, assembling a number of separately running wikis in one website, docs.indymedia.org. Today it is one of the largets TWiki installations on the World Wibe Web. The Documentation Project wiki is divided into sections made up of topic-based webs that contain links to the various working groups, documents and materials needed to understand, navigate and participate in the Indymedia multiverse. The Tech section is the home of the IMC Global Tech Team and features a variety of working groups focused on the numerous technical aspects of the Indymedia project, including system administration, IRC, security, mailing lists, and so on. There is also an FAQ, and information about Indy software and how to get involved in the tech team. Logs from past meetings, as well as drafts of policy proposals, are also stored here. The wiki's usefulness as a forum for discussing technical issues of varying degrees of importance to the smooth running of the network also becomes clear, with policy documents, proposals and meeting logs creating an invaluable store of cumulative knowledge.

While the Docs Project wiki has opened up a new mode of communication for IMC volunteers, and the tech activists that maintain the global site, it is not without challenges. A common concern about the openness of the software is the fear of vandals who delete or deface content, either in sport or from spite. Indeed, the open philosophy does not protect the site from ill-intentioned users. But wikis are designed to make it easy for users to correct mistakes (rather than making it difficult to make them), thereby providing ways to insure the validity of content despite the ease of modifications. Most wikis have a 'recent changes' page that records the latest edits, or all changes made within a specific timeframe. 'Revision history' shows previous page versions, and the 'dif feature' highlights the changes between two versions. This allows users to deal swiftly with attacks such as wiki spam or insults, correcting and malicious changes or restoring older, more appropriate content. On a small wiki, it typically takes more effort to vandalise a page than to revert it to an acceptable version. On a large installation like the IMC Docs Project, vandalism can be more of a nuisance, creating daily, tedious work. From March to September, 2006 Indymedia was unable to keep up with regular maintenance of the wiki, and the tech team disabled the editing function, rendering the site read-only. This, however, had more to do with deeper problems plaguing IMC as a globally distributed, volunteer-run collective, including activist burnout, limited resources and conflict over best practices, than shortcomings in wiki technology itself. In any case, the 'infinite undo' function offers technical insurance that no modification is ever permanently destructive (Lih, 2004: 10).

14. The emancipatory power of wikis?

What, then, are the implications of wikis for tech activism in today's global justice movement? Ebersbach-Markus Glaser (2004) assesses the emancipatory power of wikis, concluding that participating in a wiki is a political act with consequences that extend beyond cyberspace. The egalitarian structure of the wiki is based on decentralisation of authority and horizontal self-organisation. Much like Indymedia, wherein the gatekeeping power of editors and news producers to control the flow of information is obliterated, 'wikis are administered by a group of people with equal rights who control each other and whose work and decisions are subject to all users' discussion' (*ibid.*: 4). This egalitarian structure is characteristic of the GJM, which eschews formal leadership and is configured rhizomatically in loose networks of autonomous nodes. Decentralisation of power is critical for undermining the social hierarchies that define modern capitalist societies, where the few rule over the many. In modern Western capitalism, this elite minority typically dominates the production of information (as well as technology), with the majority of citizens relegated to the passive, disempowered role of perpetual consumer. In a wiki, there are no access barriers: as with Indymedia, producers of content are its consumers, and *vice versa*.

The elimination of access barriers facilitates participation in wikis as does the purposely designed ease-of-use. 'As you edit there is very little to get in the way of clear thinking and writing...The easier we can make a wiki to use, the more participants we can attract and the greater the value of the system' (Why Wiki Works, nd). Participation is further enhanced by the self-organisation that wikis require, which in turn leads to empowerment. 'Everybody feels that they have a sense of responsibility because anybody can contribute' (*ibid.*). A community grows up around well-used wikis, and users are invested in keeping their wiki relevant and functional. As discussed above, this is largely due to the collective production of content. In the process of organising their wiki, users discover shared interests and begin work on common projects that reflect the concerns and needs of the community, and that promote social cohesion in the virtual environment. Key to this collaboration is the feedback generated through the wiki's interactivity. Unlike the dominant communication technologies of radio and television, the *Internet* is highly interactive. Building upon this functionality, wiki software enables more than adding comments to existing content, as in a weblog, chatroom or email exchange, it facilitates the complete restructuring of the entire website, including its deletion. If modifications are not deemed an improvement, however, they are easily 'undone' by other users. This interaction of users with each other (via content changes) for the broader good of the wiki contributes to the community model of the Internet as a space for democratic practice.

The wiki is a social and organisational phenomenon that contrasts modern western society and prefigures alternative conceptions of social organisation, making their subversive political implications clear. The process of refining and defending views in a collaborative context leads to a deeper understanding of complex ideas, an understanding with the potential for application in the 'real world'. As Ebersbach-Markus Glaser (2004: 7) observes, 'the recognition of this might lead some people to take the organisation of work in a wiki as a model that could succeed in the real world as well'. The 'wiki way' of self-organisation and collaboration produces high quality work without capitalist incentives like competition or money, revealing other ways to live with and value technology not currently promoted by the dominant social order. The 'two-sided significance' of technology – its Janus nature of innovation and transformation – is thus evident in the wiki. The Platonic logos – the rationale for the good – as well as the Baconian ethos are realised in the design process, which in turn informs technological use. In technical terms, the wiki represents an advancement in digital communication; but in social terms, it both models and facilitates new modes of social organisation. Feenberg conceptualises Marcuse's call for a new rationality in just this way – democratically, as customising technology to fit human needs.

15. Conclusion

The Internet remains an unfinished and contested technology in that it is still subject to intervention and transformation by users. Tech activists in the global justice movement bridge the divide between geek and activist communities, creating and maintaining the digital infrastructure that supports progressive activism on a planetary scale. Through their free software development, tech activists deliberately oppose the commercial take-over of cyberspace and adapt it to democratic purposes. In the case of Indymedia, tech activists redeployed wiki software to facilitate movement goals – by creating a public space for online collaboration and by challenging inherent power inequities reflected in the broader society. The wiki's open and decentralised structure mirrors that of the GJM (and the Internet, for that matter) and remains in direct opposition to dominant societal norms based on capitalist hegemony. It is social software that prefigures progressive social change, hinting at more egalitarian, humane ways of organising our modern industrial world. It is also free software, and as such, it is indicative of how tech activists are working at the level of technical design to 'open up' Internet technology to a wider range of interests and concerns.

Viewed from a critical constructivist perspective, tech activists comprise a relevant social group that is but one node in the Internet actor-network. Through their free software development, activist geeks are contributing to the reconstruction of the Internet from a 'communication medium [to] a lever of social transformation' (Castells, 2001: 143). Indeed, a battle lies ahead for control over this virtual frontier. As such, the Internet displays interpretive flexibility – that is, it is used and understood differently by a variety of relevant social groups, as the case of tech activists demonstrates. Further, the work of tech activists may be considered an attempt to address the duality of science and technology - the internal tension between social transformation and technological invention that together comprise the modern notion of 'progress'. In their work, tech activists strive to reconnect technology with its logos – the rationale for the good served. In doing so, they remind us that technology matters, that it is political, and that it is a scene of constant struggle. Does this indicate, or contribute to, a radical reform of the technical sphere? It remains to be seen. But it certainly offers hope that another world is possible.

References

Abbate, J. (1999). *Inventing the Internet*. Cambridge, MA: MIT Press.

Arnison, M. (2002). Open publishing. *Sarai Reader 2002: The Cities of Everyday Life*. pp. 329-333.

Bennett, W.L. (2004). Communicating global activism: Strengths and vulnerabilities of networked politics. In: *Cyberprotest: New media, citizens, and social movements* (Eds. W. van de Donk, B.D. Loader, P.G. Nixon and D. Rucht), London and New York, Routledge, pp. 123-146.

Castells, M. (2001). *The Internet galaxy: Reflections on the Internet, business, and society*. New York, Oxford University Press.

Ceruzzi, P. (2003). *A history of modern computing*. 2nd edn. Cambridge, MA, MIT Press.

Coleman, B. (2004). Indymedia's Independence: From Activist Media to Free Software. *PlaNetwork Journal* 1(1). Available at: http://journal.planetwork.net/article.php?lab=coleman0704.

Cunningham, W and Leuf, B. (2001). *The Wiki way*. Boston, MA, Addison-Wesley.

Deibert, R.J. (2000). International plug 'n play? Citizen activism, the internet and global public policy. *International Studies Perspectives* 1: 255-272.

Feenberg, A. (2005). Critical theory of technology: An overview. *Tailoring Biotechnologies*, 1(1): 47-64.

Feenberg, A. (1999). *Questioning technology*. London and New York, Routledge.

Feenberg, A. (1991). *Critical theory of technology*. New York, Oxford University Press.

Feenberg, A. and Bakardjieva, M. (2004). Consumers or citizens? The online community debate. In: *Community in the digital age: Philosophy and practice* (Eds. A. Feenberg and D. Barney), Lanham, Rowman & Littlefield, pp. 1-28.

Ebersbach-Markus Glaser, A. (2004). Towards emancipatory use of a medium: The wiki. *International Journal of Information Ethics* 2: 1-9. Available at: http://www.i-r-i-e.net/inhalt/002/ijie_002_09_ebersbach.pdf.

GNU. (nd). Overview of the GNU system. Available at: http://www.gnu.org/gnu/gnu-history.html. Retrieved 22 November 2005.

Gonzolez-Reinhart, J. (2005). Wiki and the wiki way: Beyond a knowledge management system. Available at: http://www.uhisrc.com/FTB/Wiki/wiki_way_brief%5B1%5D-Jennifer%2005.pdf. Retrieved 5 December 2005.

Henshaw-Plath, E. (2002). Proposal to reform www.indy by highlighting local IMCs. Available at: http://internal.indymedia.org/front.php3?article_id=538. Retrieved 29 April 2003.

Henshaw-Plath, E. (2001). IMC-Tech summary for November 16th 2001. Available at: http://archives.lists.indymedia.org/imc-summaries/2001-November/000028.html. Retrieved 28 November 2005.

Hill, B.M. (2003). Software, politics and Indymedia. Available at: http://mako.cc/writing/mute-indymedia_software.html. Retrieved 25 November 2005.

Hughes, T.P. (1987), The evolution of large technological systems. In: *The social construction of technological systems*. (Eds. W.E. Bijker, T.P. Hughes and T.J. Pinch), Cambridge, MA, MIT Press.

Indymedia (nd). Welcome guest. Available at: http://docs.indymedia.org/viewauth/TWiki/WelcomeGuest. Retrieved 2 December 2005.

Kahn, R., and Kellner, D. (2004). Virtually democratic: Online communities and Internet activism. In: *Community in the digital age: Philosophy and practice* (Eds. A. Feenberg and D. D. Barney), Lanham, MD, Rowman & Littlefield, pp. 183-200.

Kim, E.E. (2005). The brilliant essence of wikis. Available at: http://www.eekim.com/blog/2005/09/. Retrieved 21 September 2005.

Klein, N. (2001). World Social Forum - A fete for the end of history. *The Nation*, 19 March 2001. Available at: http://www.nadir.org/nadir/initiativ/agp/free/wsf/fete.htm.

Lakhani, K.R. and Wolf, R.G. (2005). Why hackers do what they do: Understanding motivation and effort in free/open source software projects. In: *Perspectives on free and open source software*. (Eds. J. Feller, B. Fitzgerald, S. Hissam and K.R. Lakhani), Cambridge, MA, MIT Press.

Langman, L. (2005). From virtual public spheres to global justice: A critical theory in internetworked social movements. *Sociological theory* 23(1): 42-74.

Leiss, W. (1990). *Under technology's thumb*. Montreal and Kingston, McGill-Queen's University Press.

Leiss, W. (2003). The dual role of science. Paper delivered at the Canada-United Kingdom Colloquium 'Science and public policy,' November 2003, South Glouscestershire, England. Available at: http://www.leiss.ca/index.php?option=com_content&task=view&id=89&Itemid=48. Retrieved 3 January 2007.

Leiss, W. (2005). The dual role of science. Available at: http://www.leiss.ca/images/stories/Articles/dual_role_of_science.pdf Retrieved 3 January 2007.

Lemley, M.A. and Lessig, L. (2004). The end of end-to-end: Preserving the architecture of the Internet in the broadband era. In: *Open architecture as communications policy*. (Ed. M.N. Cooper), Stanford, CA, Stanford Law School. Available at: http://cyberlaw.stanford.edu/blogs/cooper/archives/openarchitecture.pdf.

Lih, A. (2004). The foundations of participatory journalism and the Wikipedia project. Paper presented at Association for Education in Journalism and Mass Communication, Toronto, Canada, 7 August 2004. Available at: http://jmsc.hku.hk/faculty/alih/publications/aejmc-2004-final-forpub-3.pdf. Retrieved 10 November 2005.

Marcuse, H. (1964). *One dimensional man*. New York, Oxford University Press.

McCaughey, M. and Ayers, M.D. (2003). *Cyberactivism: Online activism in theory and practice*. New York, Routledge.

Meikle, G. (1999). *Future active: Media activism and the Internet*. New York, Routledge.

Morris, D. and Langman, L. (2002). Networks of dissent: A typology of social movments in a global age. Paper presented at International Workshop on Community Informatics, Montreal, Canada, 8 October 2002. Available at: http://www.is.njit.edu/vci/iwci1/iwci1-toc.html. Retrieved 2 December 2005.

Obscura, V. (2005). From free software to street activism and vice versa: An introduction. Availabale at: http://garlicviolence.org/txt/drkvg-fs2sa.html. Retrieved 25 November 2005.

Open Source Initiative. (nd). FAQ. Available at: http://www.opensource.org/advocacy/faq.php.

Rheingold, H. (1993). *The virtual community: Homesteading on the electronic frontier*. Reading MA, Adison-Wesley.

Schroeder, A. (2005). Comment: The brilliant essence of wikis. Available at: http://www.eekim.com/blog/2005/09/. Retrieved 21 September 2005.

Seattle IMC. (nd). Press pass policy. Available at: https: //docs.indymedia.org/view/Local/SeattleIMCPressPassPolicy. Retrieved 5 January 2007.

Smith, J. (2001). Cyber subversion in the information economy. *Dissent* (spring 2001): 48-52.

Stallman, R. (1999). The GNU Project. Available at: http://www.gnu.org/gnu/thegnuproject.html. Retrieved 22 November 2005.

Van Aelst, P. and Walgrave, S. (2004). New media, new movements? The role of the Internet in shaping the 'anti-globalization' movement. In: *Cyberprotest: New media, citizens, and social movements* (Eds. W. van de Donk, B.D. Loader, P.G. Nixon and D. Rucht), London and New York, Routledge, pp. 123-146.

Why wiki works. (nd). Available at: http://www.c2.com/cgi/wiki?WhyWikiWorks. Retrieved 25 November 2005.

Tailoring rights regimes in biotechnology: introducing DRIPS next to TRIPS

Niels Louwaars

1. Introduction

Many articles in this book deal with the potential and actual contributions of biotechnologies, including genetic modification to development, notably the reduction of poverty, hunger and malnutrition. Key elements are the empowerment of the poor in the agenda setting of formal research, and the effective linkages of technology in local innovation systems. This is extremely relevant, but one aspect has almost systematically been ignored in the contributions so far; the impact of rights over these technologies and the materials that they use and produce over the legal use of the products of this research by the underprivileged.

We argue that tailoring biotechnologies cannot succeed when the rights regimes accompanying biotechnologies are not tailored at the same time. This article intends to provide an overview of issues involving various rights systems, including private, community and national rights over technologies and genetic resources, being either inputs into or products of research.

This will not go into details of law and country specific legislation. We rather broadly make an inventory of legal bottlenecks to reaching the potential contribution of biotechnologies to the development goals, what options states have in tailoring their legal systems towards development and what opportunities there may be at the implementation level, including institutional policies of public research institutions that support the fulfillment of their roles in this respect.

2. Biotechnologies: opportunities and risks, and rights and obligations

The biotechnology debate is dominated by genetic modification. Opportunities are framed in terms of widening the genetic base of plant breeding through overcoming natural crossing barriers in plant breeding, and opening the black box of mendelian plant breeding through the use of intimate knowledge of gene identification and functioning, and the unraveling of complex metabolic pathways. Molecular biology can both help in understanding the limiting factors in crop production at the genetic level, and in solving such problems.

Risks are commonly associated with food safety and environmental issues, which indeed are complex scientific problem. This is particularly so when the modified plants are released in smallholder farming systems where the sharing of seed is culturally embedded and the crops enter complex ecologies. This means that once released, such introductions cannot be recalled. However, these are still just scientific problems that can in principle be solved with scientific means.

A potential problem of a different order is a legal one: who owns the technology, and how does that affect the potential impact on the development goals. There are several sides to this question. Most obviously, there is the patent system. Biotechnology introduced the patent system in the plant breeding sector. A court case on a patent application in 1980 on a modified bacterium led to an ever wider interpretation of the patent system in the USA, which now allows the protection of almost any new invention in the sector, including genes, research tools, diagnostics and plant varieties (also when bred conventionally). Many countries followed, except for this last category – almost all countries exempt plant varieties from the patent system and have a softer protection regime instead, called plant breeder's rights or plant variety protection. Patents and other intellectual property rights provide for an exclusive right, i.e. that the right holder can decide on the commercialisation of the invention for a fixed time period of time. This is commonly exercised through the granting of licenses to producers or marketing organisations to use the protected subject matter against the payment of a royalty or the provision of other benefits to the right holder. Obligations are few, and mainly involving measures that avoid misuse of the monopoly rights in the market.

Other rights in biotechnology are derived from national sovereignty over genetic resources as laid down in the Convention on Biological Diversity (1992) and communal rights over traditional knowledge, including Farmers' Rights. These rights over the building blocks of biotechnology, the genes and associated knowledge that the biotechnologist uses and tries to understand and manipulate may be called genetic resource rights. Such rights may also lead to a kind of license contracts, called Material Transfer Agreement. Biotechnologists and breeders alike have to obtain prior informed consent over the use of the genetic resources and the contract includes the mutually agreed terms which commonly specify a sharing of benefits. A nation can thus grant access to genetic resources to some and exclude others. The main obligation of the CBD is that the states should conserve their genetic resources and promote their sustainable use.

It is important to note that the various rights systems that the biotechnologist has to deal with 'reach through' to users down the line. Scientists, breeders, seed producers, and finally farmers have to take such rights into account and are commonly not allowed to share the technologies or the seeds with other scientists, breeders, seed producers and farmers without the consent of the right holder. Rights over biotechnologies thus mean that the advances of science will not easily flow from research labs to application when the right holder has the intention of obtaining benefits (both on intellectual property rights and genetic resource rights). This will involve license negotiations and users who can promise benefits are likely to get a license.

3. Origins of these rights systems

3.1. IPRs

On the one hand, IPRs have a moral basis, which is laid down in Article 27 of the Universal Declaration of Human Rights: 'the right to the protection of the moral and material interests resulting from any scientific, literary or artistic production of which he is the author'. The moral grounds date back to the principle of natural law by John Locke that according to Jeremy Betham need specific protection by the state that should secure the inventor a fair share of the reward (Andersen, 2004). Current thought is though that this moral right needs to be balanced with the rights 'to take part in cultural life' and 'to enjoy the benefits of scientific progress and its applications' laid down in the International Covenant on Economic, Social and Cultural Rights of 1976 (Chapman, 2000).

The economic approach is on the other hand that IPRs are a means to increase welfare in society. Legal rights should provide incentives for inventors and authors to invest in their work and produce useful products or insights. This aspect is reflected for example in the 'industrial application' or 'use' requirements for new inventions in the patent system. The US constitution: 'Congress... promote the progress of science and useful arts by securing for limited times to authors and inventors the exclusive right to their respective writings and discoveries'. This phrase illustrates that in order to increase welfare, society needs to put limitations to the rights. In this sense, IPRs can be considered a contract between the inventor/author and society (Hardon, 2004) in which the rights are granted under particular conditions, e.g. the obligation to publish the invention for the benefit of the further advancement of science, and for effective use in the public domain after the expiry of the right, and the right of society to retaliate misuse of the exclusive right in the market through compulsory licences.

An important practical argument behind Intellectual Property Rights is that creative products tend to be non-rivalrous and non-excludable (Commission on IPRs, 2002). 'Non-rivalrous' means that the consumption by one person does not prohibit another person also using the same product. 'Non-excludable' means that others cannot easily be stopped from consuming the product. This is particularly true for biological products like genes and plant varieties that are self-replicating through seeds.

The balance between the rights of society on the one hand and those of the right holder on the other is very difficult to determine and subjective. Davis (2004) questions whether the current IPRs contribute to a social optimum in research and development (R&D). Andersen (2004) critically discusses the different economic arguments from a costs and social benefits perspective: there are administration and enforcement costs, monopoly or anti-competition costs, opportunity costs in depriving others from using the most effective solutions, which is specifically aggravated by the broad scope of patents, social costs by increasing the cost price of products through royalties, and finally costs that are incurred when patents divert investment in socially less productive channels just because protection can more easily be obtained in certain fields. This latter argument may be particularly relevant in plant breeding in developing countries, where significant social benefits can be derived from access to good varieties by the poor.

IPRs in agriculture have been irrelevant in developing countries until very recently. The WTO Agreement on Trade Related Aspects of Intellectual property Rights (1993) requires all members of the trade organisation to develop IPRs in their national laws (WTO, 1993). More recently, the requirements of trade agreements between the US or EU on the one hand and developing countries on the other significantly increase the strength of the rights compared with the minimum standards of TRIPS (GRAIN, 2004). Trade benefits are a great incentive for countries to accept these strong IPRs even though they themselves may not optimally serve development goals.

3.2. Genetic resource rights

The perception that plant species can be a strategic resource developed in the early colonial period when the emerging global powers were keen on containing valuable crops within their colonies (Plucknett *et al.*, 1987). In more recent days there have been formal (Smale and Day-Rubinstein, 2002) and informal (Fowler and Mooney, 1990) embargoes on the export of genebank materials of crops. However, among farming communities the concept of free exchange is commonly held high.

The different values of genetic resources (Brush, 2000; Birol, 2002; Smale, 2006) and the realisation that diversity eroded (Harlan and Martini, 1936; Bommer, 1991) triggered international debates about the conservation and availability of genetic resources in the agricultural sector in the late 1950s (Esquinas-Alcazar, 2005). This debate came to the conclusion that genetic resources are a heritage of mankind. The 'enormous contribution that farmers of all regions have made to the conservation and development of plant genetic resources, which constitute the basis of plant production throughout the world' was internationally recognised in the voluntary International Undertaking on Plant Genetic Resources for Food and Agriculture (IU PGRFA: http://www.fao.org/ag/cgrfa/IU.htm). This recognition was the basis of the concept of Farmers' Rights, 'vested in the International Community, as trustee for present and future generations of farmers, for the purpose of ensuring full benefits to farmers, and supporting the continuation of their contributions'.

The debate on biodiversity in the environmental sector culminated in the UN Conference on the Environment and Development in 1992. Through its (binding) Convention on Biological Diversity (CBD), biodiversity became a natural resource under the sovereignty of nations. Parties may set conditions to access to genetic resources and make it subject to agreed terms that provide for prior informed consent and which may include some form of benefit-sharing. Countries are free to negotiate and design such bilateral access agreements. The CBD thus explicitly overrode the 'heritage of mankind' principle of the IU PGRFA. The special nature of agricultural genetic resources (Stannard *et al.*, 2004) and the vast numbers of exchanges led to a multilateral system of access and benefit sharing for a number of major crop species under the International Treaty on PGRFA (2004). This Treaty also spells out the Farmers' Rights as the right to protect Traditional Knowledge, to benefit sharing, to participation in decision making at the national level, and it refers to the right of farmers to save, use, exchange and sell farm-saved seed (FAO, 2004). This concept thus links with both the CBD (benefit sharing) and the debate within the World Intellectual property Organisation (WIPO) on traditional knowledge in relation to genetic resources and folklore.

3.3. Rights on indigenous and traditional knowledge

An international agreement has not been concluded on the protection of the rights on indigenous and traditional knowledge (ITK). The Intergovernmental Committee on genetic resources, Traditional Knowledge and Folklore of WIPO is debating ways and means of protecting such rights. It is, however, difficult to determine whether such rights should be exclusive rights, equivalent to IPRs or based on recognition and benefit sharing only. The first option creates tensions because IPRs are meant to take innovations to the public domain (e.g. patent rights expire after 20 years), whereas such a temporary protection would not recognise the contribution of many generations in the generation of ITK. Secondly, the governance of such community rights needs to be clear, particularly to what extent members of the community who emigrated to cities or other regions would retain their involvement and rights to benefits. Making ITK protection an exclusive right could severely limit access by the global community to such knowledge, particularly when the exploitation of such rights might most favourably be exploited through the patent system. The limitation to recognition and benefit sharing has the risk that benefits may be small and that it will be easy to 'patent around' the ITK.

The debate on ITK is very relevant for biotechnology, but particularly to its uses in pharmacology and other industrial applications. Whereas medicinal knowledge is often kept secret in the community and handed from healer to the next generation, ITK in agriculture is commonly shared among all farmers in the community (and beyond). This means that exclusive strategies are extremely difficult to apply.

4. Impact of these rights systems: confusion and hyperownership

Even though the international agreements may be legally coherent, inconsistencies arise at the implementation level when the requirements are to be translated in national law that affect seed systems. The main reason is that they are based on unrelated goals in the agricultural, environmental and trade sectors (Leskien and Flitner, 1997; Drahos and Blakeny, 2001; Sampath and Tarasovsky, 2002). Where the CBD grants national sovereignty over genetic resources and promotes the rights of local and indigenous communities over their genetic resources and associated knowledge, TRIPS has the effect that individual IP-holders obtain control over particular genetic resources and technologies. Similarly, there is conceptual tension between the national sovereignty principle of the CBD and the multilateral approach of the International Treaty, between the promotion and the 'taxation' of intellectual property rights in TRIPS and IT PGRFA respectively, and between private IPRs and communal rights over traditional knowledge. Finally, the concept of Farmers' Rights collides with intellectual property rights principles. (Louwaars and Visser, 2004).

Another outcome of the negotiation processes is that although the different parallel discussions are not linked, still they do appear to influence each other, unfortunately not to reach an agreed optimum of rights over genetic resources and associated knowledge for individuals, communities and nations (Louwaars, 2006). Butler *et al.* (2002) claimed that strong breeder's rights have resulted in claims for Farmers' Rights. Safrin (2004) in turn called the outcome of this process 'hyperownership' a term that describes the reduction of the public domain or, as she describes it, 'the legal enclosure' of genetic resources through a spiral of increasing levels of both intellectual property and other genetic resource rights. Gepts (2004) confirmed in turn that the commoditisation of biodiversity has led to the active pursuit of IP protection on genetic resources (both in agriculture and pharmacology), leading to claims of biopiracy when appropriation is achieved without authorisation, and in turn to tighter rules. Similarly, 'thickets' of intellectual property rights create barriers to access to technologies and genetic resources (Bobrow and Thomas, 2001) and increase costs (Barton, 2000).

5. Tailoring rights

Regulatory frameworks should not be seen as a fixed 'given externality'. Intellectual property rights systems have shown to adapt to changing situations. Most changes in the patenting of life forms since the Chakrabarty case in 1980 are the result of new interpretations of existing law. This high level of dependence of the IP-system on judiciary rather than on democratic processes and the importance of case law is an excellent way to respond to the quick technological developments. Even though the general trend during the last decades has been to gradually strengthen the rights of the inventor, also clear indications can be observed that there is a way back – or 'a better way forward'.

The plant breeding sector has always been a good example of the flexibility of IPR systems. Plant varieties have been exempted from patent protection all over the world until in 1985 in te USA and soon after in Japan and Australia it became possible to apply for patents. Instead, plant breeder's rights systems were developed that are more in line with the culture and practice of farming through some important exemptions. Breeders are allowed to use any protected variety for further breeding without permission of the right holder (which would not be possible under patent regimes) and farmers are allowed, within certain restrictions, to multiply the seed for their own use and in some countries also for exchanging and selling to other farmers. The patent system doesn't normally provide opportunities for such exemptions.

The patenting of genes and other components of plant varieties, which is possible in Europe, would create ambiguity since farmers would be allowed to save seed under the breeder's rights system, but not when a component is patented; and a patent holder could control not only the gene but the whole genetic background if other breeders would not be allowed to use the variety in further breeding (Louwaars, 2007). Recent decisions by Germany and France to apply the exemptions also to varieties with patented components is both an excellent example of how developing countries would create openings in the patent system, and proof that a legal system like IPRs is not cast in stone. Another important example is the decision that put a halt on the patenting of expression sequence tags (ESTs), strands of DNA that do not have an apparent function (Kintisch, 2005). This decision followed an unprecedented run on patents as a result of the first sequencing work in the early 1990s.

At a more general level, the development of WIPO may provide additional grounds for opening up the patent systems in many developing (and industrialised) countries (Gerhardsen, 2007). Finally, the adoption of the International Treaty can also be considered a 'better way forward' within the general framework of the CBD, for example by improving access and reducing transaction costs for many major food and feed crops.

An important trend that runs counter to the possibilities to tailor rights regimes to the development needs of a country is international harmonisation of rights and systems. IPRs and genetic resource rights are territorial and based on national law. Harmonisation intends to results in transparent rights across borders and to reduce transaction costs, thus facilitating international trade in innovations and goods derived from these. It is therefore logical that the push for international harmonisation of laws is mainly based on the trade agenda. However, the trade-related aspects referred to in the TRIPS Agreement are only one group of aspects relevant to rights over genetic resources. In the implementation of TRIPS and in negotiating the Free Trade Agreements it is insufficiently realised that the primary reasons to introduce intellectual property rights are related to stimulating investments in innovation. When developing countries would be allowed to make the development-related rather than the trade-related aspects of intellectual property rights (DRIPS) leading, they would come to very different approaches, with more space for specific rules to deal with country specific objectives and a more detailed approach (Shen, 2005). Such approaches would benefit from harmonisation of implementation systems, such as joint examination of patent applications in the PCT system (Patent Convention Treaty) or standardised variety testing procedures developed by UPOV. The World Bank calls for differential rights for commercial (export) crops and subsistence crops in the scope of protection of varieties (World Bank, 2006), which is an explicit example of tailoring rights to the development needs of a sector. The conclusion of the IT PGRFA may be considered in the same line – its multilateral system is then a specially designed system to facilitate access and benefit sharing for major food and fodder crops.

6. Tailoring implementation strategies

IPRs and genetic resource rights provide rights; it is the way that they are exercised that determines their effect on farming. For example, the CBD provides national sovereignty on access to genetic resources that give states the right to make access subject to terms. The Nordic countries in Europe use this right to make their genetic resources widely available without asking for monetary benefit sharing (Evjen,

2003), whereas the countries of the Andean Community consider genetic resources and important national heritage that need to contribute to development (Louwaars *et al.*, 2006a).

Similarly, there are ways to use the patent system for making technologies widely available to the public. Patent holders may offer so-called humanitarian use licenses for the use of their protected technologies for development. This requires extensive negotiations, but these regularly lead to liberal offers by patent holders that not only include a freedom to use the technology but also training to use it properly and effectively. The greatest challenge for researchers in developing countries is to know who the owners are, and then to get them to negotiate. Organisations have been established to mediate in this field and to reduce transaction costs, e.g. AATF (www.aatf.africa.org) and ISAAA (www.isaaa.org). This shows the weaknesses of depending on such licenses: few transactions are actually concluded and this requires a good knowledge of IPRs, liabilities and contract negotiations.

A second step in this direction has been taken by the Generation Challenge Programme that developed a standard licensing agreement for all partners, which automatically makes technologies available for use for the benefit of the poor in developing countries (Barry and Louwaars, 2005). If such language could be accepted as a standard it could be used much more widely. An important limitation to such approach is the recent introduction of liability clauses in the Cartagena Protocol, which puts a significant responsibility to the developer of a technology on problems that arise out of its use (Sullivan, 2005). The result of this decision is that technology providers are less likely to accept broad licenses which mean that they don't have control over the use of their technologies.

The third step is the creation of open-source licenses for biotechnologies. This idea is particularly pursued by the Biological Information for Open Society (BIOS) initiative (Anonymous, 2004). This approach is built on the development of open-source biotechnologies, similar to methods used in software (e.g. Linux) and copyrighted text materials (Creative Commons, CopyLeft). The scientific basis will be provided by new transformation technologies developed by CAMBIA (Broothaerts, 2005; Constans, 2005) based on microbial processes other than those mediated by *Agrobacterium tumefaciens* on which hundreds of patents rest. The open-source concept uses patents to make sure that the information can be licensed freely to all, with only one major condition, i.e. all users of the patent will provide the same liberal access to all subsequent inventions derived from it. It remains to be seen how this concept may work in biotechnology where many scientists will use additional proprietary technologies in trying to expand upon the open source potentially creating barriers to the grant-back obligation. BIOS also operates a database with information from over 70 patent offices, which facilitates initial analysis of IP by researchers (see www.cambia.org).

Unfortunately, the discussion on increasing the public domain is ongoing in the field of IPRs and not in genetic resource policies (Louwaars, 2006). It is high time that the proponents of strict access regimes to genetic resources consider exclusions for the use of the resources for development objectives (both in agriculture and health).

The decision to patent or not, or to commercially license or provide free access is made by the rights-holder. This can be the individual researcher or in most cases, the organisation that employs the inventor(s). Institutional policies thus determine to a large extent whether the rights indeed create blockages to innovation downstream. Such policies are particularly important for public research and education organisations. IPRs are designed to create benefits in the market and are particularly designed to support commercial investments in R&D. Public research institutions thus have to decide to what extent they want to commercialise products and whether a focus on markets is in conflict with their public task. A watershed decision in this respect was the Bayh-Dole Act in the USA which allowed or even promoted public universities to protect and commercialise their inventions.

Claims have been made that access to information is delayed after policies were introduced at universities in the USA to seek protection (David, 2004); access to technology was also considered reduced (Zheng *et al.*, 2006); secrecy and the use of patents as blocking tools disturbed public research (Cohen *et al.*, 2000), start-up companies were hindered (Wright *et al.*, 2006), and the role of lawyers in research significantly increased (Maurer *et al.*, 2001). This decision has thus contributed to the 'anti-commons' (Heller and Eisenberg, 1998) that according to Runge and Defrancesci (2006) may lead to socially suboptimal access to the resource and inhibit innovation and development. An analysis of the impact of the Bayh-Dole Act indicates that as a result of the high costs of managing IP, very few schools make a net profit.

Also in developing countries the question is relevant. Louwaars *et al.* (2005) report that reduced public expenditure in agricultural research is a major reason to embrace IPRs as a revenue-maker, but also the expectations of public-private partnerships in agricultural research. Some respondents in their research realise that this approach will have impact on the research priorities. Focusing on revenue likely leads to an increased focus on profitable crops (like hybrids and market crops) and commercial farmers (Louwaars *et al.*, 2006b). When public policy also intends to use agricultural research for poverty reduction, agro-biodiversity management, and rural food security strategies, such institutes may have to deal with IPRs differently.

7. Conclusions

Tailoring biotechnologies to development objectives has to go hand in hand with tailoring of the rights systems that determine whether, how and by whom the technologies can be accessed. Such aspects should be taken into account throughout the technology development process, from priority setting and design of research methodologies and programmes to the (further) dissemination of products. This aspect has been under-investigated in the literature on tailoring biotechnologies.

Different types of rights affect the access to (the products of) biotechnologies by farmers: private rights in the form of intellectual property rights; communal rights in the form of rights on traditional knowledge and other Farmers' Rights, and rights (over genetic resources) based on national sovereignty. These are guided by international agreements, but granted at the national level.

Analysing impact of such rights systems on the application of biotechnologies, and proposing solutions to limitations has to take into account both the regulations themselves and the implementation through licensing strategies. Opportunities exist at both levels to tailor the rights to development objectives, and where policy space is reduced due to stronger demands from trade negotiations, openings are forged through private and NGO initiatives. It appears that such openings are not being developed in developing countries in the field of genetic resource rights.

References

Andersen, B. (2004). If Intellectual Property Rights' is the answer, what is the Question? Revisiting the patent controversies. Econ. Innov. New Techn. 13(5): 417-442.

Anonymous (2004). Open-Source Biology. *Nature* 431: 491.

Barry, G. and Louwaars, N. (2005). Humanitarian Licenses: making proprietary technology work for the poor. In: *Genetic Resource Policies and the Generation Challenge Programme* (Ed. N. Louwaars), Mexico DF, CIMMYT, pp. 23-34.

Barton, J.H. (2000). Intellectual property rights: reforming the patent system. *Science* 287: 1933-1934.

Broothaerts, W., Mitchell, H.J., Weir, B., Kaines, S., Smith, L.M.A., Yang, W., Mayer, J.E., Roa-Rodriguez, C. and Jefferson, R. (2005). Gene Transfer to Plants by Diverse species of bacteria. *Nature* 433: 629-633.

Birol, E. (2002). Applying environmental valuation methods to support the conservation of agricultural biodiversity. In: *The economics of conserving agricultural biodiversity on-farm* (Eds. M. Smale, I. Már and D.I. Jarvis), Rome, IPGRI, pp. 19-29.

Bobrow, M. and Thomas, S. (2000). Patents in a genetic age. *Nature* 409: 763-764.

Bommer, D.F.R. (1991). The historical development of international collaboration in plant genetic resources. In: Th.J.L. van Hintum, L. Frese and P.M. Perret (eds.). Crop Networks: Searching for New Concepts for Collaborative Genetic Resources Management. Papers of the EUCARPIA/IBPGR symposium held in Wageningen, The Netherlands, 3-6 December 1990. Rome, IBPGR, International Crop Network Series No. 4. Pp. 3-12.

Brush, S.B. (2000). The issues of in situ conservation of crop genetic resources. In: *Genes in the Field. On-farm conservation of crop diversity* (Ed. S.B. Brush), Ottawa, IDRC and Rome, IPGRI, pp. 3-28.

Butler, L.J. (1996). Plant breeders' rights in the US: Update of a 1983 study. In: *Proceedings of a Seminar on the Impact of Plant Breeders' Rights in Developing Countries, held at Santa Fe de Bogota, Colombia, March 7-8* (Eds. J. van Wijk and W. Jaffe). Amsterdam, University of Amsterdam.

Chapman, A.R. (2000). Approaching Intellectual property as a Human Right: Obligations related to Article 15(1)c. UN Economic & Social Council document E/C 12/2000/12.

Cohen, W.M., Nelson R.R. andWalsh, J.P. (2000). Protecting their Intellectual Assets: appropriability conditions and why US manufacturing firms patent (or not). Cambridge MA, National Bureau of Economic Affairs, Working Paper 7552. www.nber.org/papers/w7552.

Commission on Intellectual Property Rights (2002). Integrating Intellectual Property Rights and Development Policy. London, DFID, 195 pp.

Constans, A. (2005). Open Source initiative circumvents biotech patents. Researchers develop workaround for *Agrobacterium*-mediated gene transfer. *The Scientist* 19(8): 32.

Convention on Biological Diversity (1992). Available at: www.biodiv.org.

David, P.A. (2004). Can 'open science' be protected from the evolving regime of IPR protections? *Journal of Institutional and Theoretical Economics* 160(1): 9-34.

Davis, L. (2004). Intellectual property Rights, Strategy and Policy. *Economics of Innovation and New Technology* 13(5): 399-415.

Drahos, P. and Blakeney, M. (2001). *IP in biodiversity and agriculture*. London, Sweet & Maxwell.

Esquinas-Alcázar, J. (2005). Protecting crop genetic diversity for food security: political, ethical and technical challenges. *Nature Reviews Genetics* 6: 945-953.

Evjen, G.(2003). A Nordic Approach to Access and Rights to Genetic Resources. Report to the Nordic Genetic Resources Council. Available at: http://www.norden.org/jord_skog/sk/Nordic-approach.pdf.

FAO (2004). International Treaty On Plant Genetic Resources For Food And Agriculture. Available at: ftp://ftp.fao.org/ag/cgrfa/it/ITPGRe.pdf.

Fowler, C. and Mooney, P. (1990). *Shattering: Food policies and the loss of genetic diversity*. Tucson, Univ.Arizona Press, 278 pp.

Gepts, P. (2004). Who owns Biodiversity and how should owners be compensated. *Plant Physiology* 34: 1295-1307.

Gerhardsen, T. (2007). WIPO Development Agenda Meeting Snags On Norm-Setting. *Intellectual Property Watch*, 14 June 2007. Available at: www.ip-watch.org/

GRAIN (2004). Bilateral agreements imposing TRIPS-plus intellectual property rights on biodiversity in developing countries. Available at: http://www.grain.org/rights/TRIPSplus.cfm?id=68.

Hardon, J. (2004). *Patents beyond control; biotechnology, farmer seed systems and intellectual property rights*. AgroSpecial 2, Wageningen, Agromisa, 80 pp.

Harlan, H.V. and Martini, M.L. (1936). Problems and results in barley breeding. In. *Yearbook of Agriculture*, U.S. Department of Agriculture, pp. 303-346.

Heller, M., and Eisenberg, R. (1998). Can Patents deter Innovation? The anticommons in biomedical research. *Science* 280: 698-701.

Kintisch, E. (2005). Intellectual Property: court tightens patent rules on gene tags. *Science* 309: 1797-1799.

Leskien, D. and Flitner, M. (1997). Intellectual Property Rights and Plant Genetic resources. Options for shaping a Sui Generis system. Rome, IPGRI, Issues in Genetic resources No.6.

Louwaars, N. (2006). Ethics Watch: Controls over plant genetic resources – a double-edged sword. *Nature Reviews Genetics* 7: 241.

Louwaars, N.P. (2007). *Seeds of Confusion*. Wageningen University, PhD Thesis, 151 pp.

Louwaars, N. and Visser, B. (2004). Inter-regional conference on international agreements. Rome, IPGRI Newsletter for Europe. Available at: http://www.ipgri.cgiar.org/cmd/Europereadnews.asp?IDNews=347.

Louwaars, N.P., Tripp, R., Eaton, D., Henson-Apollonio, V., Hu, R., Mendoza, M., Muhhuku, F., Pal, S. and Wekundah, J. (2005). Impacts of Strengthened Intellectual Property Rights Regimes on the Plant Breeding Industry in Developing Countries, A Synthesis of Five Case Studies. Wageningen, Centre for Genetic Resources & London, Overseas Development Institute, 197 pp.

Louwaars, N.P., Thörn, E., Esquinas-Alcazar, J., Wang, S., Demissie A. and Stannard, C. (2006a). Access to plant genetic resources for genomic research for the poor: from global policies to target-oriented rules. *Plant Genetic Resources* 4(1): 54-63.

Louwaars, N., Tripp, R. and Eaton, D. (2006b). Public research in plant breeding and intellectual property rights: a call for new institutional policies. Washington, The World Bank. Agricultural & Rural Development Notes issue 13, June 2006, 4 pp. Available at: http://siteresources.worldbank.org/INTARD/Resources/Note13_IPR_PublicResrch.pdf.

Maurer, S.M., Hugenholtz, P.B. and Onsrud, H.J. (2001). Europe's Database Experiment. *Science* 294: 789-790.

Plucknett, D.L., Smith, N.J.H., Williams J.T. and Anishetty, N.M. (1987). *Genebanks and the Worlds Food*. Princeton, Princeton University Press, 147 pp.

Runge, C.F. and Defrancesco, E. (2006). Exclusion, inclusion, and enclosure: historical commons and modern intellectual property. *World Development* 34(10): 1713-1727.

Safrin, S. (2004). Hyperownership in a time of Biotechnological Promise: the international conflict to control the building blocks of life. *American Journal of International Law* 98: 641-685.

Sampath, P.G. and Tarasovsky, R.G. (2002). Study on the Inter-relation between Intellectual property Rights Regimnes and the Conservation of genetic Resources. Report, prepared for the European Commission, DG Environment. Berlin. Eco-Logic, 148 pp.

Shen, X. (2005). A dilemma for developing countries in intellectual property strategy? Lessons from a case study of software piracy and Microsoft in China. *Science and Public Policy* 32: 187-198.

Smale, M. (2006). Concepts, Metrics and Plan of the Book. In: *Valuing Crop Biodiversity; on-farm genetic resources and economic change.* (Ed. M. Smale). Wallingford and Cambridge, MA, CABI Publishing, p. 4.

Smale, M. and Day-Rubinstein, K. (2002). The demand for crop genetic resources: international use of US National Plant germplasm System. *World Development* 30(9): 1639-1655.

Stannard, C., Van der Graaff, N., Randall, A., Lallas, P. and Kenmore, P. (2004). Agricultural biological diversity for food security: shaping international initiatives to help agriculture and the environment. *Howard Law Journal* 48: 397-430.

Sullivan, S. (2005). The evolving international regime of liability and redress relating to the use of genetically modified organisms: a preliminary report. In: *Genetic resource policies and the Generation Challenge Programme* (Ed. N. Louwaars), Mexico DF, generation Challenge Programme, pp. 35-61.

World Bank (2006). Intellectual Property Rights. Designing regimes to support plant breeding in developing countries. Washington DC, World Bank Agriculture and Rural Development. Report # 35517, 77 p. Available at: http://siteresources.worldbank.org/INTARD/Resources/IPR_ESW.pdf.

WTO (1993). Agreement on Trade Related Aspects of Intellectual property Rights. Available at: http://www.wto.org/english/docs_e/legal_e/legal_e.htm.

Wright, B.D., Pardey, P.G., Nottenburg, C. and Koo, B. (2006). Agricultural innovation: incentives and institutions. In: *Handbook of Agricultural Economics Vol. 3.* (Eds. R.E. Evenson, P. Pingali and T.P. Schultz), Amsterdam, Elsevier, pp. 32.

Zheng, L., Juneja, R. and Wright, B.D. (2006). Implications of intellectual property protection for academic agricultural biologists. University of California, Berkely, Department of Agricultural and Resource Economics, mimeo.

About the authors

George Owusu Essegbey is the Director of the Science and Technology Policy Research Institute (STEPRI) of the Council for Scientific and Industrial Research (CSIR) in Ghana. His main research areas include innovation studies in agriculture and small and medium enterprises, biotechnology policy and new technologies. Geographically, his works have focused mainly on Ghana, Africa and the developing world. He holds an M.A. in international affairs (1994) and a Ph.D. in development studies (2005).

Shuji Hisano is Associate Professor of the Graduate School of Economics at Kyoto University, Japan. Between 2002 and 2004, while holding his permanent post at the Graduate School of Agriculture of Hokkaido University, he did his research as a visiting fellow at the Technology and Agrarian Development group of the Social Sciences Department at Wageningen University. His main research interest is in the political economy of reconstruction of agri-food systems and biotechnologies in the context of globalisation and localisation. He holds a master's degree in economic policy (1993) and a doctoral degree in agrarian political economy (2001).

Joost Jongerden is Assistant Professor at the Critical Technology Construction (CTC) research group of the Social Sciences Department of the Wageningen University & Research (WUR). He is managing editor of *Tailoring Biotechnologies*, a peer-reviewed journal in the field of critical theory, development studies and science & technology-studies. His main research interest is in the relation between the shaping of technology, the production of space and the construction of identities. His region of interest is the Middle East, in particular Turkey and Kurdistan. He holds a M.Sc. in rural sociology (1991) and a Ph.D. in social sciences (2006).

Niels Louwaars is senior scientist at the Centre for Genetic Resources, The Netherlands of Wageningen UR. His main research interest concerns biopolicies that have an impact on genetic resource management with emphasis on those in developing countries. He holds a PhD from Wageningen University based on a thesis '*Seeds of Confusion; the impact of policies on seed systems*'. He is manager of the major development oriented research and capacity building program at Wageningen UR, and account manager for CGIAR relations.

Les Levidow is a Senior Research Fellow at the Open University, UK, where he has been studying agri-environmental issues since 1989. A long-running case study has been the safety regulation, innovation and controversy over agbiotech. This research has focused on the European Union, USA and their trade conflicts. The case study has taken up concepts such as sustainability, regulatory science, precaution, European integration, governance, transnational civil society and organisational learning. He is also Editor of the journal *Science as Culture*. He holds a PhD in technology policy from the Open University. Details of research projects and publications are available on the Biotechnology Policy Group webpage, http://technology.open.ac.uk/cts/bpg.htm.

Ezio Manzini is Professor of Design at the Politecnico di Milano, Director of the Unit of Research Design and Innovation for Sustainability, coordinator of the Doctorate in Design. He has been Director of the Domus Academy in Milano, Chair Professor of Design at the Hong Kong Polytechnic University and visiting lecturer at the Tohoku University in Japan. In the 2006 he has been nominated Honorary Doctor of Fine Arts at The New School of New York. His works are based on strategic design and design for sustainability, with a focus on the design for social innovation. He is now coordinating CCSL, Creative Communities for Sustainable Lifestyles: a research on grassroots social innovation in China, India and Brazil. See: http://www.sustainable-everyday.net/.

William Munro is Associate Professor of Political Science and Director of the International Studies Program at Illinois Wesleyan University. He has conducted extensive research on state formation and agrarian change in sub-Saharan Africa, and is the author of *The Moral Economy of the State: Conservation, Community Development and State-Making in Zimbabwe* (Ohio University Press, 1998). His research interests include democracy and development, science and agricultural policy, and social movements. Currently he is working with Rachel Schurman on a book about the effects of social activism on the development of agricultural biotechnology, government regulatory policy, and the life sciences industry.

Elibariki Emmanuel Msuya got his BSc (2001) and MSc (2003) in agricultural economics from Sokoine University of Agriculture, Tanzania, and has worked as lecturer in the Department of Agricultural Economics and Agribusiness of the university. He is now undertaking PhD studies at the Graduate School of Economics, Kyoto University, Japan, with a JICA scholarship. His research interest is in the value chains for maize/legume-based farming system and the feasibility of different smallholder institutional setups in Tanzania.

Guido Nicolosi is researcher at University of Catania (Italy), where he is professor of Sociology of Cultural and Communicative Processes at the Faculty of Political Science and scientific coordinator at 'Centre for Technological Innovation in Socio-linguistic and Territorial Systems in the Mediterranean Area' (Braudel Centre). His research program is 'The body between social practices and cultural changes: food, communication and symbolic integrity'. He published in Italy *Corpi al limite. Linguaggio, natura e pratiche sociali* (2005) and *Lost food. Comunicazione e cibo nella società ortoressica* (2007).

Guido Ruivenkamp is professor at the Vrije Universiteit Amsterdam and head of the Critical Technology Construction (CTC) research group of the Social Sciences Department of the Wageningen University & Research (WUR). He is chief editor of *Tailoring Biotechnologies*, a peer-reviewed journal in the field of critical theory, development studies and science & technology-studies. He conducted extensive research on biotechnology, division of labour and agro-industrial food chains. He is the author of *The introduction of biotechnology into the agro-industrial chain of production: changing over towards a new form of labor organization* (PhD thesis, 1989). His main research interest are biotechnology and genomics regarding sustainable agriculture and food production in underdeveloped countries and critical theory and rural development.

Franz Seifert is Biologist and Social Scientist with a research focus on the global politics of agricultural biotechnology. He has been, amongst others, with the Institute of Advanced Studies in Vienna, the Austrian Academy of Science, and the United Nations University in Yokohama/Japan. Currently, he is independent researcher and lecturer at the University of Vienna.

Rachel Schurman is Associate Professor of Sociology and Global Studies at the University of Minnesota-Twin Cities. Her primary areas of interest include the political economy of food and agriculture, social movements and technological change, and political sociology. She is co-editor of *Engineering Trouble: Biotechnology and Its Discontents* (University of California Press, 2003), and has also written extensively on neoliberalism in Chile. Her current research focuses on social resistance to agricultural biotechnology and the myriad ways in which organised social activism has shaped the life sciences industry, government regulatory policy, and the development trajectory of agricultural biotechnology. This research, carried out in collaboration with William Munro, is slated to be published by the University of Minnesota Press in 2009.

Wietse Vroom is a PhD candidate at the Athena Institute of the Vrije University Amsterdam (VUA) and is associated to Critical Technology Construction (CTC) of the Social Sciences Group of Wageningen University & Research (WUR). In November 2004 he started his PhD research, studying the role of genetic technologies in international agricultural modernisation. His main interests are in issues of access to genetic technologies, and levels of reconstruction to make technologies 'appropriate' for developing world agriculture. He holds a M.Sc. in Molecular Sciences (2004).

Index

Printed in the United States
by Baker & Taylor Publisher Services.

Printed in the United States
by Baker & Taylor Publisher Services